CENTRAL NERVOUS SYSTEM PATHOLOGY

A New Approach

CENTRAL NERVOUS SYSTEM PATHOLOGY

A New Approach

Academician G. N. Kryzhanovsky

Academy of Medical Sciences of the USSR
Director, Institute of General Pathology and Pathological Physiology
Moscow, USSR

Translated from Russian by
Nicholas Bobrov

CONSULTANTS BUREAU • NEW YORK AND LONDON

Library of Congress Cataloging in Publication Data

Kryzhanovskiĭ, G. N. (Georgiĭ Nikolaevich)
  Central nervous system pathology.

  Rev. translation of: Determinantnye struktury v patologiĭ nervnoĭ sistemy.
  Bibliography: p.
  Includes index.
  1. Central nervous system—Diseases—Etiology. 2. Neurophysiology. I. Title. DNLM: 1. Nervous System Diseases—physiopathology. WL 100 K94d
RC361.K7913 1985   616.8′047   85-12779
ISBN 978-1-4684-7872-3   ISBN 978-1-4684-7870-9 (eBook)
DOI 10.1007/978-1-4684-7870-9

The original Russian text, published by Meditsina in 1980,
has been revised by the author for the present edition.
This translation is published under an agreement with the
Copyright Agency of the USSR (VAAP)

DETERMINANTNYE STRUKTURY V PATOLOGII NERVNOI SISTEMY
G. N. Kryzhanovsky

© 1986 Consultants Bureau, New York
Softcover reprint of the hardcover 1st edition 1986

A Division of Plenum Publishing Corporation
233 Spring Street, New York, N.Y. 10013

Preface to the English translation

The past decade has witnessed an explosion of new knowledge in clinical neurology and the neurosciences. Insights have been achieved into the nature of pain, epilepsy, "deafferentation" disorders, and a variety of functions and dysfunctions of the human nervous system. Indeed, the multiplicity and magnitude of the recent developments present a formidable task to both clinical practitioners and scientists who try to follow all the new threads of thought that will lead to a better understanding of the nervous system.

It is virtually impossible to keep up with all the new information that has been generated in recent years. It has been particularly difficult to integrate it into a coherent, integrated body of knowledge that will lead to better clinical practice as well as to further research necessary to understand the riddles of neural function that still confront us.

Professor Kryzhanovsky's new book therefore represents a major contribution by providing an integrated approach to neurological problems. His concepts of "determinant structures" and "generators of enhanced excitation" are an important step beyond the traditional concept of static, hierarchical systems which operate in a rigid predetermined fashion. Instead, he emphasizes the dynamic interactions among structures at multiple levels. Information processing in the nervous system is determined by the delicate balance of inhibitory and excitatory influences. Pathological activity is not represented exclusively by organic lesions but also includes interruptions in the balance of excitation and inhibition. New modes of therapy are suggested by ways of modulating inputs and functions so that generators of pathologically enhanced excitation are kept in check.

These are exciting new concepts and represent a valuable contribution to knowledge. Professor Kryzhanovsky has proposed important new ideas, presented in a variety of articles and chapters, for an understanding of pathological pain processes. The present book indicates that his ideas are far-reaching and have clear implications for a wide variety of problems in the field of neurology.

This book is certain to stimulate new research and new ideas. It thereby represents a major advance in a field that is urgently in need of concepts capable of integrating a variety of seemingly unconnected facts into a meaningful whole.

Ronald Melzack
McGill University
Montreal, Canada

Preface to the Russian edition

One hardly always receives due credit for working out a general theory. It is more enticing to be successful in discovering new phenomena of tangible facts even if they are particular ones. The great technical opportunities which now exist and the new methods of investigation which recently seemed to be fantastic have made that even more enticing. Therefore, the establishment of general regularities seems to be an abstract occupation whose significance is not always clear. Nevertheless, a general theory should be elaborated. Factual material should be systematized and generalized as it accumulates. As an increasing amount of data is obtained, it becomes increasingly urgent to ascertain the general regularities which determine the typical mechanisms of individual phenomena. Researchers approach these aspects of scientific activity in different ways. Some of them are interested in analyzing the specifics of a phenomenon, while others try to find a general regularity in an individual fact and look for its confirmation in new material. Both of these indispensable forms of research have been reflected in this book.

As experience is gained and research develops, it becomes more and more urgent not only to make theoretical generalizations, but also to practically implement the results of the work done. It becomes indispensable for a researcher to bring some tangible benefit. The painstaking requirements of practice have always acted as stimuli for theorizing experimenters. Pasteur wrote: 'Those who have dedicated their life to science are extremely happy when the number of discoveries increases, but a scientist is filled with delight when the results of his investigations are immediately implemented in practice. There are no two kinds of sciences. There is science and the application of science, and these two types of

activity are interconnected just as fruit is with the tree on which
it grows.' These words reflect the essence of a researcher's whole
life, which is filled with the arduously gained delights of a constant quest.

This book mainly deals with two problems: the role of the
determinant structures in the pathology of the nervous system and the
generator mechanisms of the neuropathologic syndromes. Both problems
are closely interconnected. However, each of them has its own
specifics in conformity with the role which the determinant structures and the generators of excitation play in the activity of the
nervous system. Earlier, these problems were not raised, and they
were not worked out clinically, experimentally or theoretically.
They were raised for the first time in the course of our investigations. The concepts of the determinant and the generator of excitation were formed when it became possible to generalize the material
which had accumulated. It became clear that the phenomena which were
discovered reflect the essential regularities of the activity of the
nervous system, and that they should be specially defined. Eventually, the problems were worked out with regard to other aspects
besides the initial tasks, and many results became of greater significance. Some of these results are of definite interest in the
general neurophysiological field, while others are a contribution to
the general pathology of the nervous system. Clinical practice is
taken into account also when various neuropathologic syndromes are
modeled on the basis of theoretical concepts and when the basic
principles of treatment and the mechanisms of recovery are considered. This book deals with aspects which involve other medical
sciences as well as problems of general biology and pathology.
Hence, it is intended for various researchers, experimenters and
clinicians.

It took more than two decades to do all the work whose results
are presented in this book. This work could not have been done
without my associates' assistance. Work was especially fruitful
during the last few years, when a theory was outlined and its basic
principles were formed, and when other researchers and groups of
workers began to collaborate with us in our endeavors. Consequently,
much data were obtained, work was summed up, and the necessary generalizations were made.

I am very grateful to everyone who took part in producing this
book at one stage or another, especially my laboratory coworkers A.A.
Polgar, V.N. Grafova, V.K. Lutsenko, F.D. Sheikhon, S.I. Igonkina,
Ye.I. Danilova, M.V. Dyakonova, M.B. Rekhtman, B.A. Konnikov, and
R.N. Glebov, and also my associates in other institutions: M.N.
Aliev, R.F. Makulkin, A.A. Shandra, A.A. Gun, B.A. Lobasyuk, V.V.
Russev, and Yu.I. Pivovarov. I am thankful to them and other
associates of mine for their criticism, without which the theory
presented in this book could not have been elaborated and the re-

searches could not have been carried out. I am also grateful to B.I.
Khodorov and especially A.M. Gurvich for their constructive remarks
and advice. I wish to thank everyone who has helped me to produce
and publish this book.

G. N. Kryzhanovsky

Foreword

Pathologic processes caused by various damages to the nervous system are not merely based on the disturbances of the anatomical structure's integrity and the disorganization of the functional links. Several other mechanisms are pathogenetically important in this respect. One such mechanism is the origin of a pathologic system, i.e. a peculiar functional organization formed from parts of the changed physiological systems. This functional organization is characterized by new regularities and is of disadaptive, biologically negative significance. The pathologic system itself is formed under the influence of the determinant, a hyperactive structure that produces a pathologically enhanced functional effect. The determinant plays the key role not simply in the formation of the pathologic system. It establishes both the nature and the result of the system's activity. The pathologic system's specificity, being clinically expressed in a certain neuropathologic syndrome, is connected with the origin of the determinant in certain areas of the central nervous system.

These theoretical views were the starting point in experimentally reproducing neuropathologic syndromes pertaining to diverse spheres of nervous activity. They were a prerequisite for elaborating therapeutic principles based on knowledge of the determinant's neurophysiological and neurochemical nature. These principles are connected with the view that recovery is a process of the elimination of the pathologic system.

Such is the gist of this book, the result of 25 years of work. The theory expounded in it is intended to create the general basis for studying many clinically different but pathogenetically similar disease states of the nervous system.

It is a source of joy for the physician to restore his patients' health. This is the purpose and meaning of his life. The experimenter and the theoretician are denied this joy, which makes up for the physician's failures and setbacks. They find happiness in new ideas, and it is the moment of revelation that gives them their greatest joy. Then comes the tormenting search for ways to test an idea. This search is all the more tormenting when testing and exploration subsequently lead to a general theory and a new trend in science. Gone are the days when one researcher could do all this himself. Today, only a team can cope with the tasks necessary for a theory to emerge. That is why researchers cherish opportunities to present their ideas. It is hoped that this book will attract the attention of both the experimenter and the clinician and will promote their teamwork.

I am grateful to Plenum Publishing Corporation for their kind offer to bring out this book in English, which will make it accessible to a wider audience.

I am much obliged to Professor Melzack, with whom I have been in touch on scientific matters with fruitful results, for his assistance which proved instrumental in the appearance of this English edition.

Compared with the previous edition (in Russian, 1980), this book has been greatly worked over and includes new material obtained in recent research.

G. N. Kryzhanovsky

Contents

Abbreviations

ACh	acetylcholine
ACTH	adrenocorticotropic hormone
AP	arterial pressure
ASFC	artificial stable functional connections
CN	caudate nuclei
CNS	central nervous system
CSPT	complex specific pathogenetic therapy
DA	dopamine
DCP	dorsal cord potential
DDS	determining dispatch station
DRP	dorsal root potential
DS	determinant structure
EA	electric activity
ECoG	electrocorticograph
EI	effectiveness index
EMG	electromyogram
EP	evoked potential
EpA	epileptic activity
EPR	enhanced physiologic response
EPSP	exciting postsynaptic potential
ES	electric stimulation
GA	glutaminic acid
GABA	gamma-aminobutyric acid
GE	generator of excitation
GPEE	generator of pathologically enhanced excitation

IID	interictal discharge
i.m.	intramuscular or intramuscularly
i.p.	intraperitoneal or intraperitoneally
IPSP	inhibitory postsynaptic potential
i.v.	intravenous or intravenously
ʟ-dopa	dihydroxyphenylalanine
LGN	lateral geniculate nucleus
LVN	lateral vestibular nucleus
MRF	mesencephalic reticular formation
NA	noradrenaline
PAD	primary afferent depolarization
PD	prolonged depolarization
PDS	paroxysmal depolarization shift
PG	pale globe
P-5-P	pyridoxal-5-phosphate
PS	pathologic system
RCPN	reticular caudal pontine nucleus
TT	tetanus toxin
VC	visual cortex

Introduction

Before expounding the facts, it is necessary to briefly define the concepts which had originated when the problem under consideration was being worked out and which reflect the most general regularities and mechanisms of the processes that were studied.

The <u>determinant structure</u> (DS), or the <u>determinant</u>, is a structure of the central nervous system which determines both the pattern of the activity of other parts of the system and the result of its activity. The determinant is one of the operant mechanisms of the system's functional organization. The principle of the determinant is a principle of intrasystemic relationships that determines the functional hierarchy of the system's parts.

The <u>generator of pathologically enhanced excitation</u> (GPEE) is a neuronal population which produces excessive excitation and works with a certain extent of autonomy. GPEE is the neuropathophysiological basis of the hyperactive determinant structure's activity. Unlike the generator, the determinant is a systemic category: it exists only in the system and disappears when the system is eliminated. The generator of excitation, which acts as a local neurophysiological mechanism, may then remain instead of the determinant.

This book deals with neuropathologic syndromes. We use the summary term 'neuropathologic' syndromes to cover the syndromes which relate to neurology and psychiatry and which apply to various spheres of the nervous system's activity. The term can also cover experimental syndromes which are reproduced under special conditions and syndromes which originate as a result of diagnostic and therapeutic interventions in someone's nervous system. In this book, a study was made of only the syndromes which have a central origin and which are characterized by systemic hyperactivity. Accordingly, the field of application of the theory being evolved is also limited, covering mainly the given forms of the pathology of the central nervous system.

Investigations began with the detection of a unique phenomenon of functional asymmetry in the activity of the spinal cord which was locally poisoned by tetanus toxin (TT). The essence of this phenomenon is that afferent flow to the affected spinal segments on the side of poisoning causes generalized seizure activity, which involves the whole spinal cord. Similar afferent flow from other parts of the body does not produce a convulsive reaction, and the appropriate reflexes originate instead of it. Further investigations, which were carried out on a comparative plane, have shown that the phenomenon can be reproduced in animals of different species, ranging from the frog to the monkey. It became clear that the phenomenon reflects a certain regularity of the activity of the spinal cord under the given conditions and that one of the principles of the nervous system's work is involved.

At first, we described the given phenomenon as the 'phenomenon of irradiation' or the 'phenomenon of the universal dispatch station'. This term reflects the specific fact that the functional structure, which originates in the spinal cord when it is locally poisoned by TT, can excite all the spinal and brainstem motoneurons. It has also been shown that such a structure can not only excite motoneurons and interneurons at different spinal cord levels, but also impose the pattern of its activity on them. This aspect of the phenomenon was reflected in the terms 'determining dispatch station' (DDS), the 'determinant structure' (DS) or the 'determinant'. The last two terms are the most general ones, and they are used most widely.

Several general regularities of the nervous system's activity under pathologic conditions were established by studying the mechanisms of the DS's origin and activity, and the formation and functional organization of GPEE in various areas of the central nervous system. This part of the work has become the theoretical basis.

The modeling of neuropathologic syndromes is just as important. The reproduction itself of various forms of the pathologic state on the basis of single concepts by using a single method, i.e. the creation of GPEE in various areas of the central nervous system, is of great interest with respect to the nervous system's general pathology. This part of the investigations is of special significance in this book: it is an indispensable link of the theory of the generator mechanisms of the neuropathologic syndromes and experimentally bears out its principles. An experimental model becomes really significant when it pathogenetically corresponds to the form of human pathology being reproduced, reveals its mechanisms, and makes it possible to both study the processes of its elimination and recovery and, consequently, to elaborate the principles of rational therapy. The potentialities of a theory are revealed and it is confirmed as these requirements are satisfied. Obviously, the creation of models of not only neural disturbances, but also the most common and serious forms of the pathologic state of the nervous system in man, is of great importance.

Ample space is devoted to the methodological aspects of investigations. The methods of the investigations are expounded in the works cited in the text.

A different technique of producing hyperactive structures in the central nervous system was used. To form GPEE, we locally injected chemical agents which disturb the inhibiting-exciting relationships of neurons. The neuropathophysiological peculiarities of GPEE thus produced depends on the functional and neurochemical organization of the population of neurons which are drawn into the pathologic process. The use of TT, the main effect of which is the long-lasting disturbance of various types of inhibition, was especially fruitful in these investigations. The formation of GPEE under the action of TT models the endogenous mechanism of its origination.

This brief introduction sheds only some light on the problem under consideration. It acquaints the reader with new concepts, the direction in which researches are being carried out, their stages, and certain features. It is to be hoped that the book will give an idea about the new field of the pathology of the nervous system and the prospects held out in this respect.

Part I
General theory

1
Determinant structures as functional formations: The determinant principle

DETERMINANT STRUCTURES IN THE SPINAL CORD

The determinant phenomenon was first discovered when the functional asymmetry in the activity of the spinal cord was studied as the lumbosacral segments were locally poisoned by TT (Kryzhanovsky, 1957-1968). Clinically, this phenomenon is expressed as follows: a general seizure occurs in animals when the hind leg on the side of the poisoning is stimulated. This reaction is followed by a powerful burst of electric activity (EA) in all the muscles of the back, the tail, legs, the neck, the face, and the head (Figure 1 A). The respiratory muscles are drawn into the reaction at the late stages of the process (Figure 2). A seizure does not occur when similar stimuli are applied to the opposite leg or any other site of the body, and when distant stimuli (light, sound) are used (Figure 1 B). At the late stages, when the process is generalized, sufficiently strong nociceptive stimuli applied to the opposite leg and some other sites of the body also cause a seizure. However, the stimuli applied to a leg on the side of TT administration are the most effective ones. The animal is weakened at the last stage of a disease, when spontaneous convulsions disappear. At this stage, the animal lies prostrate, its temperature is low, and its reflexes are suppressed. Then, only the stimulation of the leg on the side of TT administration can produce a general motor reaction. Stimuli applied to the given leg can produce a reaction even during agony, when the corneal reflex is absent. This phenomenon was reproduced in animals of different species, ranging from the frog to the monkey (Kryzhanovsky et al., 1961; Kryzhanovsky, 1966a, 1966b). Hence, it is not conditioned by the species specificity of the structural and functional organization of the central nervous system of animals or the species dependence of

3

the reaction to TT, but reflects the general regularity of the nervous system's activity under the given conditions.

Further analysis has shown that the phenomenon has a central origin. It is engendered by the formation of a hyperactive structure in the ventral horns of the spinal lumbosacral segments. A microinjection of TT into this region produces that phenomenon (Kryzhanovsky, 1968) (Figure 3).

The illustrations show that the burst of EA in various muscles, produced by the stimulation of a leg on the side of TT administration, has several distinguishing features. One such feature is

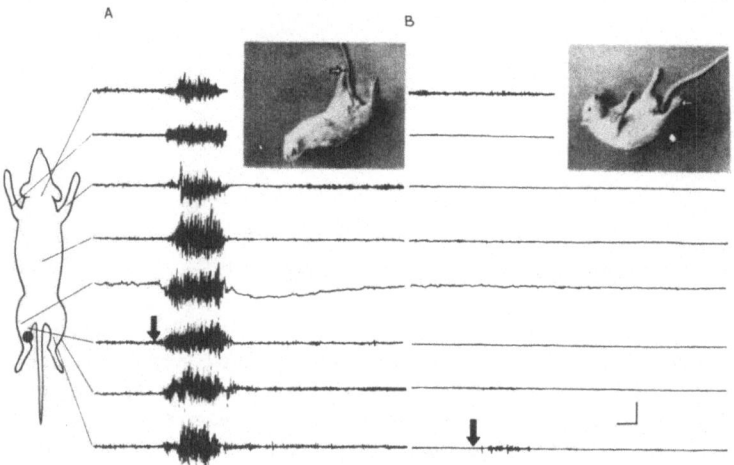

Fig. 1.   Functional asymmetry in the activity of the spinal cord
          during local poisoning of the lumbosacral segments by TT
          (the 'universal dispatch station' phenomenon). To locally
          poison the lumbosacral segments on the left, TT was injected
          into the left gastrocnemial muscle (marked by a circle).
          From there, TT goes via the sciatic nerve and the ventral
          roots into the ventral horn of the lumbosacral segments
          (Kryzhanovsky, 1966). The stimulation of the left hind leg
          (clamping of the left foot, indicated by an arrow) (A)
          causes general tonic seizure activity (as illustrated above)
          and a powerful burst of electric activity (EA) in all the
          muscles. Such stimulation of the right hind leg (B: indi-
          cated by an arrow) does not cause seizures. In this case,
          an ordinary flexor reflex occurs, and the enhancement of EA
          in the muscles of the right leg is recorded during it.
          Third day after injecting TT (4 DLM for the rat; 200 g).
          Antitoxin (0.025 IU) was injected i.v. simultaneously with
          TT to prevent TT from spreading via the blood circulatory
          system. Amplitude: 0.5 mV; time: 1 s.

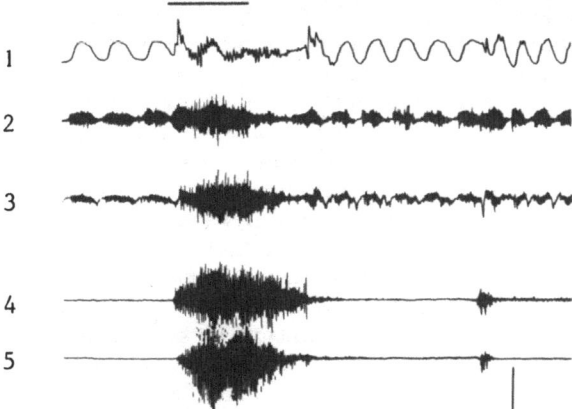

Fig. 2. Involvement of the respiratory muscles in a general seizure
in the 'universal dispatch station' phenomenon. 1: Mechano-
gram of the respiratory movements of the thorax. Recording
of EA; 2: diaphragm; 3: intercostal muscles; 4: back
muscles; 5: left gastrocnemial muscle. Generalized seizure
activity produced by squeezing (horizontal line) the foot on
the side of TT injection. Amplitude: 0.2 mV; time: 0.5 s.

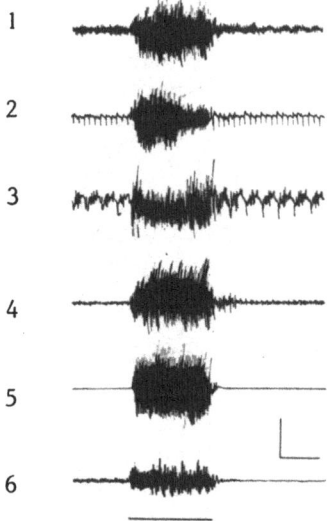

Fig. 3. Generalized seizure activity during the stimulation of the
hind leg on the side of TT injection into the right ventral
horns of the spinal lumbosacral segments. EA recordings: 1:
right neck muscles; 2: right intercostal muscles; 3: right
diaphragm; 4: right back muscles; 5: posterior group of
right femoral muscles; 6: posterior group of left femoral
muscles. The squeezing of the left foot is indicated by a
horizontal line. Amplitude: 0.5 mV; time: 1 s.

that EA originates simultaneously in all the muscle groups and simultaneously disappears. Thus, the EA burst has virtually the same
duration. Its intensity and pattern as a whole are also approximately the same. Hence, EA bursts in various muscles are due to the
propagation of excitation from the same source. Since EA in the
muscles reflects the excitation of the corresponding motoneurons of
the spinal cord, the given data show that the spinal motoneurons and
the motoneurons of the brainstem, which innervate the head muscles
(in which EA also bursts out simultaneously), are excited by impulses
from the same source and become a single functional pool which is
subordinated to the influence from that source and which uniformly
reproduces this influence. The excitation source engendered by TT
action thus has the significance of a structure which determines the
behavior of the whole system with motoneurons as the terminal links.
This source acts as the 'determining dispatch station', or the determinant.

In this case, the impulses produced by the determinant structure
propagate into all the segments of the spinal cord in descending and
ascending ipsilateral and contralateral directions reaching all the
spinal as well as supraspinal motoneurons. This nature of the propagation of excitation has given rise to the initial name of the given
phenomenon, i.e. the 'universal dispatch station' (Kryzhanovsky,
1965-1975).

The determinant effects are clearly manifested in experiments
involving the production of two hyperactive structures which work in
different modes but which have the same effectors, namely, the spinal
motoneurons (Kryzhanovsky et al., 1977a). Figure 4 shows that EA
bursts out in all the muscles when the hyperactive structure on the
left side of the lumbosacral segments is activated. This burst is at
first tonic, and then clonic. When stimulation is applied to the
right side, EA recorded in the same muscles is only tonic. Clonic
activity shows especially clearly that the synchronized discharges
originate and disappear virtually simultaneously, and each discharge
is recorded in all the muscles and has the same form.

In this case, both determinant structures function as formations
which are independent of one another. This can be seen also from the
experiment involving the administration of glycine into a region of
a hyperactive structure. Glycine is known to hyperpolarize the
neuronal membrane and produce the effect of postsynaptic inhibition
(Curtis, De Groat, 1986; Curtis et al., 1968; Werman et al., 1968;
Gushchin et al., 1970; Hösli, Hass, 1972; Curtis, Johnston, 1974;
Evans et al., 1976). Therefore, it was used to suppress the activity
of DS and eliminate it as a functional unit. Figure 4 C shows that
glycine injected into the ventral horns of the lumbosacral segments
on the left completely suppresses the activity of the appropriate DS,
and the stimulation of the left hind leg no longer produced a reaction typical of the activation of this DS. However, hyperactive DS

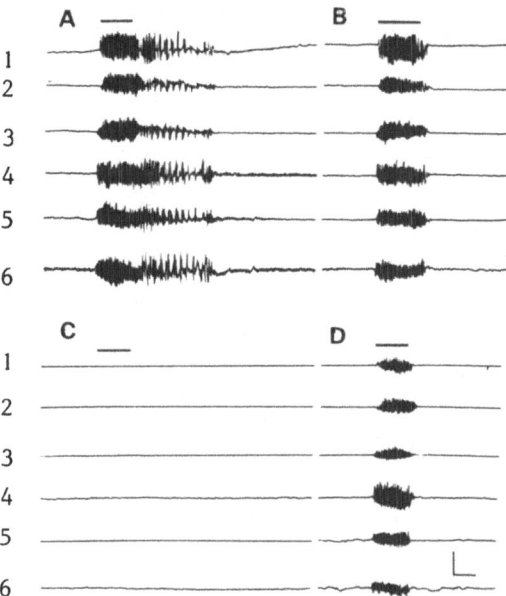

Fig. 4. Effect of the activation of various determinant structures
in the spinal cord on EA of the same muscle. A,C: EA after
activation of the 'left' determinant structure; B,D: EA
after the activation of the 'right' determinant structure;
nociceptive stimulation of the respective hind leg (the
squeezing of the toes is indicated by a horizontal line); C:
absence of evoked EA during nociceptive stimulation after
the microinjection of glycine (1.0^{-4} ml, 20% solution) into
the ventral horns L_4-L_6 on the left. Recordings of EA: 1,2:
right and left neck muscles; 3,4: right and left back
muscles; 5,6: posterior group of femur muscles on the right
and left. Duration of stimulation is indicated by a horizontal line. Amplitude: 0.5 mV; time: 1 s.

remains on the opposite side, and when it is activated by appropriate
stimulation, it produces its characteristic pattern of activity which
is, as formerly, reproduced uniformly by spinal and brainstem motoneurons (Figure 4 D).

Hence, it can be concluded that the determinant not only establishes the activity pattern of the populations of spinal neurons
which it activates, but also acts as a system-forming factor. It
produces a new system by subordinating the populations of various
neurons to itself, functionally uniting them and transforming them
into a single functional pool regardless of their initial working
characteristics and the physiological systems to which they belong.
Such a new system functions as an entity in a mode which is set by

DS: it originates only when DS is activated and disappears when the latter is either suppressed or naturally eliminated.

ROLE OF THE DETERMINANT IN THE FORMATION AND ACTIVITY OF
THE COMPLEXES OF EPILEPTIC FOCI IN THE CEREBRAL CORTEX

The spinal cord systems, especially afferent and efferent ones, are characterized by relatively rigid functional connections due to the specifics of the stationary, anatomically preformed relationships (Gaze, 1970). Although the above-mentioned determinant structure in the spinal cord is apparently localized in the system of proprio-spinal neurons (see Chapter 2), it seemed important to trace the given regularity in the operation of the area of the central nervous system in whose activity the stochastic relationships play a con-siderable role. Such functional relationships are expressed most clearly in the cerebral cortex (Burns, 1968). This part of the central nervous system was selected for studying the regularities of the formation and behavior of the functional complexes which orig-inate under the influence of the hyperactive determinant structure. The process was studied in model experiments with the local action of various agents, which cause epileptic activity, on different areas of the cerebral cortex (Kryzhanovsky et al., 1977c, 1978b, 1978c).

Foci with different levels of epileptiform activity were pro-duced, in various areas of the cerebral cortex by using diverse con-centrations of epileptogenic substances. Subepileptic foci with an amplitude of spike potentials reaching 400 μV and with a relatively low frequency of their origin were produced when relatively weak strychnine solutions were applied (Figure 5 zones 2 and 3). These foci generated potentials asynchronously and independently of one another. The given data coincide with the results of the researches carried out by other authors, who showed that foci with asynchronous activity originate when strychnine is injected into areas of the cerebral cortex that are sufficiently far away from one another (Beritov, Gedevanishvili, 1945; Roitbak, 1955; Batuev, Bogoslovsky, 1963).

Potentials whose amplitude reached 1.5-2.0 mV were recorded in a new focus (Figure 5 zone 1) when it was produced either by a concen-trated strychnine solution or by a strychnine crystal. The amplitude of the discharges in other foci (Figure 5 zones 2 and 3) also grew as the new focus was formed and the activity increased in it. Control observations have shown that this was not due to the further action of strychnine in zones 2 and 3 (Figure 5), to which the agent was no longer applied when the first epileptiform potentials appeared. The intensifications of the activity in other zones was temporally con-nected with the formation of a new focus of epileptic activity in zone 1 (Figure 5). At the same time, the discharges in the first two foci (zones 2 and 3) were synchronized with the discharges in the new

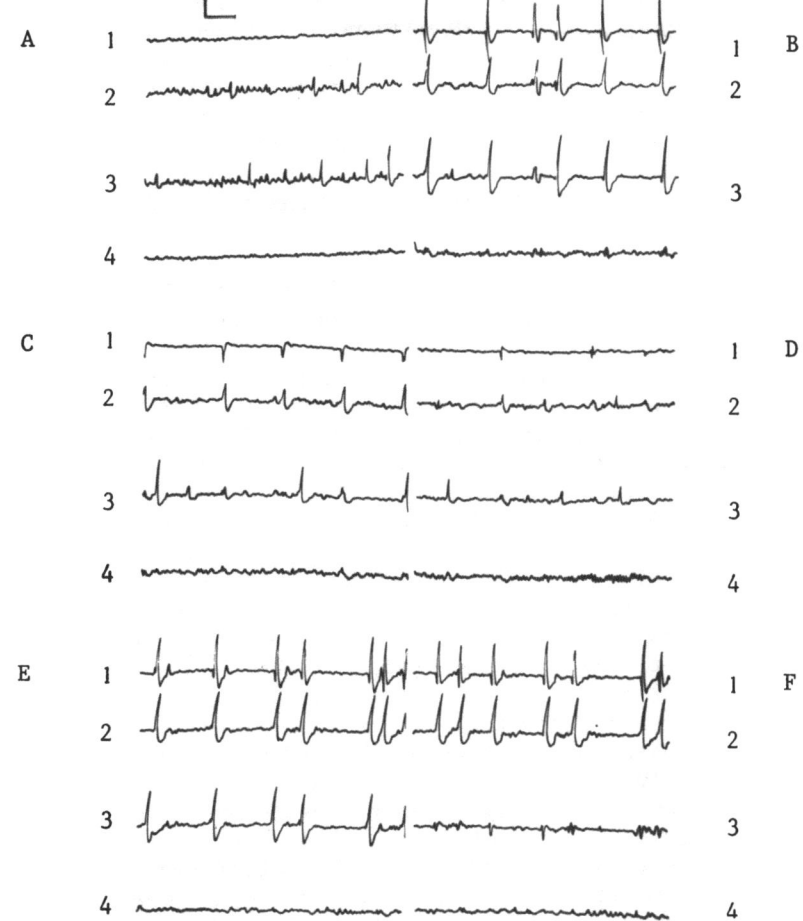

Fig. 5. Role of the determinant focus in the organization and
activity of the complex of epileptic foci in the cerebral
cortex. A: Formation of subconvulsive foci in the coronary
gyrus (zone 2) and in the posterior sigmoid gyrus (zone 3)
after 0.01% strychnine application. Strychnine application
was ceased after the appearance of activity (paper with
strychnine was removed). B: Formation of the determinant
focus in the orbital cortex (zone 1) after crystal strychnine application, and the synchronization of epileptic
activity in all the foci. C: 7 min after the application of
6% nembutal to zone 1; D: 12 min after its application;
suppression of the determinant focus and the breakdown of
the complex of epileptic activity. E: Repeated formation of
the determinant focus in the orbital cortex when crystal
strychnine was applied after nembutal had been washed out
from it, and the repeated formation of the complex (synchronization of epileptic activity in all the foci). F:
(continued overleaf)

foci (zone 1). As the process reached its peak, the appearance of discharges in zones 2 and 3 was synchronized with the generation of discharges in the new focus (Figure 5 B). Thus, a single functional complex of epileptiform activity was composed of three foci in which the new focus acted as both an organizer and a determining factor. These data are in accord with the observations made by Roitbak (1955), who showed that a weak focus is subordinated to a stronger one when excitation foci are produced in the cerebral cortex by strychnine.

Experiments involving the pharmacological suppression or the ablation of a new focus or other foci confirm the fact that the new focus acts as a determinant in the formation of the given complex and in the establishment of the pattern of its work. The amplitude of the potentials in other foci was reduced and the whole functional complex fell into separate foci and finally disappeared when the activity of the determinant focus (zone 1) was suppressed by the local application of nembutal (Figure 5 C D) or when this focus was ablated (Figure 6 C D). In this case, other foci began to function independently of one another. When the activity was suppressed in any other focus (Figure 5 F) or any other focus was ablated (Figure 6 F), the complex survived and all the remaining foci functioned in the same mode imposed by the focus in zone 1. Thus, the last focus really acted as the leading, determinant focus.

The determinant properties of the focus in zone 1 cannot be associated with its localization in the orbitofrontal cortex, although the prefrontal cortex apparently plays an important role in generating epileptic activity (Sidney, 1972; Bancaud et al., 1974; Karlov, 1974; Ludwig, Ajmone-Marsani, 1975). After the extirpation of the determinant focus in zone 1 and the breakdown of the whole complex, a new powerful focus was produced by applying a 3 per cent strychnine solution or a strychnine crystal to another zone of the cortex where one of the dependent foci was located. This engendered a new complex of epileptic activity which operated in the mode imposed by the new determinant focus (Figure 6 E).

In other investigations (Kryzhanovsky et al., 1978b, 1978c), a focus with a high level of excitation, which acted as a determinant, was produced from the very beginning not in the orbitofrontal region, but in other zones, such as the temporal cortex (Figure 7). In this case, the same relationships were observed: the determinant focus united other foci into a single functional complex and established the pattern of its activity. When the determinant focus was sup-

---

5 min after applying 6% nembutal to zone 3; suppression of the dependent focus in that zone, maintenance of the complex. Strychnine was not applied to the anterior sigmoid gyrus. Amplitude: 500 µV; time: 1 s.

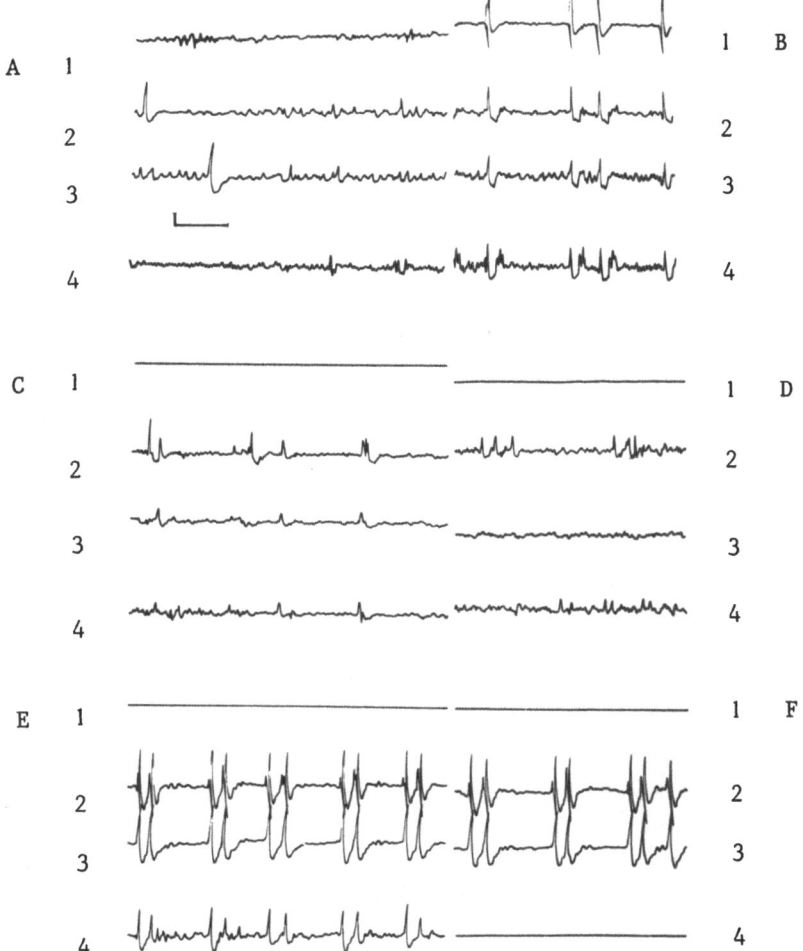

Fig. 6. Effect of removing the determinant and dependent foci on the complex of epileptic activity. A: Formation of the foci of enhanced excitability in zones 2-4 by subconvulsive strychnine application (0.01% solution). B: Application of strychnine to zones 2-4 was ceased, a determinant focus was produced in zone 1 during the application of crystal strychnine, and activity was synchronized in all the foci. C: 2 min after the extirpation of the region of the determinant focus 1; D: one hour after its extirpation; breakdown of the complex of epiletic activity. E: Formation of the determinant focus in zone 2 during the application of crystal strychnine and the repeated formation of the complex of synchronized epileptic activity (zones 2,3,4). F: 10 min after the extirpation of the focus in zone 4; preservation of the complex. The indications are the same as in Figure 5.

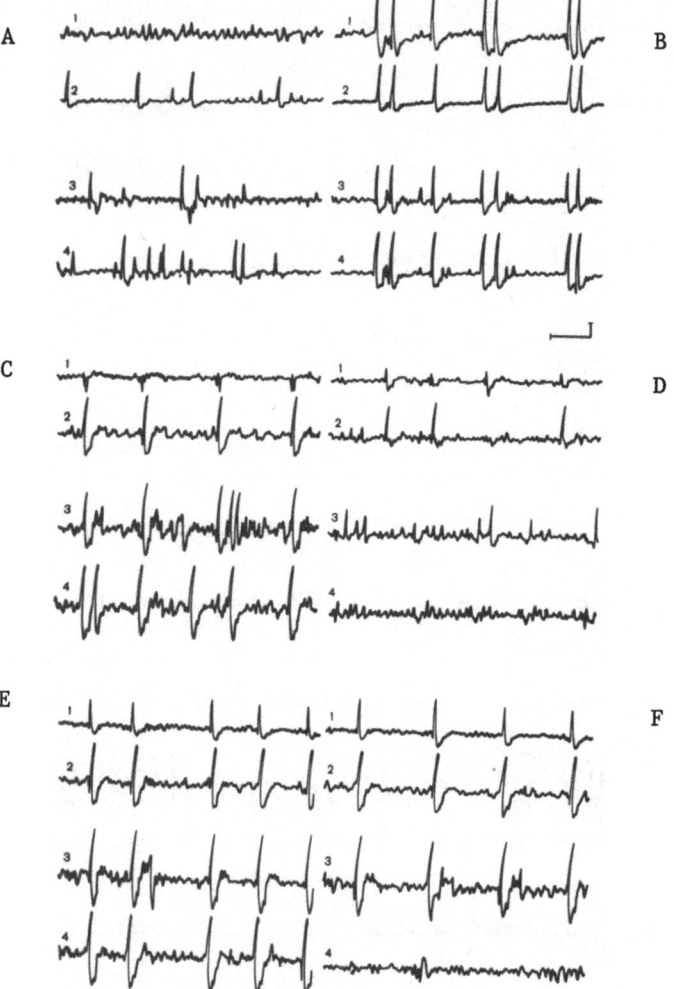

Fig. 7.   Influence of the determinant focus produced in the cortex on
          the formation and behavior of a set of foci in various parts
          of the cortex. A: Formation of subepileptic foci in zones
          2-4 during the application of weak (0.01%) strychnine;
          strychnine application was ceased after the appearance of
          activity. B: Formation of the determinant focus in zone 1
          when 3% strychnine was applied, and the synchronization of
          epileptic activity in all the foci. C: 1 min after applying
          6% nembutal to zone 1; D: 5 min after nembutal application;
          suppression of epileptiform activity in the determinant
          focus and the breakdown of the focal complex. E: Repeated
          formation of the determinant focus in zone 1 during the
          application of 3% strychnine after nembutal was washed out
          from it, and the repeated formation of the complex; increase

pressed, the whole complex broke up and its foci became autonomous. These foci began to work independently of one another in their own way.

The conditions of the formation of the complex were made more complicated by producing a determinant focus in the cortex of one hemisphere and subordinated foci in the other cortex of the other hemisphere (Makulkin et al., 1978). The formation and behavior of such a system with foci in both hemispheres were established by the determinant focus, while the suppression of its activity led to the breakdown of the whole complex. An analysis has shown that the determinant focus exerted its influence on foci of the contralateral hemisphere through the 'mirror' focus in the symmetric zone of this hemisphere. The whole complex broke up and its foci became autonomous when the appropriate part of the callosus body was transected. Such transection of this body disconnected the determinant focus from the 'mirror' and other foci in the other hemisphere (Kryzhanovsky et al., 1976a). A noteworthy fact is that when the complex as a whole exists for a rather long time before the callosus body is transected, it does not break up at once after transection. The 'mirror' focus played the role of a determinant focus for some time. But afterwards the complex broke up. Such functional-temporal relationships between the determinant focus and the whole complex were observed also when all the foci were localized in one hemisphere (Kryzhanovsky et al., 1977c, 1978c).

The properties of the determinant focus as the leading functional structure which establishes the origin of the complex and its behaviour were clearly expressed in that fact that the complex which disappeared after the determinant focus was suppressed had reappeared when the determinant focus was reactivated (Figure 5 E). In this case, the whole complex reappeared when only the determinant focus was reactivated without the additional enhancement of excitability in other foci, and all the foci in the complex functioned in a single mode.

Investigations have also shown that the activity of the epileptic complex became less intense with time and that the complex disappeared with special influences being exerted on it, passing into a latent state; but when the excitability of the brain was generally enhanced (e.g. by systematically administering bemegride), the deter-

---

and synchronization of epileptiform activity in all the foci. F: 5 min after applying 6% nembutal to zone 4. Suppression of the epileptic focus in that zone and the maintenance of the complex of epileptic activity. 1: Anterior ectosylvian gyrus; 2: coronary cortex; 3: anterior signmoid gyrus; 4: posterior sigmoid gyrus. Amplitude: 500 µV; time: 1 s.

minant focus was the first to originate and become activated, and
then synchronized activity appeared also in other foci, as a result
of which a single functional complex was formed again (Figure 8).
Conversely, the determinant focus was the last to be inhibited when
the excitability of the brain was generally suppressed (e.g. in the
case of narcosis). At first, other foci of the complex disappeared.
As a rule, the foci which were influenced least of all by the determinant focus disappeared earlier (Figure 8 H I; see also Chapter 13.)

    Investigations involving an isolated cortex have shown that the
above regularities of the of the relationships between the foci of

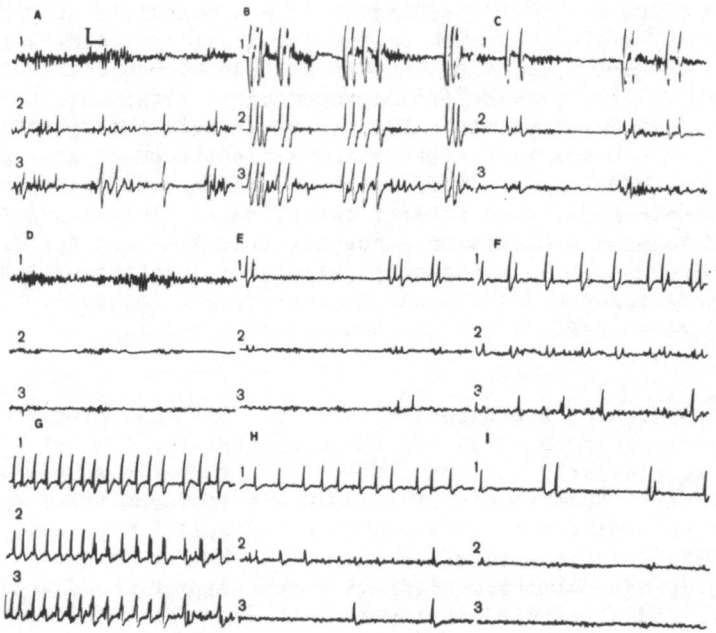

Fig. 8. Effect of bemegride and narcotan on the latent and active
         complexes of epileptic activity. A: 20 min after applying
         0.01% strychnine to zones 2,3 (foci of subconvulsive ac-
         tivity); spindle activity in zone 1. B: 2 min after applying
         3% strychnine to zone 1; formation of the determinant focus
         in this zone; formation of the epileptic complex with syn-
         chronized epileptic activity in all the foci. C: 12 min
         after washing out strychnine from zone 1; D: one hour after
         washing it out; epileptic activity is absent; instead, there
         is spindle activity. E: 5 min after i.v. injecting 0.6 ml of
         bemegride. F: 8 min after the initial administration of
         bemegride. 0.8 ml of it was injected. G: 10 min after its
         initial administration. H: 1 min after the inhalation of
         narcotan. I: 2 min after its inhalation. Other indications
         are the same as in Figure 5.

enhanced activity and DS could be realized by their own cortical
mechanisms. The subcortical structures modulate these mechanisms
(Kryzhanovsky et al., 1979b, 1980b).

Thus, the formation and elimination of the functional complexes
show that the determinant structure plays an essential role in organizing a dynamic system in the cerebral cortex and in conditioning
both its behavior as a whole and the pattern of the activity of its
constituents.

REALIZATION OF THE EFFECTS OF THE DETERMINANT AND
ITS INTERACTIONS WITH THE SUBORDINATE STRUCTURES

The question of whether the above-mentioned relationships
between the determinant focus and subordinated foci depend on their
nature inevitably arises when the effects of the determinant structure are being studied. In the above investigations, all the foci
were produced by the same agent, i.e. strychnine, which is known to
disturb glycine postsynaptic inhibition (see Chapter 5). Will such
relationships exist between the determinant focus and other foci if
their nature will be different, i.e. if they will be produced by
different agents? To answer this question, experiments were carried
out. Besides strychnine, use was made of penicillin, which disturbs
inhibition caused by gamma-aminobutyric acid (GABA) and which probably depolarizes the neuronal membrane (see Chapter 5), and also
acetylcholine, which is known to depolarize the cortical neurons.

Investigations have shown that the above-mentioned relationships
between the determinant foci and the subordinate foci are observed
also when the foci are of a different nature, i.e. when they are
produced by different agents and are based on different pathogenetic
mechanisms. For instance, the 'strychnine' determinant focus enhances and synchronizes the potentials in the 'penicillin' foci,
unites these foci into a single complex, and determines the nature of
its activity (Figure 9 A B). Such an effect is produced also when
the determinant focus is formed by penicillin, and the subordinate
foci, by strychnine (Figure 9 C D). Subordinate foci can be produced
in various regions of the cortex alternately with penicillin and
strychnine, while the determinant focus, alternately with strychnine
and penicillin. In this case, the determinant focus exerts potentiating and determining influence on the subordinate foci. Such
influence is ultimately realized in the form of a single functional
complex which is made of all foci and whose work is determined by the
determinant focus (Figure 9 E F).

Especially interesting relationships originate between the
determinant focus and the subordinate foci when the former is produced by the application of acetylcholine (ACh) with proserine on the
cortex. When ACh with proserine is applied to the cortex, including

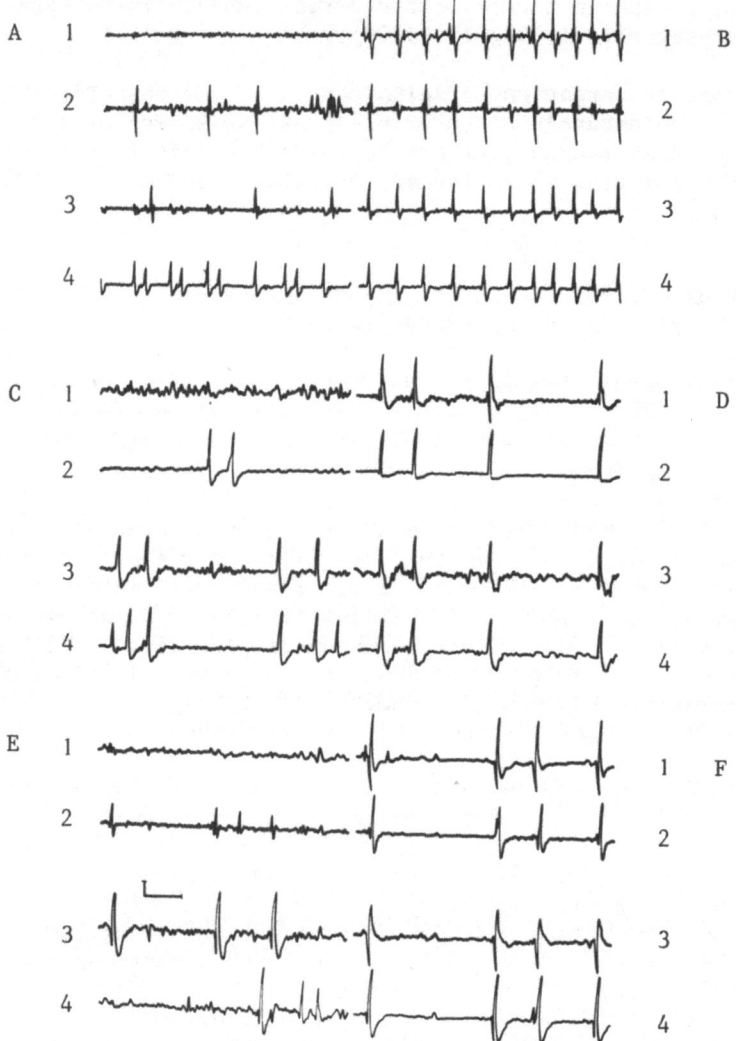

Fig. 9. Functional relationships between foci produced by strychnine
         and penicillin in various areas of the cerebral cortex.
         A,B: Experiment No. 1; A: formation of subepileptic foci in
         zones 2-4 by applying 0.25% penicillin sodium salt; penicil-
         lin application was ceased after the appearance of activity;
         B: 10 min after the formation of the determinant focus in
         zone 1 when crystal strychnine was applied; synchronization
         of epileptic activity in all the foci; C,D: experiment No.
         2; C: formation of subepileptic foci in zones 2-4 when
         0.025% strychnine was applied; D: formation of the determi-
         nant focus in zone 1 when penicillin powder was applied;
         synchronization of epileptic activity in all the foci. E,F:

an isolated strip of the cortex (Kristiansen, Courtois, 1949), a characteristic reaction occurs in the form of at first a high-amplitude hypersynchronized discharge, and then prolonged low-amplitude asynchronized afterdischarges. This activity greatly differs from the activity which is produced by strychnine and penicillin (Penfield, Jasper, 1954). Thus, the differences in the activity of the foci can be clearly seen, the influence of the determinant focus can be established, and the specifics of the formation of the complex and the interaction between the foci can be observed by using ACh for producing the determinant focus and strychnine or penicillin for creating subordinate foci (and vice versa).

Investigations (Kryzhanovsky et al., 1979c) have shown that the determinant focus, which is produced by the application of ACh and proserine to the cortical regions, behaves typically with respect to other foci which are induced by the application of strychnine or penicillin: it establishes the pattern of their activity and unites them into a complex, for which it sets a definite mode of activity. This effect is produced regardless of the cortical zones in which the determinant focus and subordinate foci are created (Figure 10). In the case of complete determination, the subordinate foci not only act in the mode set by the determinant focus, but also reproduce the activity patterns which are characteristic of ACh. Then, strychnine or penicillin activity which characterize these foci either disappears or greatly changes. When the origin of the complex is unknown, it may be assumed that all its foci are produced by ACh. However, characteristic strychnine or penicillin activity is restored in the formerly subordinate pseudoacetycholine foci when the determinant ACh-focus is either pharmacologically suppressed or ablated (Figure 10 D). Hence, 'strychnine' or 'penicillin' activity in the subordinate foci either had been suppressed or had substantially changed when the foci began to constitute the complex. It follows that the determinant focus not only imposed a new type of activity on the subordinate foci, but also suppressed and changed the pattern of their work. This new type of activity was realized even though the effect of penicillin or strychnine remained in the foci.

The dynamics of both a change in the activity of the foci and the formation of a complex is noteworthy. It can differ from case to case, but common regularities are also revealed. At first, 'strychnine' or 'penicillin' activity is replaced by the imposed 'acetyl-

Experiment No. 3; E: formation of subepileptic foci in zones 2 and 4 when 0.01% strychnine was applied, and in zone 3 when 0.25% penicillin was applied; F: formation of the determinant focus in zone 1 when crystal strychnine was applied, and the formation of the complex of foci of epileptic activty. Amplitudes: 500 μV (A,B), 100 μV (C-F). Other indications are the same as in Figure 7.

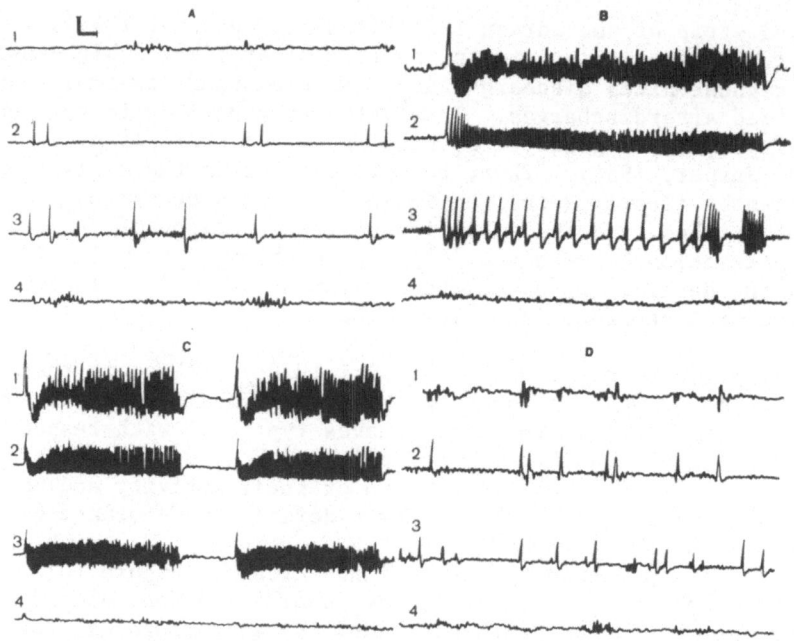

Fig. 10.   Formation of the 'acetylcholine' complex of drug-induced
foci in various parts of the cerebral cortex under the
influence of the determinant 'acetylcholine' focus.   A:
Formation of subconvulsive foci in zone 2 (application of
0.5% penicillin sodium salt) and zone 3 (application of
0.1% strychnine); drugs were no longer applied when activity appeared; no drugs were applied to zones 1 and 4. B:
1 min after applying 0.5% prozerine and 10% acetylcholine
to zone 1; initial stage of the formation of the determinant 'acetylcholine' focus in zone 1. C: 3 min after applying acetylcholine and prozerine; formation of a combined
'acetylcholine' complex. D: 2 min after applying 6%
nembutal to zone 1; suppression of the determinant acetylcholine focus and the breakdown of the complex.   Other
indications are the same as in Figure 5.

choline' activity in the focus, which is, for some reason, influenced
more by the determinant focus (Figure 10 B 2). At the same time, the
focus which is more resistant retains its specific ('strychnine')
activity, but it changes quantitatively, i.e. the potential generation frequency increases (Figure 10 B 3). At this stage, the determinant focus acts as a nonspecific activator with respect to the
given focus: it does not impose its own mode of activity, but merely
enhances the activity in that focus. Later, the specific components
of the influence of the determinant focus become manifest: strychnine
or penicillin potentials can be generated in the form of group dis-

charges whose duration corresponds to the acetylcholine discharge in the determinant focus. Finally, strychnine activity itself disappears at the last stage in all the foci, including the relatively resistant focus (Figure 10 C 3), while prolonged afterdischarges, which are typical of the determinant ACh focus and which are preceded by a high-amplitude hypersynchronized spike, originate instead. The complex is then completely formed: all the foci operate in a single mode, and the nature of the initial hypersynchronized spike, the duration of the common prolonged afterdischarge, and the frequency of the generation of potentials in every burst is the same in all the foci. A single functional complex operating in the 'acetylcholine' mode is now formed. An interesting fact is that the complex does not contain intact cortical zones which were not treated with either strychnine or penicillin and, consequently, which do not exhibit enhanced excitability.

The above ACh properties of the focus make it possible to clearly see the manifestation of different degrees of establishment of the activity of the subordinate foci which are united into a complex under the determinant's influence. Figure 11 gives an example of such relationships between the determinant foci and other foci of a complex. The influence of the determinant foci (zone 1) is realized differently in the subordinate foci (zones 2 and 3), although all foci are united into a complex: the duration of the afterdischarge and the frequency of the generation of potentials in the afterdischarge are the same in all foci. A high-amplitude hypersynchronized spike discharge is also generated in all the foci. However, the form of the potentials differs in the afterdischarge: the potentials are monophased in the determinant ACh focus. Varying positive deviations occur with respect to the high-amplitude negative spikes of the first subordinate 'strychnine' focus, while complete two-phased spikes typical of strychnine discharges are recorded in the second subordinate 'strychnine' focus.

Thus, the extent of the establishment of various functional structures which are subordinated to the determinant can differ. A 'hybrid' activity pattern reflecting the functional peculiarities of both structures can originate in subordinate foci as a result of the interaction between the activity of the structure that realizes the determinant's influence (hence, this structure acts as the 'station of destination') and the activity imposed by the determinant. An interesting fact is that such 'hybridization' is expressed more in the focus which is further away from the determinant. It can be concluded from the experiments involving the production of a determinant focus and subordinate foci in various cortical zones that the closer is a subordinate focus to the determinant focus in the system of the same connections, the more are the influences of the determinant focus realized and the less is the 'hybridization' of activity manifested in the subordinate foci. 'Hybridization' can be expressed in the activity of also the determinant focus when the influences of

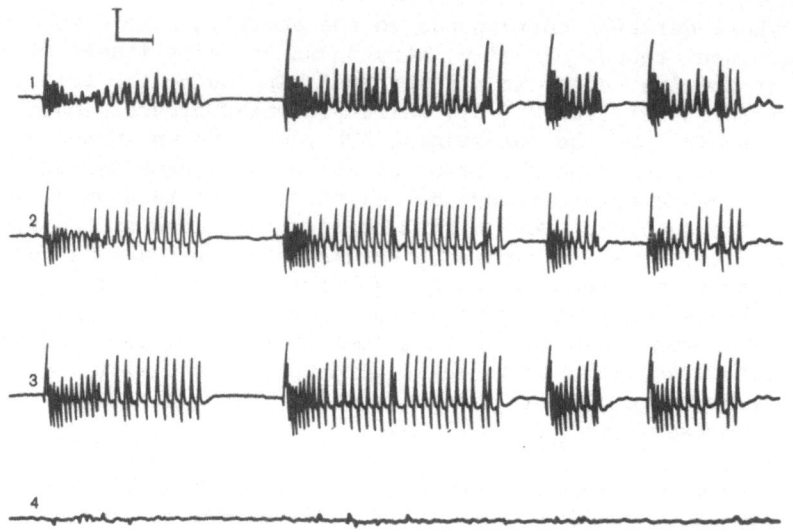

Fig. 11.   Electric activity in the acetylcholine- and strychnine-
           induced foci during the formation of complexes. Activity
           in the foci was recorded two minutes after applying 0.5%
           prozerine and 0.5% acetylcholine to zone 1.   Zones 2 and 3
           were treated beforehand with 0.1% strychnine.   Zone 4 was
           drug-free. 1: Orbital cortex; 2: coronary cortex; 3: an-
           terior sigmoid gyrus; 4: posterior sigmoid gyrus.   Ampli-
           tude: 500 μV; time: 1 s.

the subordinate focus are enhanced due to some causes. In such
cases, it is difficult to say which focus is the determinant one.
Figure 11 1 shows that such 'hybridization' is exhibited in the
activity of the determinant focus, particularly in the afterdischarge
pattern.

     The results of the investigations involving the production of
functional complexes in the cerebral cortex have shown that, under
equal conditions, excitability must be enhanced to a definite extent
(this was done by applying weak solutions of convulsive agents to the
relevant regions of the cerebral cortex) so as to allow the determinant's effects to be fully exhibited.

     At the same time, the determinant's functional message was
completely realized by motoneurons and interneurons of different
spinal cord segments and the brainstem, although they were not
treated by convulsants. Tests of monosynaptic reflex inhibition in
the efferent output system of the corresponding spinal segments which
receive impulses from hyperactive DS have shown that the state of
inhibitory control in this system is preserved: the reflex inhibition
curve is the same as the normal inhibition curve (Kryzhanovsky,
Lutsenko, 1975) (Figure 12).

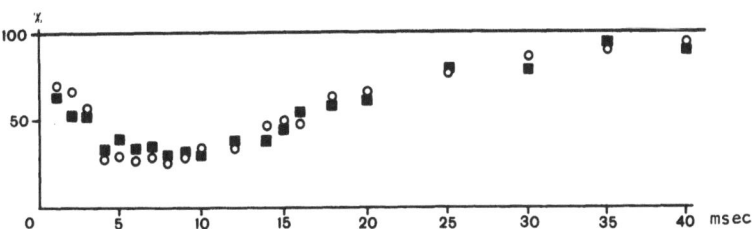

Fig. 12. Inhibition of the lumbar extensor monosynaptic reflexes in
          animals with a pathologically enhanced scratch reflex
          caused by the formation of hyperactive determinant struc-
          ture in the ventral parts of the spinal cord. Inhibition
          of the monosynaptic reflexes in the root L_6 in healthy
          animals (circles) and in animals with a pathologically
          enhanced scratch reflex (squares). Abscissa: the interval
          between conditional and test stimulation. Ordinate: ampli-
          tude of the test monosynaptic reflex (% of the control
          level).

It has already been noted that the animals in which a generalized seizure was provoked by activating the hyperactive determinant
structure in the spinal cord (the 'universal dispatch station'
phenomenon) had made the usual motions, ate, and so forth, during
interictal periods. Hence, the spinal cord structures that realize
that the message from hyperactive DS during a seizure are not damaged
and can function normally as links of the corresponding physiological
systems under ordinary conditions.

The functional state of the structures which realize the influence of hyperactive DS was studied also by taking the example of
intraspinal descending inhibition. In these experiments, hyperactive
DS was produced in the brachial parts of the spinal cord in the
scratch relief system, while the lumber flexor monosynaptic reflex,
which participates in scratching, acted as the actuating structure.
Normally, the stimulation of the radial nerve produces a two-phased
changed in this monosynaptic reflex, namely, initial inhibition and
subsequent facilitation. In principle, such a nature of the curve of
the descending inhibition and facilitation of this reflex was
recorded during the activation of hyperactive DS caused by the ES of
the radial nerve, the only difference being that both the suppression
and the facilitation of the reflex were greatly augmented (Figure
13). Thus, the system of the segmental apparatus which regulates the
lumber flexor reflex, realizing the functional message from hyperactive DS in the brachial parts, is not changed itself and does not
produce a distorted reaction to this message. The excessiveness of a
reaction is determined not by a change in the given system's functional state, but by the intensity of the functional message produced
by hyperactive DS.

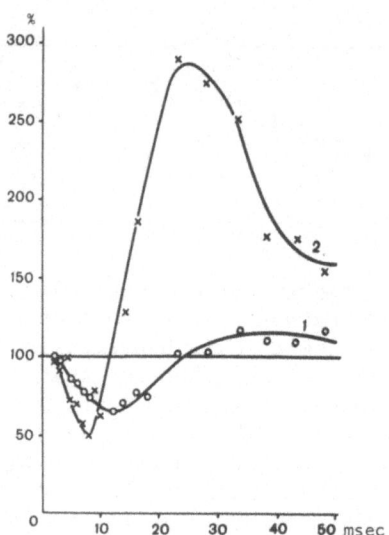

Fig. 13. Descending spinal inhibition and the facilitation of the
          lumbar flexor reflexes by stimulating the radial nerve in
          healthy animals (1) and in rats with a pathologically
          enhanced scratch reflex (2). A hyperactive determinant
          structure was formed in the ventral parts of the spinal
          cord in rats with a pathologically enhanced scratch reflex.
          Ipsilateral cordotomy – at the level S_2. The test stimulus
          was applied to the fibular nerve, and the monosynaptic
          reflex was registered in the ipsilateral root L_5. The
          strength of conditional electric stimulation of the ipsi-
          lateral radial nerve was 4 thresholds. Abscissa: the
          interval between conditional and test stimulation. Ordi-
          nate: amplitude of the test monosynaptic reflex (% of the
          control level).

     The data obtained show that the effects of the determinant can
be realized at the spinal level when the structures through which its
influence is exerted are functionally preserved with no additional
influences aimed at enhancing the excitability of their neurons. In
this case, the determinant rigidly imposes the pattern of its activity on the subordinate structures. To form functional complexes
in the cerebral cortex under the influence of the determinant, it was
necessary to weaken inhibitory control and to enhance the excitability of the neurons in the cortical regions which realize the
influence of the determinant focus, i.e. the 'stations of destination' being formed. These differences in the relationships between
the determinant and the subordinate structures in the spinal cord and
the cerebral cortex, observed in the above experiments, can be attributed to the peculiarities of the functional and anatomic relationships in the given parts of the central nervous system and to the
power of the DS which were formed.

The power of the functional message, being produced by the determinant structure, is obviously indispensable for realizing its influence. This regularity has been clearly revealed in the relationships between the foci in the cerebral cortex during the formation of functional complexes, when the focus induced by stronger epileptogenic influence became the determinant focus. In this respect, it should be noted that the unification of separate excitation foci into a single functional complex and the synchronization of their activity were clearly revealed in the case of slight nembutal narcosis, when synaptic conduction is impeded and the inhibitory mechanisms are strengthened.

The significance of the power of excitation produced by DS for realizing its effect becomes clear when the action of this excitation is taken into account. To produce a reaction with definite parameters, impulses propagating from DS should above all overcome the multisynaptic pathway. The number of synapses is very large in the system if the short-axonal propriospinal relationships in which DS originates when the above-mentioned generalized seizure (the phenomenon of the 'universal dispatch station') occurs. The short-axonal neurons, which connect the adjacent segments in the spinal cord, come into contact with many neurons, but the number of synaptic contacts with every neuron is not large (Szentagothai, 1964). The spatial and temporal summations of excitation on the neuron are especially important when such a type of relay organization exists. Hence, the excitation wave should normally become extinguished near its origin when a small amount of neurons is being excited.

That effect has been shown by taking the example of the extinction of the excitation wave within the segments which are the closest to the afferent input in the spinal cord (Gernard, Megerian 1961). For systems with probable relationships, it is necessary to excite a definite critical number of neurons so as to allow excitation to propagate throughout the system without it being extinguished near its source (Beurle, 1956). Therefore, the power of the functional message produced by a hyperactive structure, which acts as a determinant, in the realization of the whole reaction is very important.

However, the satisfaction of that requirement alone is not enough for allowing a reaction to occur in a set mode. The impulses which are generated by the DS neurons should have predominant 'synaptic weight' on the neurons of the receiving structures, on which excitations from various sources converge. As it has already been shown, these impulses should also overcome the resistance of the local mechanisms which regulate the activity of the neurons. In the case of multiple convergence and polymodal excitations on the receiving neurons, the 'synaptic weight' of every impulse plays a great role in determining a neuronal reaction. The dominant impulse action is that which has the greatest 'synaptic weight' at a given moment. This 'weight' depends on the intensity of impulse action and is realized by the mechanisms of spatial and temporal summations. The dynamic relationships between the synaptic influences coming to the neuron are decisive in the production of the synaptic input effect (Kostyuk, 1974). The relationship between the 'synaptic weights' of the converging heteromodal channels on the neuron apparently largely determines their blocking (Thompson et al., 1963; Andersen, Curtis, 1964; Stephanis, Jasper, 1964; Orlov, Pirogov, 1975; Batuev, 1978). It has been shown on the basis of neuronal models that the distribution of impulse intervals largely depends on the level and pattern of the incoming signal (Kotov, Tseitlin, 1966). Hence, the power of the incoming functional message is decisive in determining the behavior of the neurons of the receiving structures ('station of destination'). Therefore, this behavior is determined distantly, i.e. from the side of DS, when all other conditions are equal.

In this respect, it should be recalled that Sherrington (1906) was the first to formulate and demonstrate the concept of the decisive significance of the strength of the afferent reaction of the competing reflexes in the struggle for the possession of the common final pathway. Investigations involving the electric stimulation of the brain structures show that when two reactions conflict with one another, the reaction produced by stronger stimulation is realized. This concept applies also to the situation when one reaction is produced by natural stimulation, while the other, by ES (Delgado, 1969). An animal can counteract ES and suppress its effect if it is produced by a weak current (Hess, 1957; Delgado, 1969).

We have observed in the model experiments involving the ES of Deiters' nucleus that the realization of the reaction and its peculiarities depend on the intensity of the functional message produced by the structures which are determinant, i.e. weak ES causes various movements in the form of a turn of the head alone, the movement of the head and the forelegs, the movement of the head, the forelegs and the body, etc. When stimulation is strong, the animal continuously turns over lengthwise, just as in the case of vestibulopathy (see Chapter 8). Such a dependence of the graduality of a reaction on the force of stimulation of brain structures was described by many authors (Hess, 1957; Delgado, 1969).

The above-mentioned relationships between the impulses which go to the neuron from different sources and the different 'synaptic weight' being produced are realized not only in the region of the terminal pathway, but also at the intermediate stages. The functional message from DS should dominate on the neurons of all relays. Only then can its influences be realized.

However, the power of the functional message produced by DS is not always enough for realizing the determinant's influences. This is especially clearly manifested in systems with a distinct stochastic form of activity, particularly the cerebral cortex. It can be seen from the above-mentioned data that when a determinant focus is being engendered in the cerebral cortex, its effect is usually displayed in the cortical regions which were treated at first with a weak strychnine or penicillin solution, i.e. in the regions where the inhibitory mechanisms were disturbed and the thresholds of neuronal excitability were lowered; the influences of the determinant focus were not exhibited in the regions which were not subjected to such action (Figures 5-10). Activity synchronized with the discharges in the determinant focus was recorded in the previously untreated areas only when the brain's general excitability was enhanced or the determinant focus functioned for a long time.

When a complex of epileptic activity is formed in the cerebral focus, impulses from the determinant focus go to various areas of the cortex. The fact that secondary foci originate only or mainly in the cortical areas where inhibitory control is disturbed and neuronal excitability is enhanced gives an idea about the mechanisms of the 'predisposition' or 'preparedness' of the cerebral structures to be involved in producing epileptic activity and secondary foci under the influence of the determinant, and also about the factors which promote the development of a pathologic process or, to put it figuratively, its motion in one direction or another.

It follows that the determinant can fully realize its influence only when a definite relationship exists between the intensity of the flow of impulses which are produced by the determinant structure as the generator of excitation, on the one hand, and the level of excitability and the state of the inhibitory control of the neurons of the realizing structures, on the other. These relationships vary in diverse systems with a different extent of rigidity of their functional links. The functional message from the determinant structure that is not intensive can be more or less fully realized when the thresholds of excitability are greatly lowered and the inhibitory control of the neurons of the realizing structures is weakened. However, investigations have shown that the receiving structure may not be subordinated to the influences of the determinant and may function independently when the inhibitory processes are greatly disturbed and the level of its own activity is high. Moreover, such a structure can act as a determinant when it influences other formations of the central nervous system.

FORMS OF DETERMINATION

The above-mentioned experimental material and other data on both the effects of the determinant and the peculiarities of their realization show that there are three main forms of determination.

Form I (complete determination)

In the case of this form, all the parts of the system which realize the functional message from the determinant reproduce the determinant's activity pattern without any great distortions, and the whole system functions in the mode set by the determinant. This form is expressed most clearly when DS is hyperactive and functions as GPEE and/or when the mechanisms of inhibitory control are upset, the regulation of own activity is changed, and neuronal excitability is enhanced in other parts of the system which realize the message from the determinant. This form of determination can be realized in systems with relatively rigid relationships as well as systems with mainly stochastic relationships. Examples of this form of determination are both the synchronized activity of all the spinal moto-neurons, when DS in the system of propriospinal neurons (Figures 1-4) is activated, and the activity of the epileptic complex in the cerebral complex induced under the influence of the determinant focus when the control mechanisms are additionally upset and neuronal excitability is enhanced in other cortical areas (Figures 5-10). This form of determination can be realized by preliminarily suppressing or directly changing the activity of the subordinated structures. An example of such effects is given in Figure 10. Under pathologic conditions, the complete dependence of the activity of the realizing structure is clearly exhibited in systems with rigid relationships in linearly organized nuclei. Hyperactive structures in such nuclei engender a syndrome whose peculiarities are determined by the way in which the structure functions. Such relationships are clearly revealed, for instance, in the case of vestibulopathy caused by the creation of GPEE in Deiters' nucleus (see Chapter 8).

Form II (incomplete determination)

In the case of this form of determination, the system's parts influenced by the determinant realize not all the peculiarities of its functional message. The pattern of their activity and the mode of the system's behavior only partially reflect the determinant's properties. This form of the determinant is realized when hyper-active DS as GPEE produces relatively low excitation power and/or when the mechanisms of inhibitory control and the regulation of own activity are slightly disturbed and neuronal excitability is slightly enhanced in other parts of the system which realize the message from the determinant. The extent to which the determinant's effects are realized, i.e. the extent of determination, is largely conditioned by the relationships between the determinant's power, on the one hand, and the extent of the changes in the above parameters of the realizing structures, on the other. The general level of the brain's excitability and the effectiveness of integrative control also apparently play an important role in this process. Examples of this form of determination are the incomplete realization of the determinant focus's influence on other foci and the 'hybridization' of the activity patterns of the determinant focus and other foci (Figure 11). The transitional stages of the system's formation under the determinant's influence can be regarded as incomplete determination. Such a form of determination can be expressed diversely at different stages of the evolution of subordinate relations between the determinant and other structures.

Form III (system's activation without the imposition of the pattern of the determinant's activity on it)

In the case of this form of determination, the activity of the functional structures or the system influenced by the determinant merely grows without any changes in its pattern. In such cases, hyperactive DS acts as the modulator of the system's activity. Strictly speaking, it is not a determinant. Such a structure is a generator of excitation that merely activates the system. However, this activator has its own distinguishing features, e.g. it can be naturally and artificially connected with a given system and its influence is realized through the system's special mechanisms. Normally, such a form of determination is exhibited in the functions of the activating structures in various areas of the central nervous system. Such functional relationships are pathogenetically significant, as can be seen from the fact that the functions of the activating structure do not add any unusual features to the system's work and merely intensify its specific activity, but this intensification may be so great that the system's functions became pathologic. Such relationships were produced in models involving the growth of own activity in foci by applying weak convulsant solutions to the cerebral cortex. This activity grew under the influence of a more powerful focus, which enhanced excitation in the foci but did not impose a new pattern of activity on them and, consequently, did not become a determinant. These relationships are usually observed at the early stages of the formation of a complex under the determinant focus's influence.

It will be seen further that the given relations are very clearly exhibited also in the pathologic state of the brain's complex systems. For instance, the stereotyped behavior syndrome, which consists of hyperbolic fragments of species-ecologically determined behavior, is observed in animals when the dopamine system is hyperactivated in the neostriatum, one function of which is to modulate the activity of the motor sphere (see Chapter 10). When GPEE is produced in the orbital cortex, which modulates the hypnogenic system's activity, the cortex is activated, and pathologically long sleep occurs without the disturbance of its inner structure. In this case, the relationships between the phases of slow-wave sleep and paradoxical sleep remain (see Chapter 11).

The division of determinant into the given main forms is somewhat conditional. Each of these forms can be independent, but it can
also represent a stage of a process and be transitional. As it has
been shown, the significance of the types of determination or their
expression as the transitional forms of a process depends on the
peculiarities of a system in which DS originates, on the power of DS
as the generator of excitation, on the relationships between DS and
the structures which realize its influence, and on the state of
inhibitory control and the neuronal excitability level of the
realizing structures. Since the brain's general excitability and the
state of its integrative control play a big role in the determination
of the subordination relations between DS and the realizing structures, both of these factors can also considerably influence the
degree of determination. In Chapter 9, we will see the significance
of the general level of brain excitation and the balance of the
systems of cerebral electrogenesis in the realization of the effects
of hyperactive DS in generalized epileptic activity (EpA). It has
been established in the investigations involving brain sections at
different levels (cortex isolé, hemisphere isolé, cerveau isolé) that
the epileptic complexes are formed more easily and epileptic activity
in the cerebral cortex is generalized when the structures which
stabilize cerebral electrogenesis and control cortical excitability
are arrested (Kryzhanovsky et al., 1980b). Studies of epileptics
with implanted long-life electrodes have shown that the excitability
of a large number of cerebral structures increases, the conduction of
stimuli is facilitated, and the number of the functionally active
interstructural connections grows as the seizure period draws nearer
(Kambarova, 1977; Bekhtereva et al., 1978).

In this respect, an assessment can be made of the experiments
involving the ES of the cerebral structures, in which the activity of
the determinant produced by this stimulation is in essence modulated.
In narcosis, when the activity of many systems is suppressed and it
is difficult to draw them into a reaction, the stimulation of the
brain structures produces the same permanent effect which corresponds
to the physiological role of these structures. Thus, the reaction is
realized more or less in the same manner. However, when the animal
is wide awake and behaves freely, the reaction may be greatly changed
and even suppressed (Delgado, 1969). The ES of the median center
augments fear in a dangerous situation and increases relaxation in
safety (Kopa et al., 1962). Such data are especially difficult to
interpret because the thresholds of excitability of different structures in the ES region may change when the animal's state alters.
However, it is highly probable that the excitability of the realizing
structures in various systems changes and that the dominating system
is the most reactive one. In this case, we have determination of
type III, i.e. nonspecific modulation caused by the activation of the
system by GPEE.

DETERMINANT AND DOMINANT RELATIONS

The relations between the foci of hyperactivity in the cerebral cortex can develop in conformity with not only the determinant type, but also the dominant type (Kryzhanovsky et al., 1982). According to the dominant principle, which was described by Ukhtomsky as early as the 1920's, the excitation of the nerve center or the activation of a physiological system is followed by coupled inhibition of other centers and other systems. In this case, the centers which are the first to be inhibited are those that can interfere with a given excited center or a given active system. Such inhibition optimizes the conditions for the full activity of a system which is active at a given moment.

An example of the dominant relations between the foci of hyperactivity in the cerebral cortex is given in Figure 14. The formation of a new powerful focus in zone 2 suppresses the activity in the foci induced earlier in zones 1 and 3 (Figure 14 B C). The fact that the suppression of a focus in zone 2 results in the restoration of the activity of the foci in zones 1 and 3 shows that the activity in these foci weakens and disappears due to inhibition caused by the activity of a focus in zone 2 (Figure 14 D).

A very important fact is that dominant relations not only between individual foci, but also between complexes of foci can be exhibited. This phenomenon is interesting, because the functional complexes of foci operating in a unique way can be regarded as simple systems. In this case, therefore, dominant relations between systems are exhibited. Figure 15 A shows that the complex of foci created in zones 5-7 is the dominant one: it suppresses the other complex of foci created in zones 1-4. If the dominant complex (foci 5-7) is suppressed by locally applying nembutal to its foci, the formerly inhibited complex will be disinhibited, i.e. activity will appear in foci 1-4 (Figure 15 B). However, such relations are not confined to this framework. It has been discovered that the complex of foci 1-4 can become a dominant one under new conditions: if nembutal is thoroughly washed away from zones 5-7 and strychnine is applied to them, activity there either will not originate at all or will be rather weak even when the concentration of strychnine is large (Figure 15 B zones 5-7). The data of special control experiments have shown that this effect cannot be attributed to the residual action of nembutal (although such action naturally occurs): similar strychnine application to zones 5-7 after nembutal was washed away from them had engendered activity foci in the zones when foci were not produced in zones 1-4 (the origination of hyperactivity under the influence of strychnine in the cortical region after nembutal is washed away from it is illustrated also in Figures 5 and 7).

Such relations have been observed also on the spinal level: when GPEE on one of the sides in the lumbosacral segments of the spinal

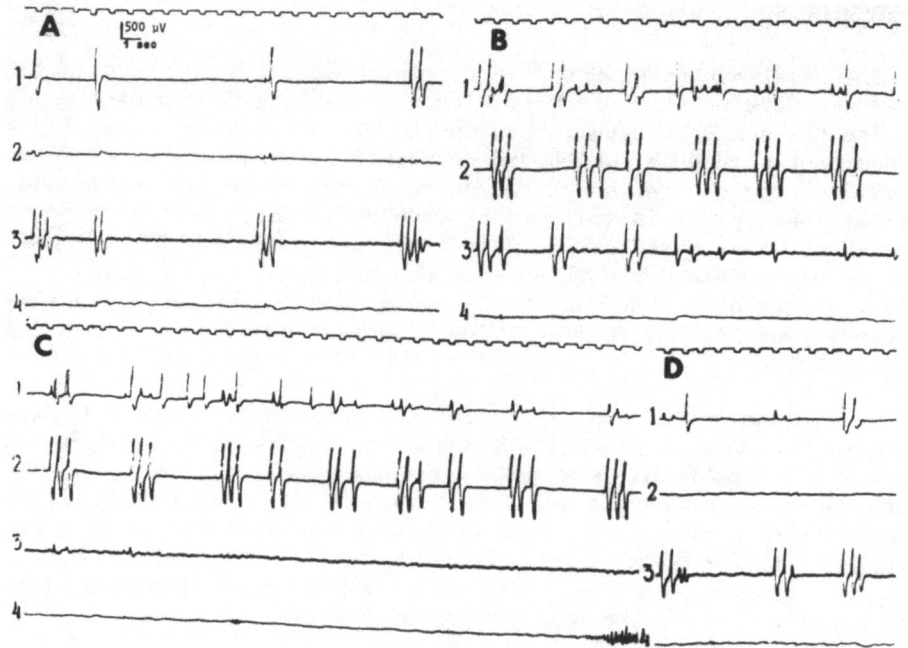

Fig. 14. Determinant relationships between the foci in the cerebral
cortex. A: Separate foci produced by the application of
0.1% strychnine to zones 1 and 3; strychnine was removed
when activity appeared. B: Origin of a focus in zone 2
after the application of 3% strychnine, diminishing activity in other foci. C: Disappearance of activity in zones
1 and 3 after the formation of a focus in zone 2 (4 min
after applying strychnine to zone 2). D: Suppression of
activity in zone 2 by applying 6% nembutal and the restoration of activity in zones 1 and 3. Zone 4 was drug-free.
1: Orbital gyrus; 2: coronary gyrus; 3: anterior ectosylvian gyrus; 4: posterior sigmoid gyrus (right hemisphere).

cord is activated, the activity of GPEE on the other side is suppressed. As a result, synchronized hyperactivity (induced by the
second GPEE) of the motoneurons of the whole spinal cord is eliminated.

Determinant and dominant relations may be exhibited simultaneously when a complex of foci is formed in the cerebral cortex (Figure
16). After the formation of separate foci in zones 2 and 3 (Figure
16 A), the induction of a more powerful focus in zone 1 enhanced the
activity in focus 3 and synchronized it with the activity in focus 1
(determinant relations). At the same time, activity in zone 2 was
suppressed, and it disappeared (dominant relations) (Figure 16 B).

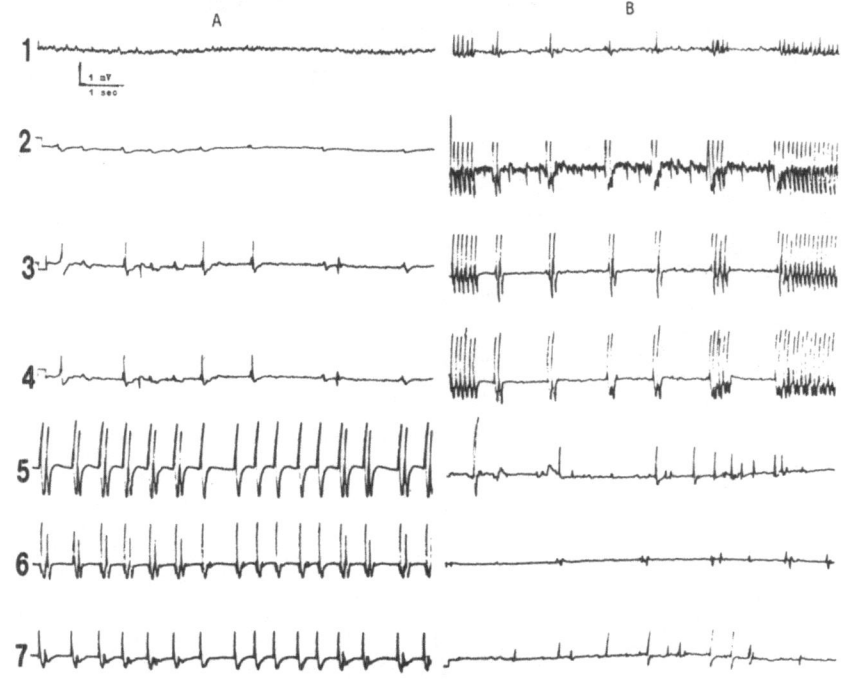

Fig. 15. Dominant relationships between complexes of foci in the
cerebral cortex. A: Complex of foci produced by applying
0.1% strychnine to zones 6 and 7, and 3% strychnine to zone
5; 3% strychnine was applied to zone 4, and 0.1% strychnine, to zones 1-3. Depression of activity in zones 1-5.
B: Suppression of focal activity in zones 5-7 of area
temporalis by locally applying 6% nembutal, and the origination of activity in zones 1-4. 1: Posterior sigmoid
gyrus; 2: anterior sigmoid gyrus; 3: coronary gyrus; 4:
orbital gyrus; 5: middle ectosylvian gyrus; 6: posterior
ectosylvian gyrus; 7: anterior ectosylvian gyrus (left
hemisphere).

In this case, the dominant effect was an expression of the transitional state: at the final stage of the process, hyperactivity that
was synchronized with the activity in the determinant focus (zone 1)
had originated in zone 2, and then a complex was formed out of three
foci with a single pattern of hyperactivity induced by the determinant focus (Figure 16 C). Thus, another way of realizing determinant
relations is exhibited in this situation, i.e. they are realized as a
result of the preliminary suppression of activity in the cortical
region which is involved in the organization of the epileptic complex.

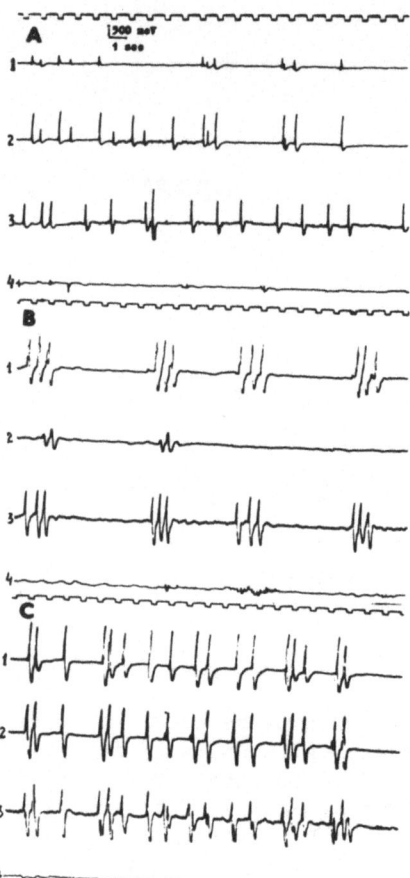

Fig. 16. Determinant and dominant relationships between the foci in
          the cerebral cortex. A: Focal activity in zones 2 and 3
          six min after the application of 0.1% strychnine. B: 2 min
          after the application of crystal strychnine to zone 1 and
          removing strychnine from zones 2 and 3. C: 20 min after B.
          1: Orbital gyrus (right hemisphere); 2: coronary gyrus; 3:
          posterior sigmoid gyrus (left hemisphere); 4: anterior
          sigmoid gyrus (left hemisphere) was free of strychnine
          treatment.

     It is assumed that dominant and determinant relations are always
combined under physiological conditions. Owing to the coupled inhibition of the interfering systems, dominant relations allow the
activated system to carry on its activity and make it possible for a
current reaction to occur without any considerable distortions.
These relations participate in the formation of the functional
channel of a reaction. The determinant relations engender the
peculiarities of the activity of the operating functional system.

Thus, each of the above principles covers various aspects of the activity of the central nervous system: the dominant principle is a principle of <u>intersystemic</u> relations, and the determinant principle is a principle of <u>intrasystemic</u> relations. The functional structure, which acts as a determinant for a given system, can be a dominant with respect to other interfering systems. The combined operation of the principles of the dominant and the determinant make it possible to achieve a programmed result in accordance with the requirements at a given moment.

Dominant relations are important also under pathologic conditions. When a hyperactive pathologic system (see Chapter 4) begins to act as a dominant, it can suppress the activity of the physiological systems, promoting the development of the pathologic start and the disorganization of the cerebral activity. On the other hand, a physiological system on the basis of which a pathologic system is formed can suppress the latter's activity when the former is sufficiently activated (antisystem relationships; see Chapter 13). This mechanism is extremely important in the maintenance of relatively stable homeostasis in recovery and is necessary for the activity of the central nervous system under pathologic conditions.

SOME THEORETICAL AND PRACTICAL ASPECTS

Both theoretically and practically, it is very important to clearly see the difference between the concepts of the 'dominant' structure and the 'determinant' structure.

In epileptology, the primary focus is often called a dominant. However, it follows from what we have seen that this focus not always can be called so. The primary focus does not suppress activity like the dominant. On the contrary, it produces secondary foci and can enhance their activity under definite conditions. At the initial stages, secondary foci can depend on the primary focus, and it can determine their activity. In such cases, the primary focus is not a dominant, but a determinant. Such a specification is not a game of words. We have seen from the description of the peculiarities of the formation and activity of the epileptic complexes that the complexes break down and the activity of the dependent foci weakens when the dominant focus is eliminated. The elimination of the dominant focus results in the activation of other foci which were inhibited until this moment due to the presence of the dominant focus. There are many clinical examples of such effects when multifocal epilepsy is surgically treated.

The terms 'primary focus' and 'determinant focus' are not always the same and may not be synonyms, The first term expresses the sequence of a process's development and the initial stage of the history of a pathologic system, while the second term expresses a

pathologic process's link which is the most important at a given
moment. The primary focus is not always a determinant one. At the
late stages of a process, other foci can act as a determinant focus.
Such a situation was simulated by the formation of epileptic complexes in the cerebral cortex, when the most powerful focus became
the determinant one, although it was produced later. It will be seen
further from the example of pathologic pain (Chapter 7), photogenic
epilepsy (Chapter 9) and other syndromes that various formations of
the central nervous system can act as a determinant structure at
various stages of a pathologic process. The 'migration' of the chief
epileptic foci is well known in the clinical picture of epilepsy. It
is very important to ascertain the real determinant focus when epilepsy is being surgically treated. Apparently, this will become
pressing in the treatment of other neuropathologic syndromes induced
by hyperactive determinant structures.

SUMMARY

     Determinant structures or simply determinants are the central
nervous system's functional formations which determine the activity
pattern of other parts of the system and, consequently, the system's
activity and its result, This property of determinant is exhibited
in various systems with the prevalence of both relatively rigid and
stochastic functional connections.

     A structure of the central nervous system becomes a determinant
when there are conditions which make it possible for its controlling
influences to be realized. If the receiving structures do not, for
some reason, realize the determinant's influence, the determinant
loses its controlling role and becomes a local generator of excitation. The determinant exhibits itself only both through the activity of the structures which receive its influence and through the
activity of the system which it induces.

     The system which the determinant has induced breaks up and
disappears when the determinant is eliminated. Such an effect is not
produced when other parts of the system are eliminated. The system
is restored or a new system is induced when the determinant structure
is reactivated or a new one is created. Thus, the determinant acts
as a system-forming factor.

     When there are several hyperactive structures, the one with the
highest level of excitation acts as a determinant. This rule is
evident when the foci of epileptic activity in the cerebral cortex
are taken as an example: such foci intensify their activity and
combine into a single functional complex under the influence of the
most powerful focus, which determines the pattern of the activity of
the whole complex. When the complex is generally suppressed, the
determinant focus is the last to disappear, and during restoration,

it is the first to be reactivated and promotes the reactivation of other foci, combining them again into a complex and determining its behavior.

A functional message produced by the determinant should be powerful enough for the determinant to realize its influences. A sufficiently powerful message helps overcome numerous synaptic inhibitions and surmount the control mechanisms in the realizing structures, and allows 'synaptic weight' to prevail on the neurons of these structures. The significance of the power of the determinant's functional message is exhibited most clearly in the systems where rigid connections prevail. The enhancement of neuronal excitability and the disturbance of the control mechanisms in the structures which receive the impulses from the determinant are another important prerequisite of the realization of the determinant's influences. This mechanism is important in determining both the course of the process and the brain structures' 'predisposition' to or 'readiness' for the origination of secondary foci under the influence of the determinant focus. This regularity is most clearly expressed in the parts of the central nervous system where stochastic connections prevail.

The extent to which the determinant's effects are realized and the peculiarities of this realization show that there are three main forms of determination: form I, or complete determination, i.e. the determinant's influence is realized completely without any distortions; form II, or incomplete determination, i.e. the pattern of the activity of the system's parts that are subordinately influenced by the determinant only partially expresses the peculiarities of the determinant's functional transmission, and form III, or the activation of the system or its parts without imposing the pattern of the determinant's activity on them. Strictly speaking, the matter in question in the case of form III is not the determinant, but the generator of excitation which merely activates the system. The relations between such generators and the system which they activate have several distinguishing features. The forms can either be significant in themselves or can represent some stages of realization of the determinant's effects.

The determinant as the working principle of neural activity greatly differs from the dominant. The determinant conditions the functional relations inside a system and is, consequently, a principle of intrasystemic intercentral relations. The dominant represents the principle of intersystemic relations, according to which other systems undergo coupled inhibition when one system is active. Both principles are not an alternative. They supplement one another, and together they allow the nervous system to carry on purposeful integrative activity under normal conditions. Dominant-determinant relations are observed also in the origination of functional complexes under the influence of the determinant focus.

Dominant relations may exist not only between individual foci, but also between complexes of foci. When a model is taken as an example, this means that dominant relations exist between systems: other systems, particularly those which interfere with an activated system, are inhibited when one system is activated. The functional structure, which acts as a determinant for a given system, can produce a dominant effect with respect to other antagonistic structures and systems.

Like determinant relations, dominant relations may be important under pathologic conditions. If a hyperactive system becomes a dominant, it can cause the activity of physiological systems to be suppressed, promoting the development of the pathologic state and disorganization of the brain. In epilepsy, it is important to distinguish between the determinant focus and the dominant focus. Their elimination can produce different effects: the elimination of the determinant focus causes other foci to weaken and the pathologic system to break up, while the elimination of the dominant focus results in the enhancement of other foci's activity.

2
Generators of pathologically enhanced excitation in various areas of the central nervous system

.It follows from the previous chapter that the definite intensity of the determinant's functional message is an important prerequisite for realizing the determinant's influence. The higher the level of excitation generated by the determinant's structures, the more fully are its influences realized when all other conditions are equal. If the determinant structure becomes hyperactive, its influences become pathologically intensive and cannot be suppressed. Investigations have shown that such a structure is neurophysiologically based on the generator of pathologically enhanced excitation (GPEE).

The main aim of this part of the book is to show whether it is fundamentally possible for GPEE to be formed in various physiological systems of the central nervous system. At the same time, it is important to ascertain individual peculiarities and the general regularities of the formation and activity of GPEE in various areas of the central nervous system. The material pertaining to this subject is expounded also in other chapters where individual neuropathologic syndromes are considered.

GPEE IN THE SPINAL CORD'S EFFERENT OUTPUT

GPEE in the spinal cord's efferent output was produced in the rat's lumbosacral segments with small TT doses, which were either injected into the hind leg (in this case, TT entered the ventral horns of the lumbosacral segments via the muscle nerves) or micro-injected directly into the ventral horns of the given segments (Kryzhanovsky, 1966, 1973).

Enhanced electric activity (EA) appeared in the muscles of the injected leg after a certain latent period (its duration depended on the amount of TT which was injected). At first, EA bursts occurred during stimulation (which was produced by squeezing the foot or fingers). In this case, the amplitude of potentials was relatively low and the duration of the burst was short. In the course of time, EA bursts became longer and more powerful, and the amplitude of potentials grew. During this period, bursts could appear in response to slight stimulation (e.g. slight tactile stimulation), and they could originate also spontaneously, i.e. without any special stimulation. These bursts, which consisted of high-amplitude and high-frequency potentials, were long and lasted even when stimulation ceased, passing into after-discharge activity, and the amplitude and frequency of its potentials gradually diminished. This activity, which could remain for an indefinitely long time, is in essence enhanced background activity (Figure 17 A). Clonic activity in the form of brief, occasionally regular EA bursts could sometimes be observed together with that tonic activity at the height of the process and especially at its late stages.

Changes in the muscular tonus corresponded to the above EA changes in the leg muscles. The muscular tonus gradually grew, while articular mobility diminished. As a result of the prevalence of the extensor hypertonus, the leg gradually stretched out and was drawn backwards. At the height of the process, the muscular hypertonus turned into muscular rigidity; the leg was as stiff as a stick, and articular motion was absent (Figure 17 B). At the late stages, clonic twitching or jerks of the affected leg were observed. These motions corresponded to clonic EA bursts.

The given electrophysiological and clinical characteristics were observed in all the animal species, beginning with frogs and ending with monkeys. This form of a pathologic state is known as local tetanus in literature.

When an electromyogram was taken with coaxial electrodes capable of recording muscular units (Buchthal et al., 1960) as the animal was in a state of relative rest, it was observed how a powerful EA burst consisting of high-amplitude and high-frequency potentials characteristic of groups of muscular units gradually passed into EA in the form of individual synchronized high-amplitude potentials which later also disappeared; low-voltage potentials, which represented EA for individual muscular units, were then recorded instead of those potentials (Figure 18).

That EA pattern in the muscles reflects the activity of motoneurons in the ventral horns of the lumbosacral segments, where GPEE is formed under the influence of TT. A large motoneuronal population becomes involved in a reaction when GPEE is activated, as a result of which high-amplitude muscular potentials originate. The amplitude of

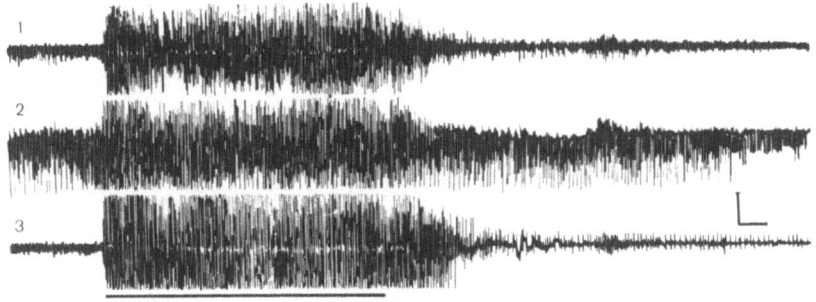

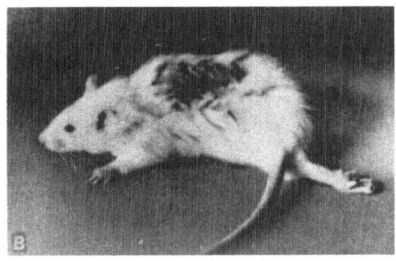

Fig. 17.   Electrographic and clinical manifestations of the activity
           of GPEE produced in the efferent output of the lumbosacral
           segments of the spinal cord. A: Electric activity in the
           left hind leg muscles; 1: EA in the posterior group of
           femoral muscles; 2: EA in the gastrocnemial muscle; 3: EA
           in the posterior group of crural muscles. The horizontal
           line indicates stimulation (the squeezing of the ipsi-
           lateral foot). Amplitude: 0.3 mV; time: 0.2 s. B: Rigid-
           ity of the left hind leg muscles. GPEE was produced in the
           ventral horns of the lumbosacral segments (L_6-S_1) by TT
           microinjection.

these potentials decreases as the activity (power) of the GPEE diminishes, indicating that individual groups of neurons have come out of
the general neuronal population which constitutes GPEE. This phenomenon shows that the neurons which constitute the given population are
functionally different: some of them become rapidly inactive, others
become inactive more slowly, and still others are highly active for a
long time. We will see further that such a situation is characteristic of the duration of GPEE also in other areas of the central
nervous system, including the cerebral cortex.

    However, motoneurons are the terminal link of the system of the
efferent output of the ventral horns. Are they the operative part of
GPEE that produces the pathologically enhanced flow of impulses, or
do other neurons perform this?

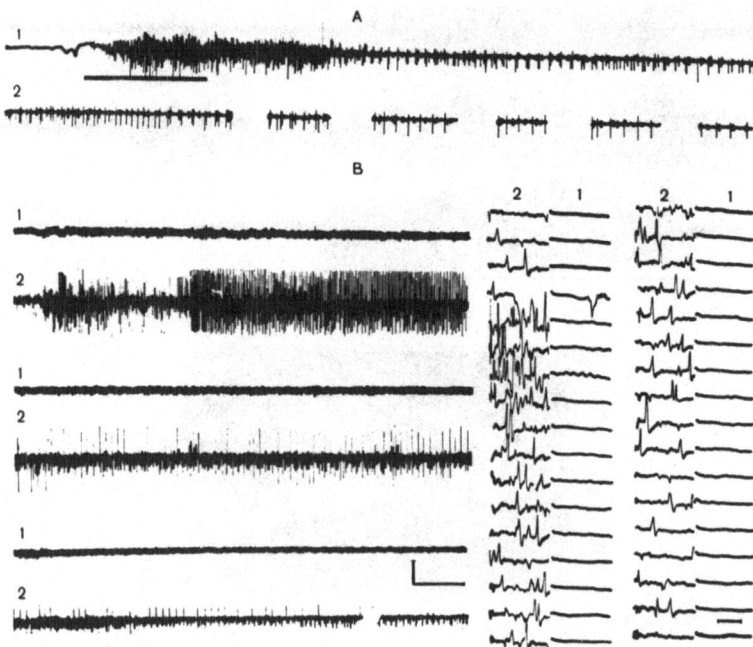

Fig. 18. Background, evoked and after-discharge EA in the
          gastrocnemial muscle of a monkey (A) and a rat (B) during
          the production of GPEE in the motor nuclei of this muscle.
          A: Burst of EA in a monkey's gastrocnemial muscle evoked by
          squeezing the ipsilateral foot (horizontal line). The
          successive recordings of after-discharge EA have been made
          at intervals of 2-5 s. B: EA in a rat's right (1) and left
          (2) gastrocnemial muscle. A burst of EA was produced by
          provoked locomotion. Recordings of discharge activity at
          intervals of 5 s. Right: EA in the right and left gastro-
          cnemial muscles presented rapidly (1 ms). In all the
          cases, GPEE was produced in the ventral horns of the
          lumbosacral segments on the left by injecting TT in small
          doses into the left gastrocnemial muscle. The potentials
          were recorded by coaxial electrodes. Amplitude: 0.3 mV;
          time: 0.5 s.

     Investigations have shown that EA in the muscles which is simi-
lar to that described above can be recorded when GPEE is produced in
the ventral horns of the lumbosacral segments and when the respective
posterior roots are cut (Kryzhanovsky, 1966, 1967, 1973). This fact
rules out the role of monosynaptic stimulation in motoneuronal hyperactivation and shows that the disinhibition of the segmental gamma-
system engendered by TT (Erzina, 1961; Takano, Kano, 1973) is of no

great significance in that phenomenon. When TT acts on the spinal
reflex mechanisms, monosynaptic reflexes are not enhanced (Brooks et
al., 1957; Sverdlov, 1960, 1969; Kryzhanovsky, Dyakonova, 1964;
Kryzhanovsky, 1966, Kryzhanovsky, Sheikhon, 1973; Kryzhanovsky et al,
1973b, 1973c), although various types of postsynaptic motoneuronal
inhibition are upset (see Chapter 5). Motoneuronal responses are not
enhanced also in the case of antidromic stimulation (Figure 19 A).
However, polysynaptic reflexes are greatly enhanced (Brooks et al.,
1957; Sverdlov, 1960, 1969; Kryzhanovsky, Dyakonova, 1964;
Kryzhanovsky, 1966, 1967: Kryzhanovsky, Sheikhon, 1973; Kryzhanovsky
et al., 1973b, 1973c; Figures 19,20,45,46). The polysynaptic discharges recorded in the anterior roots on the side of GPEE not only
are substantially enlarged, but also constitute an unusual high-
amplitude potential, which can be reproduced with a high frequency
(Kryzhanovsky, Dyakonova, 1964; Kryzhanovsky, 1966, 1967; see Figure
48). Such characteristics of the polysynaptic reflexes show that the
interneurons, which have an outlet to motoneurons, are disinhibited
to a great extent and their activity is considerably synchronized.

The intracellular recording of the responses of the lumbosacral
motoneurons in GPEE has shown that the amplitude and duration of
polysynaptic EPSP greatly increase until large potentials originate,
indicating that polysynaptic bombardment is intensive and long
(Figure 20 B) (Kryzhanovsky, Sheikhon, 1973; Kryzhanovsky et al.,
1973a). Since polysynaptic activity is enhanced, multiple motoneuronal responses originate when the structures connected with GPEE
are stimulated by a single shock (Figure 20 C). The excitability of
the motoneurons assessed intracellularly by direct pulse current, and
the physical constants of the motoneuronal membrane (resistance, time
constant, etc.) are unchanged during the seizure-free period
(Sverdlov, 1969; Kryzhanovsky et al., 1973a). These data and the
fact that the monosynaptic reflexes are not enlarged suggest that the
motoneurons are not the basic cellular substrate which forms GPEE.

It follows from all the above data that the interneurons of the
efferent output, which are directly connected with the motoneurons,
constitute the operant part of GPEE that maintains enhanced electrogenesis. These interneurons may probably be the tonic interneurons
which were described by Granit and his associates (1957) and which
are directly connected with motoneurons. The authors called these
interneurons the 'first final common path', unlike the 'second final
common path', which the motoneurons represent.

However, motoneurons are apparently not the passive part of
GPEE. When interneurons are hyperactive, intense polysynaptic bombardment causes the motoneuronal membrane to become depolarized,
greatly enhancing its ability to react to orthodromic impulses with a
high frequency (Figures 19 B; 20 A C). Thus, the given GPEE, which
originated in the efferent output of the spinal cord under the influence of TT, is believed to consist of efferent interneurons and

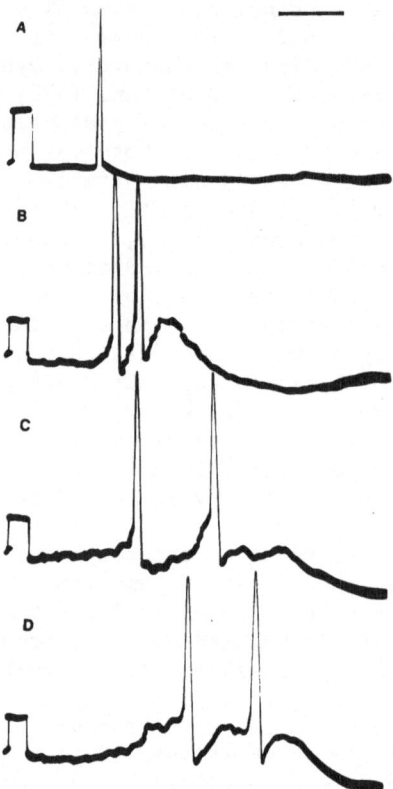

Fig. 19. Responses of the flexor motoneurons in the TT-poisoned
          segment of the spinal cord with GPEE localization there
          during ES of various medullar nuclei. Response of the
          antidromic stimulation of the PBST nerve (A); effects when
          the caudal pontis nucleus (B), Deiters' nuclei (C), and the
          raphe nucleus (D) were stimulated by a short series of
          stimuli. Amplitudes: A: 20 mV; B-D: 5 mV; time: 10 ms.

motoneurons, while the interneurons constitute the main, operant
part, which generates and maintains enhanced electrogenesis.

        A similar picture was observed by other authors when penicillin
was applied to the pial surface of the lumbosacral segments of the
spinal cord (Kao, Crill, 1972a, 1972b; Lothman, Somjen, 1976a; Crill,
1980). The only difference is that the segmental myoclonus (which
lasted 20 minutes) had originated and there was no muscular rigidity
in their experiments. This form of a pathologic state is similar to
spinal myoclonia ('universal dispatch station' phenomenon; see Chapters 1 and 6), which is due to the presence of GPEE in the system of
propriospinal neurons (see the next section of this chapter). When

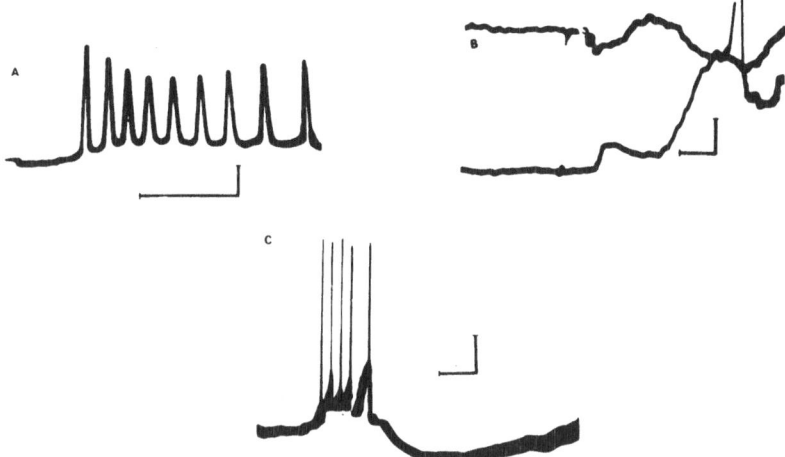

Fig. 20. Responses of three different motoneurons to rhythmic (A) and single (B) ES of Deiters' nucleus and to the single stimulation of the giant-cell nucleus (C). Amplitude: 100 μV; time: 10 ms.

penicillin was applied to the region of the motoneuron which was studied intracellularly, its substantial and prolonged depolarization (PD) was recorded. This PD is similar to the PDS of the cortical epileptic cell, but its amplitude is lower and the variability of its duration is less. Post-PD hyperpolarization has also not been observed (Kao, Crill, 1972b; Lothman, Somjen, 1976a); the monosynaptic EPSPs of the motoneurons did not change (Davenport et al., 1977; Dunn, Somjen, 1977), and the disynaptic IPSPs (Davidoff, 1972; Davenport et al., 1977; Dunn, Somjen, 1977), recurrent inhibition (Davenport et al., 1977) and apparently presynaptic inhibition (Lothman, Somjen, 1976b) were not upset, although, according to some authors (Davidoff, 1972), presynaptic inhibition has been disturbed. A sharp increase in polysynaptic EPSPs is noteworthy. This increase is observed even when the cutaneous nerve is stimulated by a single shock (Davenport et al., 1977). The interneuronal circuits are apparently responsible for the origin of PD and the hypersynchronization of seizure activity in the case of the given form of penicillin-induced myoclonus (Crill, 1980). Gradual depolarization which originates in some motoneurons suggests that the membrane properties change in the neurons. This is evident from repetitive firing that is imposed on prolonged depolarization, which originates when the applied depolarizing current stimulus ceases. The change in the membrane of the given motoneurons is probably an additional mechanism of epileptiform activity (Kao, Crill, 1972; Crill, 1980).

GPEE IN THE SYSTEM OF PROPRIOSPINAL NEURONS

The localization of GPEE and the peculiarities of its activity
were studied on rats in connection with the above-mentioned investigations (see Chapter 1) of the generalized seizure reaction of spinal
origin (the 'universal dispatch station' phenomenon), known as spinal
myoclonia (see Chapter 6) (Lutsenko, Kryzhanovsky, 1975). In this
phenomenon, a generalized seizure occurs when the hind leg is stimulated on the side of GPEE; stimulation of the opposite leg produces
only a local reflex reaction.

The recordings of electric activity (EA) in the anterior roots
have shown that prolonged high-amplitude discharges occur in them
when the phenomenon is realized (Figure 21). By their duration and
intensity, they correspond to the above-mentioned (Chapter 1) powerful bursts of EA in the muscles. The simultaneous recording of these
discharges, the dorsal cord potentials (DCP) and the dorsal root
potentials (DRP) has shown that all of them have the same intensity
and duration, being much greater than those observed under normal
conditions; DCP increased due to a positive component, namely, the P
wave (Figure 21 2). The duration of the P wave, which was recorded
on the dorsal surface of the lumbosacral segments, exactly correspond
to the duration of the discharge on the lumbar roots. This corre-
spondence was observed when activity was induced by stimulating the
ipsilateral sural nerve (Figure 21 1-4; 22 A) and when spontaneous
bursts of seizure activity occurred (Figure 22 B). These bursts
differed from one another by their intensity and duration. However,
the intensity and duration of the discharges in the anterior root
always exactly corresponded to those of the P wave.

Thus, the intensity and duration of the discharges in the
anterior root strictly corresponded to those of the P wave in an
induced or spontaneous seizure reaction. This suggested that the
discharges in the anterior root and the augmentation of the P wave
are due to the enhanced activity of a functional structure which
causes a seizure reaction.

Under normal conditions, the P wave and DRP are known to originate after an afferent volley. They express the extent of the
depolarization of the primary afferent endings in the spinal cord.
This depolarization correlates with presynaptic inhibition (Eccles et
al., 1962). However, there is no afferent volley from the periphery
when seizure activity spontaneously originates. Therefore, the
enlarged P wave cannot be due to the activation of the depolarizing
neurons by the afferent wave. When the activity was induced by
electrically stimulating the central part of the sectioned dorsal
roots, the P wave and prolonged discharges in the anterior roots
could be produced as the spinal cord was completely deafferented
below its transection (Figure 21 G).

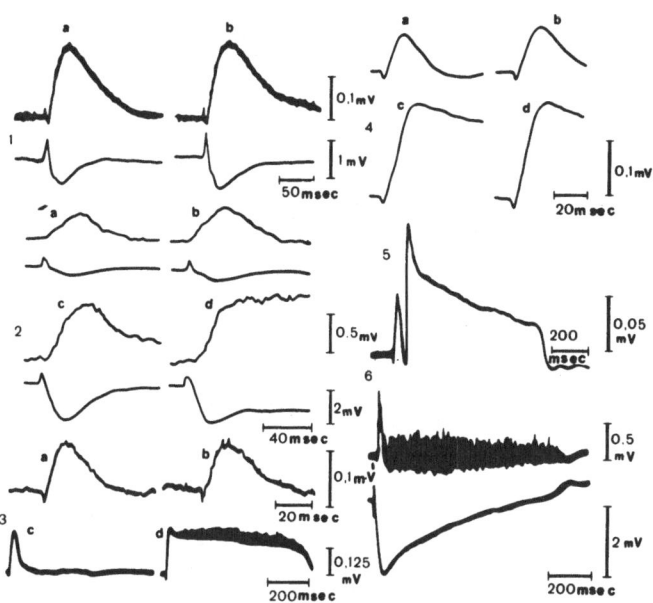

Fig. 21.   Dorsal root potential (DRP) and the dorsal cord potential
(DCP) of the spinal cord in a healthy rat and in animals
with GPEE in the lumbosacral segments of the spinal cord.
1: Ipsilateral (on the side of nerve stimulation) DPR L_5
(top) and DCP (bottom) in a healthy rat; 2: DCP L_4 and DRP
L_6 in a rat with GPEE; 3,4: ipsilateral DRP L_5 in a rat
with GPEE. In all the recordings, the potentials were
produced by nerve stimulation; a,b: on the ipsilateral
'healthy' side; c,d: on the side of TT injection; 5: DRP L_5
is recorded on the side contralateral to the TT injection
side; short DRP evoked by nerve stimulation on the
'healthy' side; long DRP induced by nerve stimulation on
the TT injection side. The left curves 1-4 (a) represent
the reaction to stimulation with a strength of 4 thresholds, and the right curves 1-4 (b), the reaction to stimulation with a force of 40 thresholds; 6: discharges in the
anterior root L_2 (top) and DCP (bottom) were evoked by
stimulating the posterior root with a strength of 4 thresholds. Complete deafferentation below the level (Th_7) where
the spinal cord was cut. Total curarization, artificial
respiration.

All the above-mentioned facts show that there is a new component
in the composition of the enlarged P wave, which appears during an
induced or spontaneous seizure, together with the components that are
homologous to the main component of the P wave under normal conditions. This new component reflects seizure activity, and not the

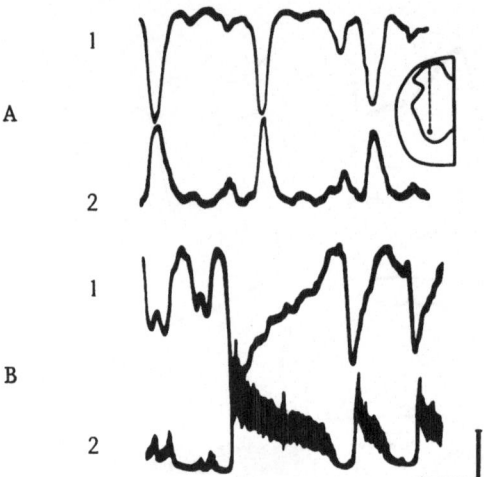

Fig. 22. Potentials of the lumbar segments of the spinal cord
          recorded on the side of GPEE after provoked (A) and spon-
          taneous (B) seizure activity. Dorsal cord potential (1)
          and focal potentials (2) in the ventral horn of segment L_6.
          The point on the cross-section scheme indicates the pos-
          ition of the focal potential recording. Amplitude: 0.25 mV
          (A), 0.5 mV (B); time: 0.2 s.

depolarization of afferents. It is due to this component that the P
wave grows in the phenomenon of generalized seizures. In other
words, the component reflects the activity of the new functional
structure which produces a seizure and acts as GPEE.

When the cutaneous nerve was stimulated on the side of GPEE,
i.e. when GPEE was activated, the P wave and ipsilateral DRP originated, and their amplitude and duration were far greater (sometimes
they were 20 times greater) than those of either healthy or the same
sick animals when the nerve was stimulated on the opposite side (i.e.
when GPEE was not activated) (Figure 22). Hence, it can be seen that
GPEE, which produces a seizure, is very powerful. The fact that
seizure discharges in the anterior root and the respective P wave
originate in the spinal cord segments, which are isolated from the
central influences (below the section of the spinal cord) and are
completely deafferented (section of all the dorsal roots), shows that
GPEE is functionally autonomous, i.e. it can develop prolonged and
self-sustaining activity without additional afferentation.

Further experimental analysis has shown that the additional
component of the P wave, which correlates with seizure activity and
reflects the activity of GPEE, is made up of two components, namely,
the early and the late ones. These components are connected with the

activation of GPEE by various afferents: when low-threshold fibers
are mechanically damaged, the early component drops out and the late
one, which is connected with the activation of high-threshold afferents (A-post-δ- and C-fibers), remains.

The enlarged P wave originates only during a generalized seizure
(the 'universal dispatch station' phenomenon). It is absent in the
case of another pathologic form, i.e. local tetanus (Figure 23),
which is characterized by the affection of inhibitory control in the
formation of GPEE in the efferent output system (motoneurons and the
interneurons connected with them, see the preceding section of this
chapter). The fact that the P wave does not grow in local tetanus,
but increases in the phenomenon of a generalized seizure reaction
means that, in the last case, GPEE consists of interneurons which do
not belong to the efferent output system. This is confirmed by
experiments involving nembutal, which is used for changing neuronal
activity. The well-known effect of barbiturates, i.e. the elongation
of the P wave and the suppression of the discharges in the ventral
root, originated when nembutal was administered in a specially selected dose to healthy animals (Figure 23 1 b). Nembutal produced a
similar effect also in animals with local tetanus: the P wave was
elongated and the discharges were suppressed in the ventral root
(Figure 23 2 d). The agent produced a complex effect in animals in a
generalized seizure (the 'universal dispatch station' phenomenon)
(Figure 23 3 f): (1) the suppression of the late component of the P
wave and prolonged discharges in the anterior root that correlate
with the late component of the P wave, and (2) the dismemberment of
the early part of the P wave into two components, namely, the component which acts, with respect to nembutal, as a homologous component in a healthy animal (i.e. it increases under the influence of
nembutal), and a component of smaller duration at the beginning of
the P wave (the conversion of this component into an ordinary one is
indicated by an arrow in the illustration). The duration of the
additional component correlated with that of the remaining part of
the discharges in the ventral root. Thus experiments with nembutal
confirmed that the enlarged P wave is heterogeneous and has a complex
structure in the case of a generalized seizure of spinal origin (the
'universal dispatch station' phenomenon). They showed that the late
component of the enlarged P wave, which appears only in this phenomenon, acts differently from other components observed in healthy
animals, too. That component also reflects the activity of the new
functional structure, GPEE, which produces a seizure.

Investigations of the focal potentials in the spinal cord, which
were carried out in order to ascertain the localization of GPEE, have
shown that its constituent neurons are in the ventral horn of the
grey substance of the lumbosacral segments (Figures 24,25). The
generated focal potentials in the ventral horn exactly corresponded
to DCP by their duration, configuration and time of origin. They
were a 'mirror' image of DCP. A study of the characteristics of the

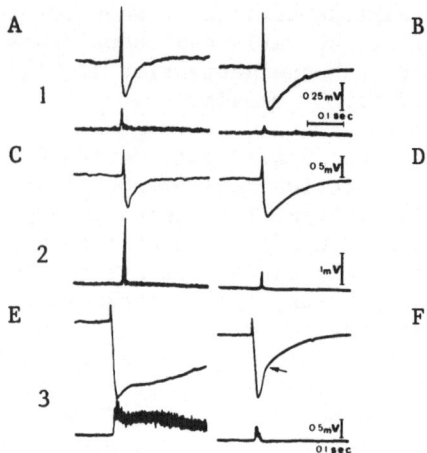

Fig. 23.   Effects of nembutal on the dorsal cord potential and the
            ventral root discharge evoked by the stimulation of the
            ipsilateral cutaneous nerve in healthy animals (1), in rats
            with local tetanus (2) and in animals with GPEE (3).   The
            top curves represent the dorsal cord potential, and the
            bottom curves, the discharges of the ventral roof L_5;
            before (a,c,e) and after (b,d,f) nembutal administration
            (14 mg/kg).   Nerve stimulation strength was 40 thresholds.

focal potentials, which were drawn off when a microelectrode was
inserted into the spinal cord (the microelectrode track and the
averaged-out curves of a change in the amplitudes of the focal potentials are represented in Figure 25), has shown that the focal potential, which corresponds by its duration to the N wave of DCP in
animals with a generalized seizure phenomenon ('universal dispatch
station' phenomenon), is not substantially changed in comparison with
that under normal conditions.   The maximum of this negative potential
is determined in the dorsal horn of the gray substance, being in
accord with the data presented by other authors (Coombs et al.,
1956).   The positive DCP potentials, which correspond to the P wave,
greatly changed: they were substantially enlarged with respect to
both the amplitude and duration when GPEE was activated (when the
nerve on the side of GPEE was stimulated), but did not greatly change
when GPEE was not activated (when the nerve on the opposite side was
stimulated).   The maximum amplitude of the P wave was recorded near
the surface of the spinal cord.   At the base of the dorsal horn, the
P wave changed its sign and became a negative wave, which reached the
maximum in the ventral horn (Figures 24,25).   This negative wave,
i.e. the focal potential which corresponds to the P wave, had the
same configuration as that in normal animals or in animals with a
'universal dispatch station' phenomenon when GPEE was not activated
(when the nerve on the opposite side was stimulated).   However, its

dimensions were much larger when GPEE was activated (when the ipsilateral nerve was stimulated). A negative potential which was
greatly enlarged with respect to the amplitude and duration and also
numerous neuronal discharges were recorded on the side of GPEE.
Figure 24 shows the successive recording of the potentials at a
different depth in an experiment. A prolonged focal potential and
numerous neuronal impulses were recorded on the side of GPEE at a
depth of 1.6-1.9 mm when the ipsilateral nerve, which activates GPEE,
was stimulated. However, the stimulation of the contralateral nerve,
which does not activate GPEE, produced a brief potential and one
impulse (the depth was 1.9 mm). An interesting fact is that the
duration of the slow negative focal potential correlated with the
duration of impulse activity, which originates when GPEE is activated. The presence of the focal potential, whose amplitude and
duration are enhanced and which is accompanied by high impulse activity, also testifies to the activity of such a hyperactive structure under consideration as GPEE.

An analysis of the waves in DCP has shown by means of the displacement of the electrode along the dorsal surface of the spinal
cord in the caudate and cranial directions (Figure 26) that the N
wave is spatially distributed in the same way as under normal conditions, while the P wave is substantially enlarged in the cranial
and caudate directions. The P wave's enlarged length reflects the
dimensions and power of GPEE, which is localized in the ventral horns
of the spinal cord's gray substance and which consists of many interneurons.

According to Coombs and co-workers (1956), the N wave of DCP
reflects the activity of the dorsal horn's interneurons, which apparently belong to the ascending systems. It has been shown that the P
wave does not pertain to the ascending systems, and it is not connected with the efferent output system, since it does not change when
the inhibitory mechanisms in the efferent output system are upset in
the case of local tetanus. It can be seen from these data and from
the site of the maximum amplitude of the focal potential and impulse
activity during the activation of GPEE that the latter is localized
in the ventral horns and consists of neurons which are in the system
of propriospinal connections. GPEE probably consists of neurons
which are in the VII and VIII laminae of the ventral horn according
to Rexed's classification. The neurons of the VII lamina are known
to respond to a volley in high-threshold afferents by a tonic discharge which produces prolonged activity of the motoneurons
(Jankovska et al., 1967). At the same time, the neurons of GPEE are
connected with the system which causes the depolarization of .the
primary afferent endings in the spinal cord, being expressed in a
change in the P wave when GPEE is activated.

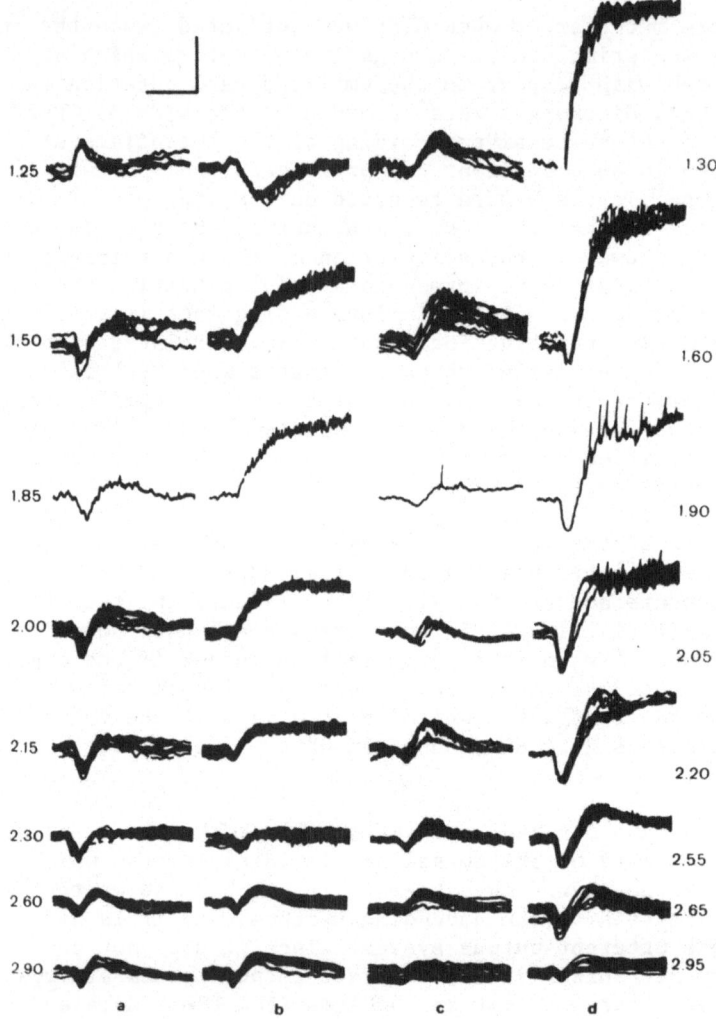

Fig. 24. Focal potentials in the intermediate zone and in the
          ventral horn of the spinal cord in rats with GPEE. a,b:
          Focal potential recordings of the intact side during the
          stimulation of the ipsilateral (a) and contralateral (b)
          cutaneous nerves with a force of 4 thresholds; c,d: focal
          potentials on the side of GPEE evoked by similar stimuli.
          The values represent the depth of microelectrode insertion
          (the dorsal cord is taken as the 'zero' level, mm). Ampli-
          tude: 0.5 mV; time: 20 ms.

GPEE IN THE BULBAR GIANT-CELL NUCLEUS

     GPEE was produced in the cat's bulbar giant-cell nucleus by
micro-injecting TT into the nucleus, using stereotaxic techniques
(Kryzhanovsky, Sheikhon, 1976).

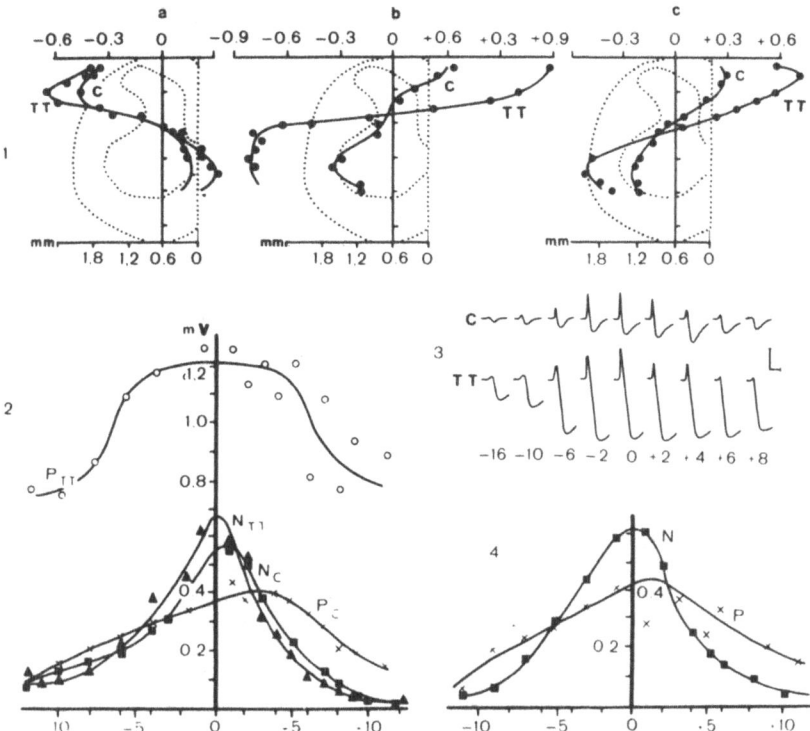

Fig. 25.   Distribution of the spinal cord potential amplitudes in
rats with GPEE (1-3) and in healthy animals (4). 1: The
influence of microelectrode insertion into the spinal cord
on the focal potential amplitude (mV) corresponding to the
N wave (a) and the P wave (b,c); a,b: potentials on the
ipsilateral part of the spinal cord; c: potential of the
contralateral part of the spinal cord. The average potential amplitude curve and a scale of the microelectrode
position are superposed on the spinal cord cross-section
(distance from the dorsal cord, mm); T: potentials evoked
by nerve stimulation on the TT injection side; C: potentials evoked by nerve stimulation on the contralateral
side; 2,3: N and P wave amplitude change during microelectrode displacement in the cranial (-) and caudal (+)
directions (mm) from the point with a maximum value of the
N wave; 4: the same as 2, but in healthy rats. P_T, P_0: P
wave amplitudes during nerve stimulation on the side of TT
injection and on the opposite side. N_T, N_0: the same for
the N wave. Amplitude: 0.5 mV; time: 5 ms.

The initial changes in neuronal activity in the giant-cell
nucleus (1.5-2 hours after the injection of TT) were expressed in an
increase in the amplitude and frequency of background impulse activity and in the number of active neurons, the appearance of neurons

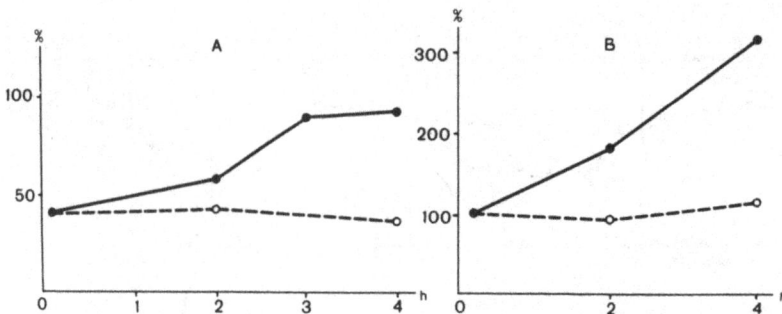

Fig. 26.   Change in background impulse activity during the formation
            of GPEE in the giant-cell nucleus. A: Number of active
            points in giant-cell nuclei after injecting active TT into
            a nucleus (continuous line) and inactivated TT into the
            contralateral nucleus (dashed line). Abscissa: time after
            TT injection. Ordinate: number of active points (% of the
            initial level). B: Change in total background impulse
            activity in the giant-cell nucleus after injecting active
            (continuous line) and inactivated (dashed line) TT.
            Abscissa: time after TT injection. Ordinate: integral
            impulse activity of neurons (% of the initial level).

with burst activity, and the growth of the amplitude and duration of
the generated potentials. These changes grew with time.

Change in Background Activity

It has been learned from the establishment of the number of
active points, i.e. points in which neuronal activity was recorded as
the microelectrode passed through the nucleus on various planes, that
their number grows with time after TT is injected. Before TT was
injected, it constituted about 50 per cent of all the points which
were studied, while three and four hours after the injection of TT,
it constituted 90 and 95 per cent of all the points, respectively
(Figure 26 A). The number of active points virtually did not change
throughout the investigation in the opposite nucleus into which
inactivated TT was injected.

As the number of active points increased, the level of the
activity being recorded in them grew. High amplitude and high-
frequency discharges were recorded at these points. They were absent
in background activity before TT was injected or after inactivated
toxin was administered in the control tests. The total intensity of
background activity (Figure 26 B), being established by an inte-
grator, had substantially increased, just as the frequency of the
discharges of the neuronal aggregates being recorded (in four hours,
it constituted 300 per cent of the control level).

Burst Activity

Group discharges of individual neurons appeared more frequently
with time after the injection of TT. In most cases, the burst consisted of several (up to ten) impulses, but a larger number of
impulses was occasionally recorded (Figure 27).

Evoked Activity

The stimulation of the cutaneous nerves of the hind leg by a
single stimulus (1-2 thresholds) caused inconsiderable changes in
neuronal activity before TT was injected or after inactivated TT was
administered in control experiments. After TT was injected, the
afferent stimulus of the same strength caused greater impulse activity. The frequency, amplitude and duration of evoked activity
substantially increased in four hours. The discharges, which could
be produced even by a slight single stimulation of the cutaneous
nerve (this is ineffective under normal conditions), occasionally
continued for 7-8 seconds and even more (Figure 28).

Induced impulse activity increased also when natural stimuli
were used (squeezing or bending the foot and the toes, stroking the
back, etc.) and when the giant-cell nucleus was activated from the
side of the other areas of the central nervous system, e.g. when the
cerebellar fastigial nucleus, which has an outlet to the giant-cell
nucleus (Magoun, 1950; Angaut, Brodal, 1967; Eccles et al., 1974;
Batton et al., 1977), was stimulated. Four hours after the injection
of TT into the giant-cell nucleus, the amplitude of the discharges in
it increased by three or four times in response to single stimulation
of the nucleus, the frequency grew by more than 20 times, and the

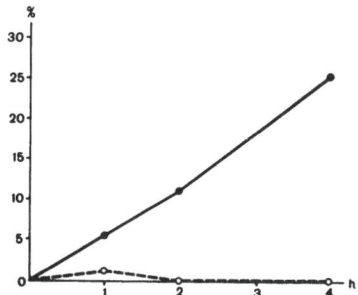

Fig. 27. Burst activity in the giant-cell nucleus after the
          injection of active (continuous line) and inactivated
          (dashed line) TT. Abscissa: time after TT injection.
          Ordinate: number of points with burst activity (% of all
          the points being studied).

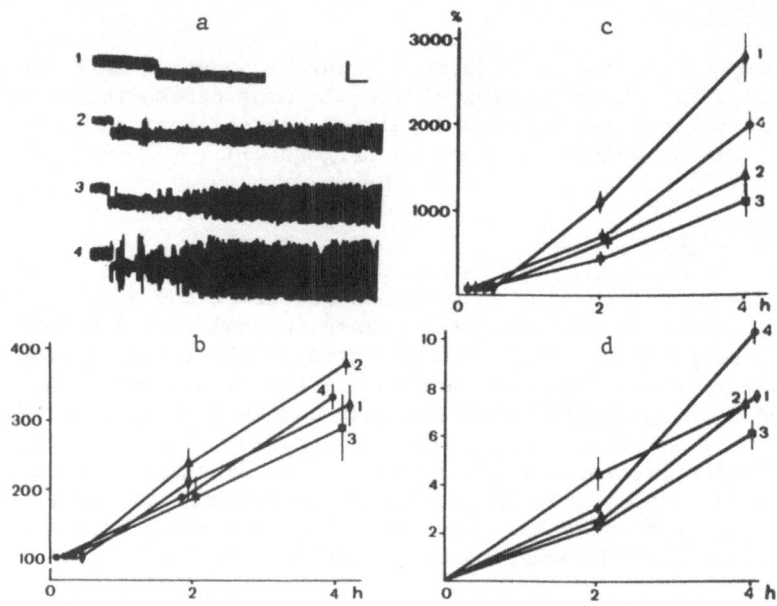

Fig. 28.   Change in evoked activity after injecting TT into a
           giant-cell nucleus. X: Activity evoked by single stimu-
           lation (0.1 ms, 2 thresholds) of the gastrocnemial nerve
           before (a) and two (b), three (c) and four (d) hours after
           injecting TT into the giant-cell nucleus. The duration of
           stimulation is marked by a drop in the levels of the curve.
           The graphs show the changes in the amplitude (II), fre-
           quency (III) and duration (IV) of the discharge caused by
           single stimulation of the (1) sural, (2) deep peroneal and
           (3) gastrocnemial nerve and the (4) fastigial nucleus.
           Ordinate: II: amplitude; III: frequency of evoked activity
           in percentage of the initial level; IV: duration of evoked
           activity, s. Abscissa: time after TT injection, h. Apli-
           tude: 100 μV; time: 1 s.

duration of induced activity reached 10-20 seconds. In this case,
the discharges were augmented and became more frequent for the second
time.

     It follows that GPEE is formed in the giant-cell nucleus after
TT is injected into it. This is expressed above all in the appearance of virtually sustained intense background activity of the
nuclear neuronal populations and in the sharp increase in evoked
activity, which could be recorded even when the stimuli were minimal
and normally ineffective. Such activity is, in essence, no longer an
evoked response. It continues for a long time after stimulation.
Following the first burst of impulses, discharges often grow and

become more frequent for the second time. This is probably connected
with the appearance of new active neurons or a new cycle of excitation of the neurons which were activated earlier. Such a form of
activity shows that it develops independently of trigger stimulation
and is an expression of the operation of GPEE. Afferent and other
stimuli merely trigger GPEE. The activity of GPEE increases with
time after the injection of TT as it spreads through the giant-cell
nucleus. Consequently, the inhibitory mechanisms are disturbed more
and more and an ever larger number of neurons are drawn into the
pathologic process, producing excitation, i.e. GPEE becomes larger
and more powerful. The experiment involving the suppression of GPEE
activity by glycine shows that the given mechanisms are based on the
insufficiency of inhibition in the nuclear neuronal population. As
it has been shown (see Chapter 1), glycine hyperpolarizes the postsynaptic membrane, engendering the effects of postsynaptic inhibition. Glycine exerts the same influence also on the neurons of the
medulla (Hösli, Tebecis, 1970; Tebecis, 1973). Enhanced background
activity was suppressed when glycine was microinjected into the
giant-cell nucleus poisoned by TT (Figure 29). This effect continued
for 30-40 minutes. It originated only in the region where glycine
was injected. Background activity remained in the adjacent tracks ar
a distance of 500μm from the tip of a microcannula. The inhibitory
effect was due to the specific action of glycine and was not connected with a trauma, since the injection of a saline in an appropriate concentration into the GPEE region did not produce such an
effect.

GPEE IN THE LATERAL GENICULATE NUCLEUS

     Like in the previous investigations, GPEE was produced by
stereotaxically microinjecting TT into the rat's and the cat's
lateral geniculate nucleus (LGN).

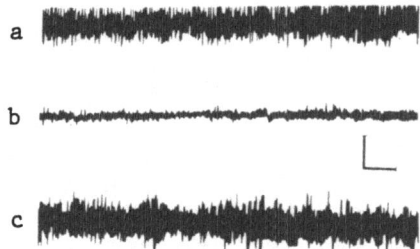

Fig. 29.   Suppression of enhanced impulse activity in the region of a
           giant-cell nucleus poisoned by TT after glycine micro-
           injection. Background activity before (a) and 5 min (b)
           and 40 min (c) after glycine microinjection. Amplitude:
           200 μV; time: 1 s.

The impulse activity of the LGN neurons, which react to light
stimulation, was processed by a computer in order to at first separate definite groups of neurons in accordance with the peculiarities
of their induced and background activity (Rekhtman et al., 1978). A
comparison of the phases of neuronal activity as a response to light
stimulation (Figure 30) has made it possible to identify the excitatory neurons, the inlet of which is directly connected with the
fibers of the visual tract and which react to the on-signals (neurons
of groups I and II) and the off-signals (neurons of group V), and the
inhibitory neurons (groups III, IV and VI).

The neuronal activity of LGN during the formation of GPEE was
analyzed at different stages of the pathologic process, i.e. photogenic epilepsy (Kryzhanovsky et al., 1978e). The neurons of all the
same groups recorded in a healthy animal were identified during the
latent period, when there were no changes in both the brain's electric activity and the animal's behavior. At this stage, the ratio
(in percentage) of the amount of neurons according to groups did not
change. The functional organization of LGN substantially changed
during the preconvulsive period, when the amplitude of the induced
potentials on LGN and in the ipsilateral visual cortex increased. As

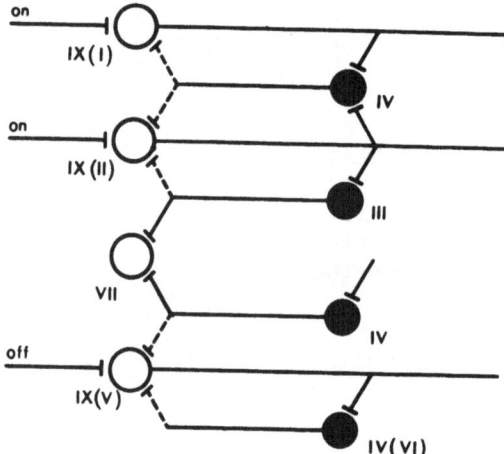

Fig. 30. Change in the functional relationships of the neurons in
          the lateral geniculate nucleus during the formation of GPEE
          after injecting TT into the nucleus. White circles stand
          for relay neurons, and black circles, for inhibitory
          neurons. Roman numerals indicate the neuronal groups;
          those in parentheses are the neurons before, and those not
          in parentheses are the neurons after TT injection and the
          formation of GPEE in the nucleus. The dashed line stands
          for synaptic links which become ineffective as a result of
          TT action.

the inhibition of the relay neurons of groups I and II was disturbed, the inhibitory pause characteristic of them had disappeared after the first discharge in response to light stimulation (Figure 31). Consequently, the number of neurons in group I decreased, and group II disappeared. Instead, group IX appeared, being made up of neurons which were no longer under inhibitory control (Figure 30). Under new conditions, excitation went as formerly along the fibers of the visual tract and reached the relay neurons of groups I and II, and then it was transmitted to the inhibitory neurons of groups III and IV. However, the neurons of groups I and II operated without inhibitory pauses as neurons of group IX. The neurons of the 'off' type are known to undergo active inhibition in response to light stimulation (Shevelev, 1971). It could be that the neurons of the 'off' type were converted into those of the 'on' type because the inhibitory interaction of the 'on'- and 'off'-systems was disturbed especially due to the disturbance of the inhibitory influence of the neurons of group IV on the neurons of group VI. This explains why GPEE is activated and why seizures occur during the convulsive stage of a process in the case of not only on-stimulation, but also off-stimulation. Thus, the disturbance of the inhibitory mechanisms caused a change in the parameters being recorded in the neurons of other groups. Consequently, a part of the neurons of some groups was converted into the neurons of other groups, and the quantitative ratio of the neurons by groups changed.

Evoked potentials (EP) with respect to a light flash in LGN and the visual cortex (VC) were studied in order to assess the changes in total electrogenesis in the LGN-visual cortex system. These changes occurred during the formation of GPEE in LGN and reflected GPEE's

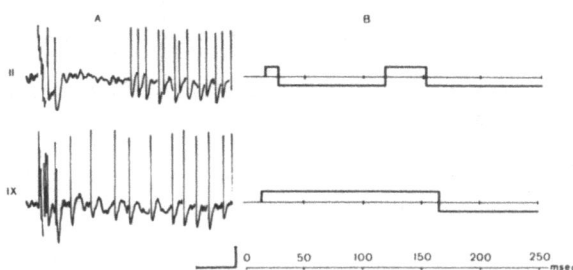

Fig. 31. Response of the neurons of groups II and IX to a light flash. A: Recordings of the neuronal reaction to a light flash (the recordings began when the flash occurred). Amplitude: 100 μV; time: 50 ms. B: Average post-stimulation histograms of the neurons of groups II and IX (the number of the group is indicated by Roman numerals on the left). Deviation upwards corresponds to an increase in neuronal impulse activity, and deviation downwards, to its inhibition. Amplitude: 100 μV; time: 50 ms.

activity. Since there is a direct monosynaptic pathway between LGN
and VC, the output parameters of GPEE in LGN could be assessed by EP
in VC. The EP changes in LGN and VC in the ipsilateral and contralateral hemispheres were compared with the stages of development of
photogenic epilepsy.

Negative EP variations increased in LGN as GPEE was formed
during the preconvulsive period. The appearance of a new negative
component in the primary response potential, $N1^c$ (Figure 32), was an
important sign of the EP change in LGN that correlated with the EP
changes in VC. This component, being in the form of additional
negative deviation, grew with time. Therefore, the $N1^c$ component was
the main negative component of the primary response at the end of the
preconvulsive phase.

Figure 33 shows that the tan α, i.e. the parameter of the rate
of increase in the anterior front of the $N1^c$ component, ceased to
monotonously grow at the end of the seizure phase, while the parameter of the rate of diminution of the posterior front of the $N1^c$
component grew during this period. This phase can be regarded as
both the GPEE formation phase and the growth of GPEE power up to the
level that is necessary for producing a seizure. The new $N1^c$ component is probably connected with the activity of the new functional
group of neurons that is engendered as a result of the disturbance of
the inhibitory connections in LGN which form the basis of GPEE.

It can be seen from Figure 34 that the forms and parameters of
EP in contralateral LGN, into which inactivated TT was injected, had
remained virtually unchanged until the convulsive phase. Thus, the
EP changes which occurred during this period on LGN, into which
active TT was injected, were due to TT's specific action, namely, the
disturbance of the inhibitory mechanisms and the formation of GPEE.

The respective EP changes occurred in ipsilateral VC. They were
expressed in the fact that the amplitude and duration of the primary
negative component (NI) gradually grew (Figure 34). By the middle of
the preconvulsive period, the amplitude of the N1 component in ipsilateral VC considerably increased in comparison with the normal one
and with the one in the contralateral cortex. These changes coincided by their duration with the changes in the primary response in
LGN, particularly with the appearance of the $N1^c$ component in it. At
the same time, the first positive variation (P1) in VC virtually did
not change throughout the pathologic process.

The primary brief latent negative variation of N1 in the primary
response in LGN was gradually replaced by the new $N1^c$ component,
which appeared later, possibly due to the disturbance of the inhibitory mechanisms, which are realized under normal conditions during
the initial period of the formation of a response (so-called early
inhibition) and the inclusion of an additional group of neurons with

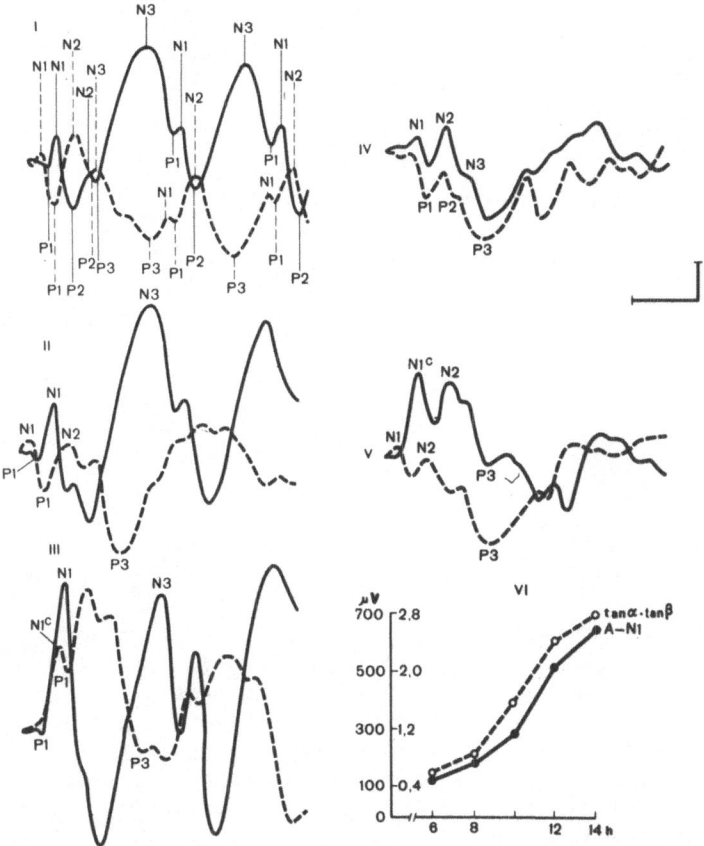

Fig. 32. Comparison of evoked potentials (EP) in the lateral
          geniculate nucleus (LGN) and the visual cortex of both
          sides under normal conditions and during the formation of
          GPEE in LGN after TT injection. Average EPs in LGN (dashed
          line) and the ipsilateral visual cortex (continuous line)
          in the normal state (I) and 10 h after TT injection on the
          contralateral side (II) and on the TT injection side (III).
          Average EPs in LGN 3 h after injecting TT into this nucleus
          (continuous line) and in contralateral LGN (dashed line)
          (IV). EPs in the same structure 12 h after TT injection
          (V). VI: Dependence of the amplitude of the first negative
          component of EP in the visual cortex (A–N1) and the multi-
          plicative slope (tan α·tan β) of the first negative
          component of EP in LGN (on the TT injection side) on the
          time which passed after injecting TT into LGN. Recordings
          were made when light stimulation (illumination of 10 luxes)
          was used. Amplitude: 100 μV; time: 100 ms.

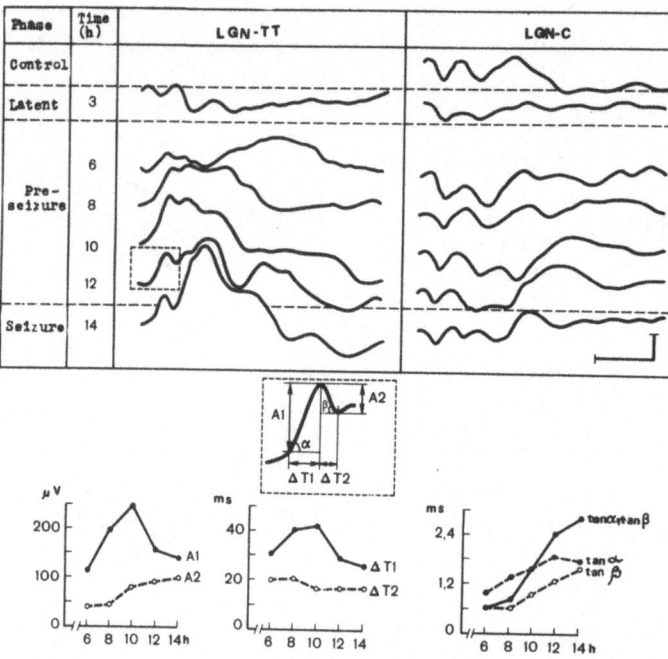

Fig. 33.   Dynamics of EP in both LGN after the formation of GPEE in
           one of them by TT injection.   LGN-TT: nucleus on the side
           of TT injection.   LGN-C: nucleus on the contralateral side.
           Time (h) after TT injection is given on the left.   The
           lower part of the diagram shows graphs of the relationships
           of the parameters which characterize the slope of the
           growth and drop of the fronts of the component $N1^c$ of the
           evoked potential in LGN on the TT injection side, depending
           on the time after the injection.   The method of measuring
           the corresponding parameters is indicated on the enlarged
           fragment of the potential (in the dashed frame).   The
           graphs show the dependence of the amplitude of the anterior
           (A1) and posterior (A2) fronts, the time of the growth of
           the anterior ($\Delta T1$) and diminution of the posterior ($\Delta T2$)
           fronts, the values of the slope angle of the anterior (tan
           α) and posterior (tan β) fronts and the multiplicative
           slope (tan α·tan β) on the duration after TT was injected
           into LGN.   Amplitude: 150 µV; time: 100 ms.

greater latency. The fact that the small latent N1 component
leveled, i.e. that it became less defined, is probably due to the
longer duration of the short latent neurons which form the component.
This duration is longer probably because the inhibitory mechanisms
are upset. Both of those neuronal groups are apparently in the
population which constitutes GPEE.

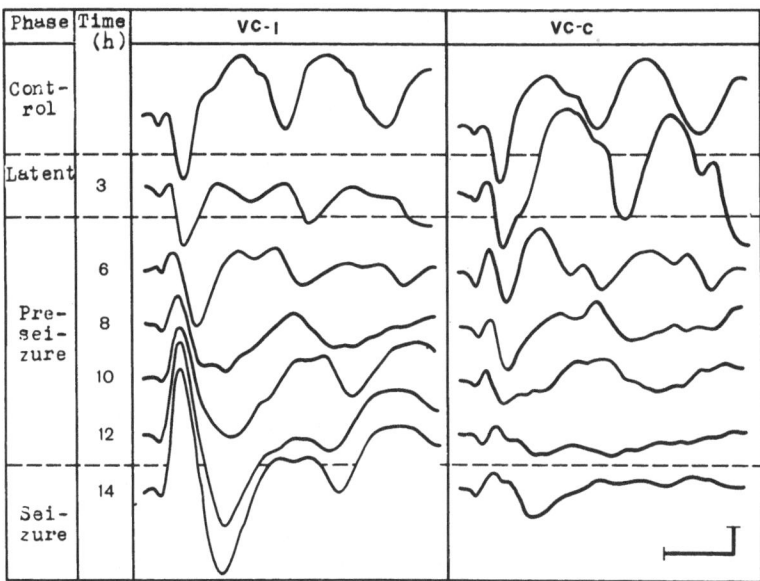

Fig. 34.   Change in EP in the visual cortex (VC) during various
           periods of the formation of GPEE in LGN. VC-I: Ipsilateral
           region, and VC-C: contralateral region of VC with respect
           to LGN into which TT was injected. The time after TT
           injection is indicated on the left. EP was recorded when
           light stimulation (illumination of 45 luxes) was used.
           Amplitude: 150 µV; duration: 100 ms.

At the same time, the late slow positive component (P3) remained
in modified EP, which was recorded in LGN when GPEE was being formed
in the nucleus. This component is connected with the prolonged
inhibition of the relay elements of the nucleus (Vastola, 1960;
Burke, Sefton, 1966). Its preservation shows that not all the forms
of inhibition in LGN are disturbed under the influence of TT and,
consequently, not all the relay elements of the nucleus form GPEE.

The peculiarities of a change in the photoreactivity of contralateral VC during the formation of GPEE in LGN and, in this respect,
the origin of the hyperactive LGN (generator)- ipsilateral VC system
are noteworthy. Investigations have shown that the parameters of the
primary response in the contralateral VC did not considerably change,
but the amplitudes of the secondary response (P2) substantially
decreased in comparison with the normal ones (Figure 34). This fact
should be taken into account when an assessment is made of the integrative mechanisms of the brain during the development of the pathologic process engendered by GPEE. It shows that the mechanisms which
control the excitation of the VC elements are switched on and are
gradually enhanced. These mechanisms can be set to suppress excess-

ive excitation, which is produced by the LGN-generator, and to counterbalance the disturbances in the neurodynamic relations in the brain that result in a seizure syndrome (see Chapter 9).

Besides the above-mentioned experiments, which were carried out with a view to studying the formation of GPEE in the LGN of rats, similar experiments were carried out with cats. These investigations showed (see Chapter 9) that the same GPEE originates in cats when inhibition is disturbed under the influence of TT in LGN (Kryzhanovsky et al, 1976c). The amplitude and duration of the focal potentials in the nucleus, being produced by a light flash, increased with time. Such an EP change in LGN testified to an increase in total electrogenesis in the nucleus, being characteristic GPEE formations. In turn, intense EP in LGN produced intense EP in ipsilateral VC, showing that the flow of impulses generated in LGN had intensified, i.e. that GPEE became more powerful as it was being formed. The changes in EP in LGN fully corresponded to those in ipsilateral VC at the early stages of a pathologic process. Thus, characteristic relations originated between the determinant structure in LGN, which forms a functional message, and the structure which receives and reproduces this message in VC. The changes in this relation were observed only at the late stages of the process. It was expressed in the fact that EP in LGN began to relatively diminish, while EP in VE, to increase. This phenomenon is apparently connected with the disorganization of GPEE on LGN at the late stages of the process and with the growth of neuronal excitability in VC, in which the secondary generator begins to be formed. The diminution of EP in LGN, which shows that electrogenesis in the generator nucleus decreases, can probably be due to the start of some mechanisms which suppress GPEE activity at this stage.

GPEE IN THE CEREBRAL CORTEX

In the cerebral cortex, GPEE's are the foci of epileptic activity that can be produced by various techniques involving diverse agents. Like in the previous experiments experiments in creating GPEE's in various areas of the central nervous system, GPEE in the cat's cerebral cortex was produced by TT (Kryzhanovsky et al., 1978d). Other researchers have produced epileptic foci in the cortex by TT in a similar way (Carrea, Lanari, 1962; Brooks, Asanuma, 1965). Investigations have revealed several peculiar features of GPEE in the cerebral cortex.

Clear electrocorticographic changes were recorded in cats under the conditions of their free behavior after a certain latent period following the injection of TT into the motor cortex. These changes became more pronounced as the process developed. They consisted in the origin of bursts of high-amplitude seizure discharges in the zone into which TT was injected (Figure 35 A 1). Such epileptic activity

was recorded also in the symmetrical cortical zone of the contralateral hemisphere ('mirror' focus) (Figure 35 A 2). These bursts
corresponded to the clinical seizure contraction of the muscles of
the extremities which were contralateral to the primary focus.
During a seizure, the frequency of the epileptiform spikes changed:
at first, it was 3-4 per second; by the middle of the process, it
increased to 5-8 per second, and at the end of the seizure, it diminished to 2 per second (Figure 35 A). As the process developed, the
seizures could last 60 seconds, while the frequency of their origin
was 1-2 per minute.

The foci of activity at the symmetrical points of both hemispheres were functionally unequivalent. These differences were
clearly exhibited during electrocorticographic recordings when the
animals were immobilized in a stereotaxic frame, being treated with
myorelaxin, and were subjected to artificial respiration: under these
conditions, epileptic activity was recorded only in the zone were TT
was injected (Figure 35 B 1) and was absent on the opposite side
(Figure 35 B 2). Electric activity was changed in the GPEE region
itself: some discharges grouped into complexes (Figure 35 B 1)
appeared instead of the prolonged high-amplitude and high-frequency
discharges (Figure 35 A 1). Hence, the disappearance of epileptic
activity in the 'mirror' focus under the given conditions is probably
due to the attenuated production of excitation in GPEE, the hindered
propagation of excitation from GPEE to the opposite hemisphere, and
the suppression of the activity of the neurons of the 'mirror' focus
when the central nervous system is generally suppressed. It will be
shown (Chapter 13) that such relations between the hyperactive determinant structure and the dependent structures are observed during the
action of various agents which suppress the central nervous system.

An analysis of the focal ECoG potentials, which are recorded in
an immobilized animal in the GPEE region, the membrane potential and

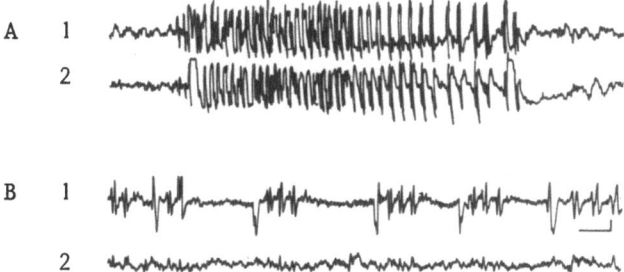

Fig. 35. Electric activity in the cat's cerebral cortex in the TT
          injection zone (1) and in the symmetrical point of the
          contralateral hemisphere (2). A: Electrocorticogram of
          freely moving animals; B: electrocorticogram of motionless
          animals. Amplitudes: 100 μV (A); 200 μV (B); time: 1 s.

impulse activity of the GPEE neurons (intracellular recordings) have
revealed several features of the electric activity of GPEE in the
cortex. There were two types of ECoG activity: type I, or spontaneous spikes, are interictal discharges (IID) whose amplitude
reached several millivolts, and their duration was 80-200 ms, and
type II, or waves of smaller amplitude (100-200 μV) without peak
potentials. During IID, continuous background impulse activity was
recorded in most of the neurons which were studied in the GPEE
region, while irregular background impulse activity was recorded in
other neurons. This activity was not recorded in some neurons.

  Paroxysmal depolarization shifts (PDS) of the membrane potential, which occur simultaneously with IID, were observed during the
activity of type I in most of the neurons (85 per cent) which were
studied (Figures 36,37,38). Only an increase in the frequency of the
discharges without any substantial shift of the membrane potential
was observed in some neurons (15 per cent). The frequency of the
neuronal discharges reached 500 per second during the PDS (Figure
36 A). Often, impulses which originated then had ceased (Figures
36 B, 38).

  The frequency of the discharges increased as the membrane was
slowly depolarized in neurons with continuous background impulse
activity during the intervals between the IID complexes (Figure 37).
The activity of these neurons was characterized by the absence of
membrane hyperpolarization after PDS (Figure 36 A B), while the PDS
amplitude varied in some neurons (Figures 36,37). During the IID

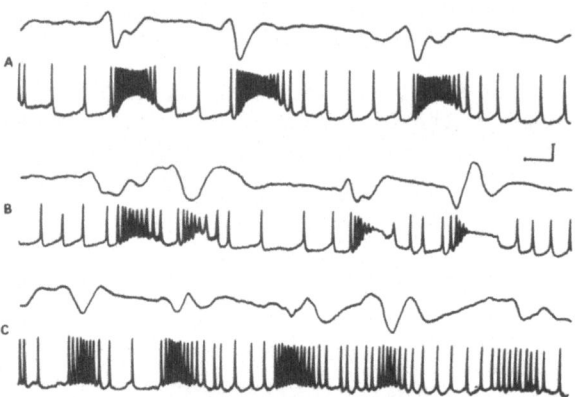

Fig. 36. Electric activity of neurons with continuous background
          impulse activity (A,B) and neurons with irregular back-
          ground impulse activity (C) during the interictal dis-
          charge. Top recordings are electrocorticograms, and the
          bottom recordings are intracellular activity. Amplitudes:
          100 μV (electrocorticogram), 20 mV (intracellular record-
          ings); time: 40 ms.

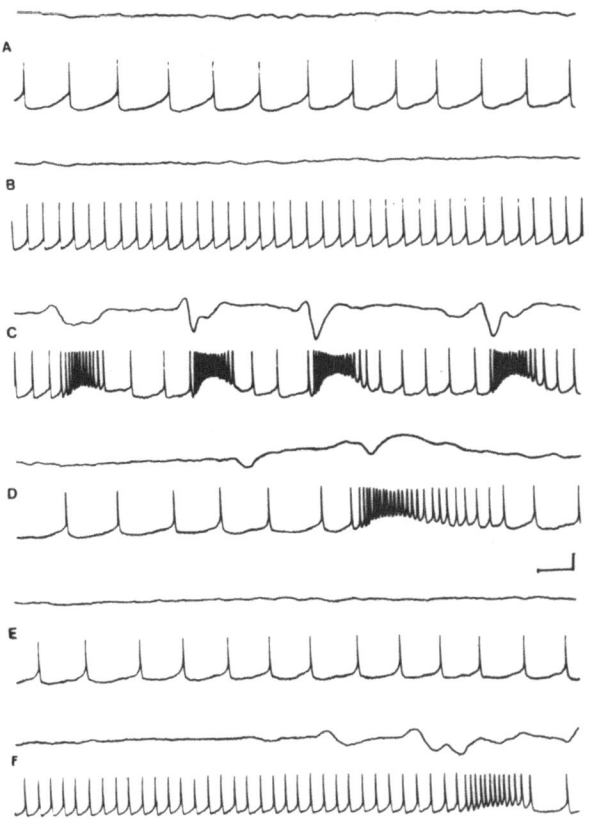

Fig. 37. Change in the electric activity of a neuron with continuous
background impulse activity during the interval between
interictal discharges. A-F: Successive recordings after
40, 45, 85, 90 and 130 s, respectively. Other indications
are the same as in Figure 36.

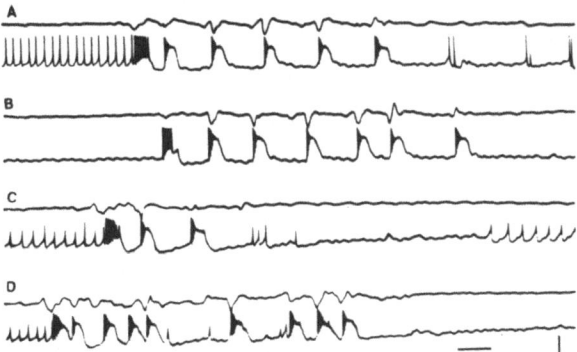

Fig. 38. Electric activity of a neuron with irregular background
impulse activity. A-D: Successive fragments of recordings.
Time: 200 ms. Other indications are the same as in Figure 36.

complexes, every successive PDS was accompanied by an increase in the interimpulse interval of background activity. The development of PDS in the neuron and IID in the cortex stopped when the interval increased to 80-100 ms. The frequency of the discharges increased without any substantial PDS in some neurons with irregular background impulse activity during IID (Figure 36 B), while PDS with a relatively constant amplitude was recorded in other neurons (Figure 38). Besides this, membranous hyperpolarization was observed after PDS in neurons with irregular background activity (Figure 38).

Irregular PDS with a frequency of 1-5 per second were recorded in neurons during the activity of type II (Figure 39). The frequency of background impulses decreased after every PDS. There was no longer any PDS, and the IID amplitude decreased when the intervals between the background impulses increased to more than 100 milliseconds.

The growth of the discharge frequency in the GPEE neurons, being both coupled with the development of PDS and connected with the origin of IID, has been observed in the cortical foci produced by various epileptogenic agents (Sawa et al., 1963; Matsumoto, Ajmone-Marsan, 1964; Prince, 1967; Okujava, 1969; Dichter et al., 1972; Kao, Crill, 1972; Ebersole, Levine, 1975). The frequency of PDS and, accordingly, IID in the zone of GPEE produced by TT averaged 3-5 per second, being greater than the frequency of IID in acute epileptic foci produced by strychnine of penicillin (1-2 per second) (Sawa et al., 1963; Matsumoto, Ajmone-Marsan, 1964; Sypert et al., 1970; Ebersole, Levine, 1975) and less than the frequency of IID in chronic foci produced by aluminum cream (9-10 per second) (Ward, 1972; Calvin et al., 1973).

The presence of neurons with continuous and irregular impulse activity (and also neurons in which background impulse activity is not recorded) in GPEE shows that the neuronal population which con-

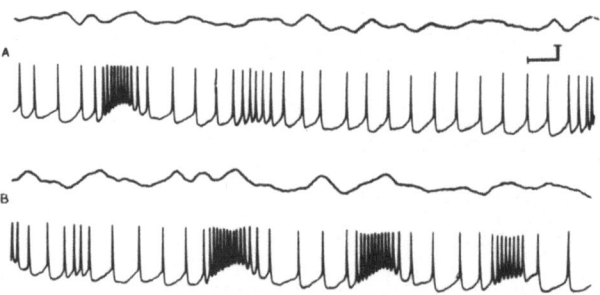

Fig. 39. Electric activity of a neuron with continuous background impulse activity during the electrocorticogram of the activity of type II. A-B: Successive fragments of recordings. Indications are the same as in Figure 36.

stitutes GPEE is functionally heterogeneous and that its neurons are
unequivalent in the activity of GPEE as an epileptic focus. The
involvement of the GPEE neurons in synchronization shows that a part
of them acts as 'leading' elements which activate other neurons, draw
them into the pathologic process, synchronize their activity, and
thus form an 'epileptic neurogenic aggregate' (Ajmone-Marsan, 1961),
i.e. GPEE. The 'leading' neurons also in other epileptic foci have
been described by several authors (see Chapter 3). Neurons with
regular impulse activity can play the leading role. Other neurons
may be effectively drawn into the generation of pathologically enhanced excitation when the percentage of the 'leading' neurons is
high. Theoretical analysis (Calvin, 1972) shows that the origin of
PDS in the normal neurons of the focus may be connected with this
process. The high degree of synchronization of the 'leading' neurons
should also promote the process, since all the presumed 'leading'
neurons increased the frequency of their impulses simultaneously with
the appearance of IID. The formation of PDS in the GPEE neurons was
clearly observed during the origination of IID. The frequency of the
discharges of the 'leading' neurons increased when the IID complex
began to develop (Figure 37). PDS originated only during the subsequent periods. In this case, gradual changes in the amplitude and
duration of PDS were revealed (Figures 36 B;37). It could be that
PDS in the 'leading' neurons change gradually, unlike those in the
neurons which are drawn for the second time into the process and in
which PDS's originate according to the 'all-or-nothing' law (Figure
38). A relatively large number of 'leading' GPEE neurons shows that
there is an epileptization mechanism which is common to all these
neurons. The disturbance of the inhibitory processes under the
influence of TT may be such a mechanism.

    Several factors participate in the formation of GPEE in the
cerebral cortex, which has a sufficiently large critical mass of
neurons and is therefore pathogenetically significant (see Chapter
3). An increase in the frequency of the discharges of the 'leading'
neurons (Prince, Futamachi, 1970; Glotzner at al., 1973; Ebersole,
Levine, 1975), the activation of the positive feedbacks (Dichter,
Spencer, 1969; Ayla et al., 1973) and other processes which augment
the synaptic bombardment of neurons are important, since PDS may
originate even in a normal neuron when a definite level of stimulating input is attained (Calvin, 1972, 1975)

    It has already been shown that hyperpolarization shifts do not
appear in neurons with continuous background impulses after the
origin of PDS and seizure discharges. The membrane was hyper-
polarized to a certain extent only in other neurons. The absence of
hyperpolarization indicates that postsynaptic inhibition is substantially reduced. This is in accord with the data on the ability of TT
to upset inhibition of various types (see Chapter 5). The inhibitory
mechanisms of neurons with continuous and irregular background activity may be considered upset to a different extent, and this fact
establishes their different role in GPEE activity. The different

frequencies of IID and PDS in acute and chronic foci are associated with a different extent of the disturbances of inhibitory processes: the hyperpolarization wave is more prolonged in acute foci than in chronic ones. The hyperpolarization potentials are regarded as an important mechanism which terminates epileptic activity (Prince, 1967, 1968; Okujava, 1969; Prince, Futamachi, 1970; Glötzner, 1974; Wyler, 1974). The disappearance of epileptic activity when the inhibitory mechanisms of most GPEE neurons are either absent or substantially disturbed can show that not all the types of inhibition are upset in these neurons. This fact is in accord with the data which show that TT disturbs not all the types of inhibition (see the previous pages) and that there are also other mechanisms which suppress this activity. In particular, the generator mechanism may be inactivated during PDS (Figure 36 B), and then the frequency of the background neuronal impulses decreases with every successive PDS. Such an inactivating mechanism has been described also by other authors. When impulse production by the neurons decreases, the effectiveness of their synaptic interaction diminishes, ultimately causing the 'epileptic aggregate' to break down, i.e. GPEE is either reduced or disappears.

SUMMARY

The data of this chapter show that GPEE is the neuronal population which produces an extremely intensive and long flow of impulses that is inadequate for the activating stimulus. Such production of excitation can occur also spontaneously, i.e. without any special stimulation. The local insufficiency of the inhibitory mechanisms in the population of neurons which constitute GPEE is the main cause of the origin and activity of GPEE.

The generator can be produced in various areas of the central nervous system, because all the areas have populations of neurons with many positive and negative connections.

The GPEEs formed in various areas of CNS have their specifics due to the preformed structural and functional connections inherent in the given area of CNS. This fact also illustrates the peculiarities of the experimental analysis of the activity of various GPEEs. At the same time, GPEEs reveal several common features in their activity due to a common property, namely, excessive electrogenesis. These features are expressed in the intensification of background neuronal impulse activity, the growth of induced activity, and the appearance of components which are new in comparison with normal conditions in the evoked responses of the populations of neurons that form GPEEs at a late stage of their formation. GPEEs can operate as autonomous structures which are capable of developing prolonged and self-sustaining activity without any additional stimulation.

3
Functional organization and properties of GPEE and the mechanisms of its formation

FUNCTIONAL STATE OF THE NEURONAL POPULATIONS OF GPEE

The change in the functional state of the neuronal populations during the formation of GPEE is characterized by two processes: (1) neurons lose their initial functional distinctions; figuratively, this process can be regarded as the pathologic 'functional homogen-ization' of the neuronal pool, and (2) neurons acquire new pathologic properties and constitute new functional groups in accordance with these properties; figuratively, this process can be regarded as the pathologic 'functional differentiation' of neurons.

Inhibitory mechanisms play an important role in the first pro-cess. The example of the organization of GPEE in LGN (see Chapter 2) has shown that, as a result of the disturbance of the inhibitory mechanisms, individual neuronal groups 'disappear', while neurons which react to the 'off'- stimulus produce reactions to the 'on'-stimulus and become trigger ones that are capable of 'starting' GPEE. All this engenders a new group of neurons which are dis-inhibited and whose reactions are of a single type. This group grows with time at the expense of neurons of other groups. The development of this process is connected with the distribution of the pathogenic agent, which upsets the inhibitory mechanisms, and with increasing synaptic neuronal bombardment, which promotes the depolarization of neurons and, consequently, the attentuation of the inhibitory effects and the enhancement of neuronal excitability.

The second process, i.e. pathologic 'functional differen-tiation', has complex mechanisms. Besides the insufficiency of inhibitory control, the process depends on both the endogenous pro-perties of the neurons and the influences to which they are subjected

when GPEE is being formed. The example of the formation of GPEE in
the cerebral cortex (see Chapter 2) has shown that the neurons which
produce epileptic activity possess various properties. The definite
extent to which these properties are expressed, their stability, and
the new nature of the activity of some neurons has made it possible
to single out the given neurons as 'leading' or 'trigger' ones.

Many authors have thoroughly studied the properties and behavior
of neurons in epileptic foci in experiments with animals and in
clinical observations of man (Matsumoto, Ajmone-Marsan, 1964; Prince,
1967; Okujava, 1969; Sherwin, 1970; Sypert et al., 1970; Glötzner et
al., 1973; Wyler et al., 1973; Wyler, 1974; Fetz, 1974; Schmidt et
al., 1976; Loeser, Howe, 1980; Ward, 1980; Wiler, Ward, 1980; Prince,
Wong, 1981). As a result, three main groups of neurons which participate in the production of epileptic activity can be singled out.

Neurons of group I have relatively stable 'epileptic' features:
the spontaneous, monotonously developing depolarization of the membrane (against the background of which PDS appears), burst activity
during PDS, a definite value of the so-called burst index, the absence of a change in activity during the action of environmental
stimuli (including conditioned stimuli) during sleep, etc. Hence,
the activity of these neurons is relatively autonomously epileptic
from the very start, and they are not subject to integrate inhibitory
control of the brain. These properties show that the neurons are
'leading' or 'trigger' ones. Their number in a focus may be relatively small, but they are the ones which draw other neurons into both
the pathologic process and general pathologically enhanced electrogenesis cause the neuronal population of GPEE to be synchronized.
They constitute the initial critical neuronal mass which is necessary
for triggering GPEE (this will be discussed further).

Neurons of group II are capable of changing the pattern of their
activity under the influence of various stimuli and also when the
functional state of the brain changes, e.g. during sleep, they are
not immediately drawn into the epileptic process, but once they are
drawn into it, the synchronize their activity with that of the neurons of group I, and then they become similar by their electrogenetic
properties to the neurons of group I (PDS, burst discharges, and the
subsequent inactivation of the generator mechanism in them). Hence,
the neurons of this group are also 'epileptic neurons', but with
variable epileptic characterization. Normally, they are subjected to
inhibitory control by the brain. This group of neurons is not the
'leading' one; it establishes to a great extent the fluctuation of
the dimensions of the active epileptic focus, namely, GPEE.

Neurons of group III are normal neurons. Just as the neurons of
group II, they are drawn into the pathologic process for the second
time, but even more slowly. Afterwards, however, their properties do
not become similar to those of the neurons of group I. The burst

activity of the neurons of this group is, as a rule, not accompanied
by a shift in the membrane potential and the subsequent inactivation
of the generator mechanism. Therefore, the neurons of this group do
not become completely epileptic ones even when they are drawn into
summary epileptic electrogenesis. They can be subjected to integrative inhibitory control more than other neurons. The mechanisms
which realize this control on the given neurons are not damaged, but
they are functionally insufficient when they are drawn into general
epileptogenesis. The population of such neurons constitutes half or
even the greater part of the cells with respect to the amount of the
neurons in the focus.

Functional epileptic heterogeneity is more pronounced in chronic
epileptic foci than in acute foci. The involvement of neurons in
epileptic activity in chronic foci is relatively low in percentage
terms (Glötzner et al., 1973), and there is virtually no correlation
between the activity of individual neurons of a focus and ECoG activity (Prince, Futamachi, 1970; Schmidt et al., 1976). These peculiarities of the chronic focus should be taken into account when an
experimental analysis is being made.

An analysis of the properties of the 'leading' neurons in
chronic epileptic foci, being based on the specifics of the so-called
structuralized discharge (Glötzner et al., 1973; Wyler et al., 1973;
Wyler, Fetz, 1974), has shown that the population is also heterogeneous and that neurons with a higher or lower degree of epileptization can be distinguished in it.

The above-mentioned features of the functional heterogeneity of
the neuronal populations of the epileptic foci are noteworthy not
merely in connection with the study of the functional organization of
GPEE. They also shed light on some mechanisms which activate the
trigger GPEEs (this will be discussed further) and on the characteristics of the effects of therapeutic influences which are intended to
suppress GPEE (see Chapter 13.).

CRITICAL MASS OF NEURONS WHICH CONSTITUTE GPEE

The critical mass of neurons which constitute GPEE is its important parameter. This concerns not only the power of GPEE, but
also the conditions and mechanisms of its formation and activity.
The matter in question is the neuronal mass that should constitute
the population which can operate as an independent generator when the
inhibitory mechanisms are insufficient, i.e. operate under the
conditions of relative autonomy, maintaining the pathologically
enhanced level of excitation.

That question can be discussed by taking the example of acute
and chronic epileptic foci in the cerebral cortex. According to some

authors (Goldenson et al., 1970), the area of the penicillin-induced
focus in which epileptic activity is recorded can be one square
millimeter. Dichter et al. (1972) produced an epileptic focus with
a diameter of 2-5 mm in the hippocampus. According to Reichental and
Hocherman (1977), stable epileptic activity has been recorded in a
focus with an area of 0.6 mm^2, while the focal area of 0.4 mm^2 is
critical for the origin of epileptic activity. Walsh (1971), who
used iontophoresis of penicillin to the cortical neurons, has recorded only an increase in the spontaneous variations of the membrane
potential when the zone of penicillin action was limited; in some
cases, impulse bursts appeared without any substantial shift in the
membrane potential. His data show that the minimum focal dimensions
are 0.4 mm^2. Leuders and others (1980) have shown that an area of
0.25 mm^2 is enough for the origin of epileptic activity in the cat's
sensorimotor cortex. In general, the area required for epileptic
activity is inversely proportional to the concentration of penicillin
being applied. According to Petshe and others (1979), the focal
dimensions can reach 300 microns. According to Walsh (1971), the
diameter of the horizontal branches of the excitatory interneurons in
the upper and lower layers of the neocortex is about 400 microns
(Endo, Araki, 1981). These interneurons probably form the basis of
the neuronal aggregates in the cortical GPEE. Investigations with
microelectrodes have shown that such aggregates correspond to neuronal columns (Gabar et al., 1979). A noteworthy fact is that the foci
of epileptic activity cannot be formed by the application of convulsants to the cerebellar cortex, which has no excitatory interneurons
(Eccles et al., 1967; Yuasa et al., 1981). This has been shown also
in our investigations (the experiments were carried out by Makulkin
and Shandra) involving the application of strychnine, penicillin and
acetylcholine to the cortex vermis of the cerebellum. Epileptic foci
originate especially easily in the cortex of the hippocampus, where
the density of the excitatory neurons is high. Hippocampal pyramidal
neurons can produce epileptic activity also under normal conditions
(Kandel, Spencer, 1961; Ogata, 1975, Schwartzkroin, 1975; Wong,
Prince, 1977, 1978; Crill, 1980). There are facilitating connections
between the hippocampal neurons. These connections are confined to a
group of 60 neurons (the aggregate is 200 microns) (Traub, Wong,
1981).

   The integral activity of GPEE is not an arithmetical sum of the
activities of its individual neurons and their aggregates because of
many reciprocally facilitating influences which originate during the
formation and activity of GPEE. This should be taken into account
when the critical mass of the GPEE neurons is being assessed. It
will be shown that the large content of extracellular K^+ plays an
important role in hyperactivating the neurons. By taking the example
of the spreading Leao depression with a hyperactive component, whose
generator is a potassium focus (Kuznetsova, Koroleva, 1978), it has
been shown that at least one cubic millimeter is the minimum amount
of the nerve tissue in which the critical concentration of K^+, being

required for the origin of both the focus and spreading depression, is attained (Zachar, Zaharová, 1963; Matsura, Bures, 1971).

Thus, when the matter in question is the critical mass of GPEE neurons, two parameters should be taken into account: (1) the minimum number of 'leading' (trigger) neurons that is needed for drawing other cells into the process of forming pathogenically significant GPEE, and (2) the minimum amount of neurons that is needed for forming pathogenically significant GPEE and for its activity. That number depends on the power of the electrogenesis of the trigger neurons, the extent of 'readiness' for hyperactivating the neurons which were drawn secondarily into the process, and on factors which modulate the course of the processes in GPEE. Some data show that the critical number of the initial epilepticized neurons is 8-10 for the penicillin focus in the cortex. The dimensions of GPEE increase further, the number of its active neurons grows, and they are epilepticized as pathologically enhanced excitation is produced. Those processes can occur in a chain reaction.

GPEE POWER

The power of GPEE depends on the final number of neurons which are drawn into summary electrogenesis and on the extent of electrogenesis of every GPEE neuron. Both of these factors are interconnected and potentiate one another: when the membrane is substantially depolarized, the neuron generates multiple discharges, while the extent of this depolarization depends on synaptic bombardment, i.e. on general electrogenesis. The generator's power depends on its dimensions when the density of the GPEE neurons is more or less uniformly distributed in the nervous tissue. The example of GPEE produced in the system of propriospinal neurons in the spinal cord (Chapter 2) has shown that the P wave, being recorded from the cord surface, and the appropriate focal potential, which is drawn from the depth of the cord where GPEE is localized, increase as the pathologic process develops. As the potentials reflecting the summary electrogenesis of GPEE increased, the region in the spinal segments where the potentials were recorded grew, i.e. GPEE became larger. The focal potentials, the number of active points in the region where GPEE is formed, and the dimensions of GPEE increased also when the generators were formed in other areas of the central nervous system (see Chapter 2). The growth of the dimensions of GPEE can be traced by observing the growth of its trigger zones (this will be discussed further).

In analyzing the pathogenetic role of GPEE, its power should be assessed by the extent of the effects which cause excitation produced by it.

Figure 40 shows the dynamics of the changes in the electric activity (EA) of the muscles of the hind leg when GPEE is being formed in the ventral horns of the lumbosacral segments, i.e. in the system of efferent output (see Chapter 2). The growth of the dimensions of GPEE, which gradually involves the nuclei of various muscles, can be traced by the EA change in the respective muscles.

The effects of GPEEs in other areas of the central nervous system, namely, Deiters' nucleus, the reciprocal nucleus, and the giant-cell nucleus of <u>medulla oblongata</u>, have been traced in the same way (Kryzhanovsky, Sheikhon, 1976). The changes in the descending inhibitory and facilitating influences on the spinal reflexes during the formation of GPEE were the criterion used in this respect.

To activate the nuclei in which GPEE was formed in these experiments, i.e. to activate GPEE itself, its structures were directly electrically stimulated. Instead, use was made of the orthodromic message from the fastigial nucleus of the cerebellum, which was stimulated by an electric current of the same strength throughout an experiment. When this nucleus is stimulated, the descending effects are realized via the bulbar reticular formation (Lindsley, 1952; Walberg et al., 1962; Eccles et al., 1974; see also Chapter 2). The giant-cell nucleus and Deiters' nucleus are relay structures in which impulses from the cerebellum are switched over (Magoun, 1950; Brodal et al., 1962; Sasaki, Tanaka, 1963; Grilliner et al., 1970). This method of indirect stimulation for activating GPEE is important in such experiments, since it is necessary to be sure that the originat-

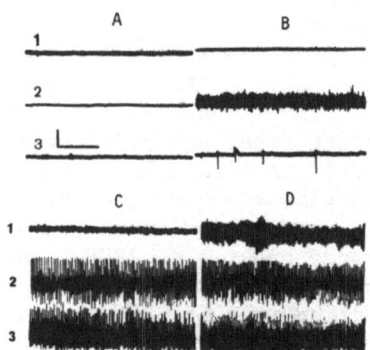

Fig. 40. Change in EA in the hind leg muscles after the formation of GPEE in the ventral horns of the spinal lumbosacral segments. Electric activity in a rabbit's anterior group of the left crural muscles (1), the left gastrocnemial muscle (2), and the left quadriceps (3) 3 h (A), 24 h (B), 48 h (C) and 72 h (D) after injecting a pain-producing dose of TT into the left gastrocnemial muscle. Amplitude: 0.3 mV; time: 0.5 s.

ing functional effects are connected with the natural activation of
GPEE and not with the electric stimulation of all the structures
which are within the range of action of the electric field near the
electrode.

Figure 41 shows that the descending facilitation of the extensor
monosynaptic reflexes from Deiters' nucleus grows when GPEE is being
produced in the nucleus. This effect increases with time as the
generator is being formed. At the height of the process, the facilitation of reflexes can be very great, being 5-10 times greater than
the normal level. A noteworthy fact is that the curve of facilitation in the activation of GPEE is similar to that under normal
conditions: the peak of facilitation is attained within 15-20 ms
after the cessation of conditioning stimulation. In many cases, the
curve was double-peaked (just as under normal conditions), especially
at the rather early stages of GPEE formation. Such a type of curve
has been noted by other authors, too (Sasaki, Tanaka, 1963). Thus,
purely quantitative distinctions were observed in comparison with the
normal level. The facilitation of reflexes was very considerable and
prolonged. It could be 10-20 times longer than the normal level. In
some experiments, the thresholds of the monosynaptic reflexes decreased and their amplitudes increased without the conditioning
stimulation of the cerebellum, showing that GPEE was spontaneously
active in Deiters' nucleus.

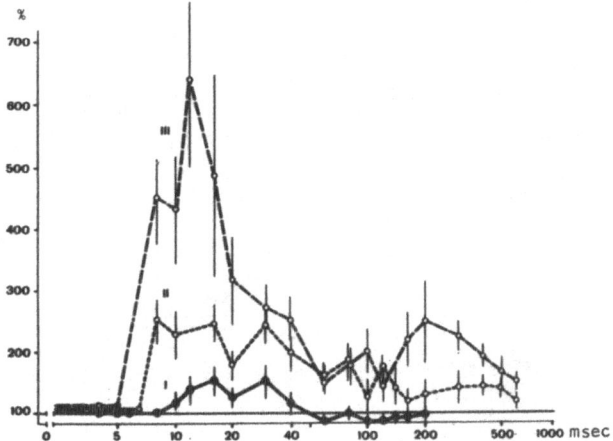

Fig. 41.   Supersegmental facilitation of extensor monosynaptic
           reflexes after the formation of GPEE in Deiters' nucleus.
           Before (I) and two (II) and four (III) hours after inject-
           ing TT into Deiters' nucleus. Abscissa: time (ms) between
           the stimulation of the cerebellar fastigial nucleus and the
           stimulation of the gastrocnemial nerve. Ordinate: reflex
           values (% of the initial level).

A similar effect was observed when GPEE was being formed in the reciprocal nucleus (Figure 42). The facilitation of the flexor monosynaptic reflexes, being observed when the nucleus is activated under normal conditions, increased and became very considerable (it reached 500 per cent) and prolonged (by 5-10 times) at the height of the process when GPEE was produced in the nucleus. Purely quantitative distinctions were observed in comparison with the normal level in this case, too: the curve of reflex facilitation when GPEE was worn out had the same nature as that before GPEE was formed.

A similar result was obtained when GPEE was produced in the giant-cell nucleus (Figure 43). The inhibition of the polysynaptic reflexes was more profound and prolonged when GPEE was activated than when the nucleus was stimulated before the formation of GPEE. The extent of inhibition and its duration increased with time as GPEE was being formed; they reached the maximum at the height of the process, when the reflexes were almost completely inhibited (the effect grew by 5-10 times) and the duration of inhibition considerably increased (by 5-10 times). In this case, the curve of the process during the activation of GPEE was similar to that of the process which occurs under normal conditions (maximum inhibition was observed 100-150 ms after the cessation of conditioning stimulation). Hence, purely quantitative distinctions were observed in comparison with the normal level.

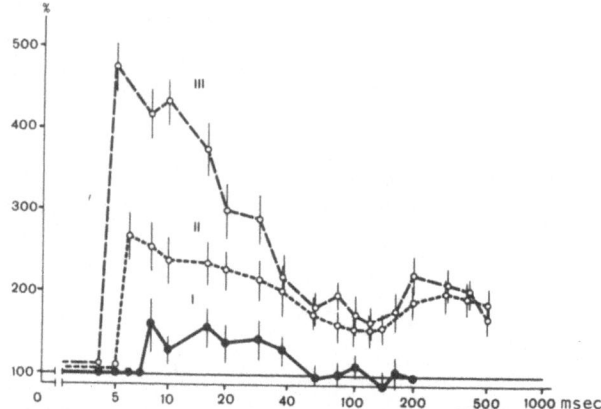

Fig. 42. Supersegmental facilitation of the flexor monosynaptic reflexes after the formation of GPEE in the medullar reciprocal nucleus. Before (I) and two (II) and four (III) hours after injecting TT into the reciprocal nucleus. Abscissa: time (ms) between the stimulation of the cerebellar fastigial nucleus and the stimulation of the gastrocnemial nerve. Ordinate: reflex values (% of the initial level).

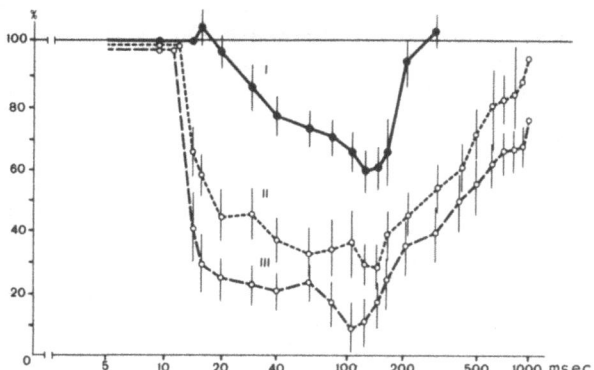

Fig. 43. Descending inhibition of the spinal polysynaptic reflexes
after the formation of GPEE in the medullar giant-cell
nucleus. Before (I) and two (II) and four (III) hours
after the injection of TT into the giant-cell nucleus.
Abscissa: time (ms) between the stimulation of the cerebellar fastigial nucleus and the stimulation of the gastrocnemial nerve. Ordinate: polysynaptic reflex values (% of
the initial level).

     Thus, the same results were obtained when GPEEs were produced in
the functionally different systems which realize the facilitation and
inhibition of various spinal reflexes, i.e. natural effects characteristic of the activity of the given nuclei were extremely enhanced
under ordinary conditions. This intensification is inadequate with
respect to activating stimulation. It grows with time as GPEE is
being formed and shows that the power of GPEE increases.

POSITIVE NEURONAL CONNECTIONS IN GPEE ACTIVITY

     Much data suggest that virtually all the cells of an animal
organism are capable of developing endogenous pacemaker electric and
other rhythmic activity (Carpenter, 1982). In classical experiments,
Alving (1968) showed that a neuron without synaptic inputs generates
regular spikes. Synaptic spikes modulate this activity: a neuron
under synaptic influence generates irregular discharges. Therefore,
neuronal activity can be regarded as the admixture of endogenous
pacemaker activity and synaptic inputs. Neuronal deafferentation
causes the neuron to become hyperactive (see Chapter 5) probably due
to the disinhibition of its ability to generate endogenous activity.
The membrane potential and resistance, with which cellular endogenous
electric activity is connected, depends on the concentration of
intracellular Ca^{2+} (Meech, Standen, 1975). Voltage-dependent Ca^{2+}
and associated slow K^+ conductance mediated by intracellular Ca^{2+}
play a critical role in generating spontaneous low-frequency spike

activity and burst firing by the CNS neurons (Wong, Schwartzkroin, 1982). Hippocampal neurons can generate Na^+-dependent Ca^{2+} spikes (Schwartzkroin, Slawsky, 1977). Such spikes were recorded in neurons of various CNS regions. Endogenous burst activity was recorded in the CA3 hippocampal neurons when chemical synaptic transmission was blocked (Wong, Prince, 1978).

The ability of the neurons to endogenously generate electric activity should play a definite role also in the mechanisms of their hyperactivity. Under in vivo conditions, however, a very complicated and largely inexplicable effect of diverse extracellular factors which modulate neuronal properties is observed. Abnormal neuronal activity seems to be the outcome of the interaction of intrinsic and extrinsic properties of nerurons and their population under pathologic conditions (Schwartzkroin, 1982).

Epileptic activity, whose pattern is pathologic, differs from that which is endogenously generated by the neuron. It is produced by neuronal 'aggregates' whose burst activity is more or less synchronized (Schwartzkroin, 1982). In short, intercellular interactions and diverse mutually potentiating effects, which are enhanced due to the insufficiency of the inhibitory mechanisms in the GPEE neuronal populations, play an important role in the formation of GPEE as a neuronal 'aggrtegate'. Such effects include positive intra- and extrapopulational neuronal connections.

Positive interneuronal connections are the basis of the activity of any formations of the central nervous system. They are especially significant under pathologic conditions. In particular, the ability of GPEE to produce an intense and prolonged wave of excitation is largely due to numerous positive neuronal connections.

Even the electrogenesis of neurons in tissue culture shows the significance of the interneuronal connections in the origin of a prolonged multiple discharge: the tendency towards a multiple response originates as the culture grows. High-amplitude waves occur two weeks after cultivation in response to a single stimulus. it is followed by a characteristic prolonged multiple discharge (Crain, 1972). Such a discharge is similar to the one which originates in the neuronally isolated cortical strip (Purpura, Housepian, 1961; Bogoslovsky, 1968; Burns, 1968; Shuranova, 1977). Complex multiple discharges can originate also under the influence of the 'pacemaker' neurons, the spontaneous firing of which can last a long time. The presence of a definite number of neurons, i.e. a definite critical mass of neurons, is a prerequisite of the duration of discharges and their spontaneous origin (Crain, 1972). The amplitude of slow waves and discharges sharply increases under the influence of strychnine. They assume the form which is similar to that recorded during epileptic activity in the cortex treated with strychnine (Crain, 1972; Crain, Bornstein, 1972). The neurons of the central nervous system, which were completely separated from one another after treatment with trypsin and which formed aggregates again in the culture, create a functional synaptic net with excitatory and inhibitory mechanisms that warrant the given features of the net when it functions (Crain, 1972). This picture of the neuronal interactions in tissue culture shows the similarity of the phenomena which are observed in the neuronal population of GPEE.

Normally, excitation in the neuronal populations of the struc-tures of the central nervous system is maintained by numerous direct positive connections in the form of branching neuronal chains (Livanov, 1965, 1975). In such systems, a stimulus activates a large number of neurons, and a reaction intensively occurs. Later, this reaction becomes limited and is maintained for a long time at a relatively constant level. Individual neurons in this system can maintain excitation for several seconds. The inhibitory mechanisms, which participate also in the abatement of the spread of excitation, limit a reaction in time and space. Reverberating chains in such a system are considered to be a mechanism which sustains excitation for a long time. Apparently, the neuronal population should be converted into GPEE when the inhibitory mechanisms are upset in it.

It is very enticing to use the mechanisms of reverberation for explaining many systemic neural processes. The presence of reverber-ating chains is theroetically postulated, but it is very difficult to obtain direct proof of their existence in physiological experiments (Kostyuk, 1974; Shuranova, 1977).

The idea of the existence and physiological significance of the reverberating neuronal chains was defined most clearly for the first time by Lorento de No (1933, 1938), who assumed that closed neuronal chains can allow excitation to circulate for a long time in such a contour. Later, this idea became the subject of special investigat-ions (Chang, 1950; Verzeano, Negishi, 1960; Burns, 1968; Bess, 1970; Shibata, Bures, 1972; Raeva, 1977; Shuranova, 1977; Kuznetsova, Koroleva, 1978). According to Asanuma and Brooks (1963), the pre-servation of excitation in the pyramidal neurons of the cortex for several tens of milliseconds following antidromic stimulation is connected with the activation of neurons via positive feedback. The origin of late discharges after brief grouped action potentials following single electric stimulation of a cat's motor cortex is regarded as proof of the existence of reverberating links (Okujava, 1969). The spontaneous activity of cortical pyramidal neurons may be due to exciting feedbacks from the cell to the branchings of its apical dendrites, which have a low threshold for generating the action potential (Arshavsky et al., 1966). Data suggesting that there are recurrent exciting connections have been obtained when various regions of the cortex, the hippocampus and the isolated cortex were being studied (Andersen, Lime, 1966; Takahachi et al., 1967; Guselnikov, Supin, 1968; Shibata, Bures, 1972; Guselnikov,

Guselnikov, 1975). Numerous direct and recurrent positive connections are characteristic of the so-called 'cyclic' formations of the
brain (Leontovich, 1978, 1980). It has been indicated (Sheibel,
Sheibel, 1961) that excitation may reverberate in the structures of
the reticular formation owing to the ability of the reticular neurons
and the collaterals of their axons to form cyclic pathways. Reverberation mechanisms in other nuclei and areas of the central nervous
system, i.e. the limbic system, the cerebellum, the pons, LGN, etc.,
have also been described (Arduini, Pompeiano, 1957; Sefton, Burke,
1965; Oniani et al., 1972; Tsukahara, 1972). They have been observed
also in the nervous system of animals (Burmistrov, 1965). According
to some researchers (Shuranova, 1977), the group pattern of the
discharges of the pyramidal cells (Stefans, Jasper, 1964; Takahashi
et al., 1967) and the neurons of Clarke's column (Kostyuk et al.,
1970) indicates (when these cells are subjected to antidromic stimulation) that the recurrent exciting collaterals are present. However, the 'recurrent EPSPs' are very weak, and are insufficient for a
discharge. They originate in neurons with other functional properties. Therefore, they supposedly cannot maintain circulating excitation for a long time. It could be that they bring one level of
synaptic excitability of a functional group of neurons into accord
with another group (Kostyuk, 1974). Other researchers also believe
that the recurrent postsynaptic influences are insufficient for
generating the action potential (Stefans, Jasper, 1964; Okujava,
1969).

Weak and poorly effective recurrent synaptic influences can
probably become functionally significant under pathologic conditions
owing to intense synaptic stimulation, which causes a sufficient and
steady shift of the membrane potential. As it has been shown in this
case, high-frequency firing with respect to even one input can cause
a neuronal discharge. When GPEE is formed, however, its neurons can
be discharged at a high frequency, as a result of which there may be
a small number of inputs that are necessary for activating the
neuron. Thus, reverberating mechanisms can be effective in establishing the power and autonomy of GPEE and the duration of every
cycle of its activity when the inhibitory mechanisms are insufficient
and synaptic bombardment is intense.

In studying the epileptic focus produced by penicillin in the
hippocampus, Dichter and Spenser (1969) drew the conclusion that the
positive feedback systems participate in the formation of PDS, which
occurs in some neurons involved in the epileptic process as the focal
connections are activated. Several researchers (Ayala, Vasconeto,
1972; Ayala et al., 1973) believe that the positive feedbacks are
important in the formation of epileptic activity. It is held that
the ring pathways and reverberation mechanisms play a certain role in
the pathogenesis of extrapyramidal disorders (Raeva, 1977).

To prove that excitation is reverberated in the neuronal net, Burns (1968) proposed that use should be made of local single electric shock of rather strong force that simultaneously excites all the neurons in the net of the given region. As a result, the neurons become synchronously refractive and their repeated excitation, being engendered by the reverberation mechanisms, ceases for a certain length of time. Burns used this method to analyze the activity of the neuronal net in a isolated cortical strip. We used this method in our investigations of GPEE produced in the giant-cell nucleus of the bulbar reticular formation (see Chapter 2). Rather strong direct electric shock of the region of the nucleus in the GPEE zone, where high background impulses were recorded, has suppressed that activity for a short time. Such an effect occurred after the appearance of a high-amplitude synchronized discharge (Figure 44). Electric activity intensified when the symmetrical region of the opposite nuceleus, in which GPEE was not produced (injection of inactivated TT), had been stimulated with approximately the same intensity. In considering the mechanisms of the effect, account should be taken (besides the simultaneous origin of refractiveness of the activated neurons) of the fact that the inhibitory mechanisms can be switched on. This was pointed out by Krnjevic and others (1966). However, it should also be taken into account that the inhibitory mechanisms in the population of the neurons which constitute GPEE were disturbed in the given experiments due to TT. Instead of the suppressing effect in the case of such stimulation, neurons were activated when the inhibitory mechanisms remained (in the opposite nucleus).

When the peculiarities of GPEE produced in the system of propriospinal connections were being considered (Chapter 2), it was

Fig. 44. Suppression of spontaneous background impulse activity in the medullar giant-cell nucleus region with GPEE after single strong ES. The top curve shows background activity in the nuclear region where GPEE is localized, and the bottom curve illustrates background activity in the symmetrical point of the opposite nucleus into which inactivated TT was injected. The moment of direct electric stimulation of both regions simultaneously is marked by an artifact. Stimulation strength: 30 V. Amplitude: 200 µV; time: 200 ms.

shown that GPEE localized on one side of the lumbosacral segments can
become very active for a long time when the spinal cord is cut (Th_7),
when the given segments are completely bilaterally deafferent, and
when the animal is completely immobilized by curare, i.e. when there
is no flow of impulses from the upper spinal segments, the supraspinal regions and the periphery. The activity of GPEE produced in
the spinal efferent output system continues similarly in the case of
decentralization and deafferentation (Chapter 2). These data show
that GPEE can develop powerful self-sustaining activity due to the
internal positive connections. Naturally, they are not confined to
the connections inside GPEE. There may be different circuits which
cover various structures on the segmental and suprasegmental levels.
A similar situation is believed to exist when GPEE is active also in
other areas of the central nervous system.

The functional organization of the CNS structures on the basis
of which GPEE originated is important in establishing the pattern of
the activity and power of GPEE, the duration of stimulus generation
by it, and the extent of its automony. The structure of GPEE and the
peculiarities of its activity will be different when it is produced
in linear formations, where long-axonic cells have no or almost no
intranuclear collaterals, and in cyclic formations, which have many
such collaterals and where intro-connections and feedbacks are multiform (Leontovich, 1978, 1980).

In discussing the mechanisms which maintain the activity of the
GPEE neurons for a long time, account should be taken of the presence
of external chains with positive feedbacks and the direct positive
connections of nuclei (in which GPEE is formed) with other areas of
the central nervous system. Such connections, which maintain tonic
influences, have been described also with respect to nuclei where
GPEEs had been created, namely, the giant-cell nucleus (Magoun, 1950;
Magni, Willis, 1964; Ito et al., 1970; Eccles et al., 1974), Deiters'
nucleus (Gernandt, Thulin, 1952; Markham et al., 1966; Markham, 1968;
Ito et al., 1969a), LG, and other areas of the central nervous system
(see Chapters 2, 7-12).

REPLACEMENT OF INHIBITORY EFFECTS BY EXCITING
EFFECTS IN THE POPULATIONS OF GPEE NEURONS

A peculiar phenomenon of exciting effects instead of inhibitory
ones originates when structures, whose stimulation normally produces
an inhibitory reaction, are activated in the event of a distrubance
of the inhibitory mechanisms which cause the formation of GPEE.

That phenomenon was discovered in an investigation of the state
of descending and segmentary inhibitions in the spinal cord when GPEE
was induced by TT in the efferent ouput system (Chapter 2).

The inhibitory postsynaptic potential (IPSP) was not recorded and the exciting postsynaptic potential (EPSP) appeared instead in almost all the motoneurons which were studied and which constituted the output of GPEE when the supraspinal structures were stimulated. Descending messages produced by the stimulation of bulbar structures, which normally inhibit motoneurons, had engendered only EPSP in all cases (see Figures 19, 20) (Kryzhanovsky et al., 1973c; Kryzhanovsky, Sheikhon, 1973). A similar effect was described also with respect to the stimulation of the efferent nerves: IPSP disappeared and depolarization potentials appeared instead of IPSP in the motoneurons of the extensors when the flexor efferents were stimulated (Sverdlov, 1969). In such cases, the extensor moto-neurons are excited by the same messages as the flexor motoneurons and produce a reaction which is synchronized with them. Thus, the substantial functional distinctions of both neuronal pools as antagonists disappear, and they operate as a single pathologic functional pool in which, as we have already seen, the 'functional homogenization' of the pools of the GPEE neurons is expressed. These effects are accompanied by intense polysynaptic activity, which makes it possible for high-amplitude and prolonged EPSP to originate. Powerful polysynaptic EPSP (Figure 20 C), being characteristic of epileptic activity, often occurs in response to a single stimulus of the bulbar structures. Such responses have been observed also in the spinal cord under the action of strychnine (Shapovalov, Arushanyan, 1965; Shapovalov, 1966). The ability of the motoneurons to produce action potentials of high frequency during rhythmic stimulation is connected with intensive polysynatpic activity (Figure 20 A).

In recording the potentials from the ventral roots of the segments of the spinal cord where GPEE is formed under the influence of TT, those effects are expressed in the fact that the inhibition of the reflex reactions, being characteristic of normal conditions, does not occur. Instead, intensive responses as a rule originate and great polysynaptic activity always occurs. Such effects are observed when either the supraspinal structures (Figure 45) or the afferent nerves (Figure 46 B) are stimulated. When electric activity in the muscles innervated by the segments at the GPEE site is being recorded, that effect is expressed in the origin of intense activity (instead of inhibition that occurs under normal conditions or on the opposite side), which is polysynaptic (Figure 47) (Kryzhanovsky, Sheikhon, 1968).

Such effects were noted for the first time by Sherrington (1906). According to him, one of the possible explanations of these effects was that the inhibition processes were inverted to excitation under the action of TT and strychnine. However, an analysis has shown that reversed EPSPs do not originate under the given conditions and that the depolarization responses in the motoneurons, which originate instead of IPSP, are true polysynaptic EPSPs (Sverdlov, 1969).

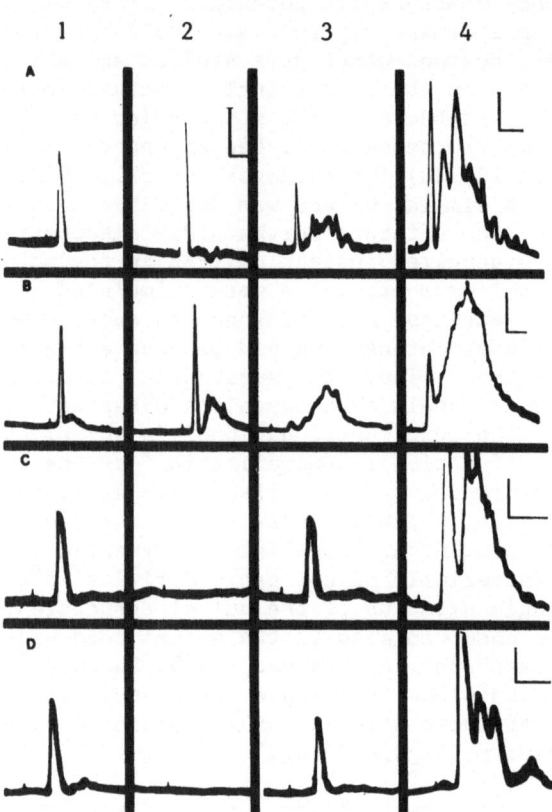

Fig. 45.   Effect of ES of the medullar structure on monosynaptic and
           polysynaptic spinal reflexes in local unilateral poisoning
           of the spinal cord with TT (formation of GPEE). 1,2:
           reflex responses on the side opposite to TT injection; 3,4:
           responses on the TT injection side; 1,3: before the stimu-
           lation of the medullar structures; 2,4: during their stimu-
           lation. A,B: effects of ES of the parvocellular nucleus.
           A: extensor monosynaptic reflexes produced by stimulating
           the gastrocnemial nerve. Amplitude: 20 µV; time: 5 ms.
           B: flexor monosynaptic reflexes produced by stimulating the
           PBST nerve. In this case, the initial monosynaptic re-
           flexes on the TT injection side were reduced (late stage of
           the process). Amplitude: 500 µV; time: 5 ms. C,D: effects
           of the stimulation of the giant-cell nucleus and the raphe
           nuclei, respectively, in the same animal. Extensor mono-
           synaptic reflexes were produced by stimulating the gastro-
           cnemial nerve. Amplitude: 250 µV; time: 5 ms. In all the
           experiments, the reflexes were tested on the 25th second of
           the stimulation of the nucleus; stimulation frequency was
           80/s.

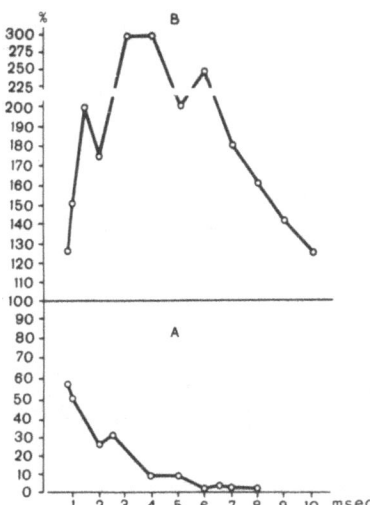

Fig. 46.   Occurrence of a stimulating effect instead of an inhibitory
one after poisoning the ventral horns of the lumbosacral
segments with TT (formation of GPEE). A: inhibition of the
monosynaptic reflex by stimulating the gastrocnemial nerve
during conditioning transmissions from the peroneal nerve
in the normal state. B: stimulating effect of the same
conditioning transmission after poisoning the lumbosacral
segments with TT (formation of GPEE). Abscissa: time
between the conditional and test stimuli. Ordinate: amplitude of monosynaptic reflexes (% of the initial level).

The given phenomenon of the replacement of the inhibitory reactions by the exciting ones can be explained by the notion that
mixed influences, i.e. inhibitory and facilitating ones, are exerted
on motoneurons under normal conditions when a stimulus is applied to
the afferent nerves or CNS structures, whose stimulation produces an
inhibitory effect (Ushima et al., 1960; Sasaki, Tanaka, 1963; Honge
et al., 1965; Shapovalov, Arushanyan, 1965; Kryzhanovsky, 1968).
Such a duality of influences can be assumed to originate from the
very beginning due to the stimulation of the formations which produce
different effects. However, it can also be assumed that the duality
of the message originates at the subsequent relays as impulses are
sent from the structures being stimulated to both exciting and inhibitory interneurons from which exciting and inhibiting influences,
respectively, converge on motoneurons. Such a form of functional
relations was defined as a principle of a double functional message
(Kryzhanovsky, 1968; Kryzhanovsky, Sheikhon, 1968). Under normal
conditions, the inhibitory effect is due to the prevalence of the
inhibitory component in the response of the reacting neurons. However, the exciting component is removed. In this case, the exciting
component can be substantially augmented due to greater polysynaptic

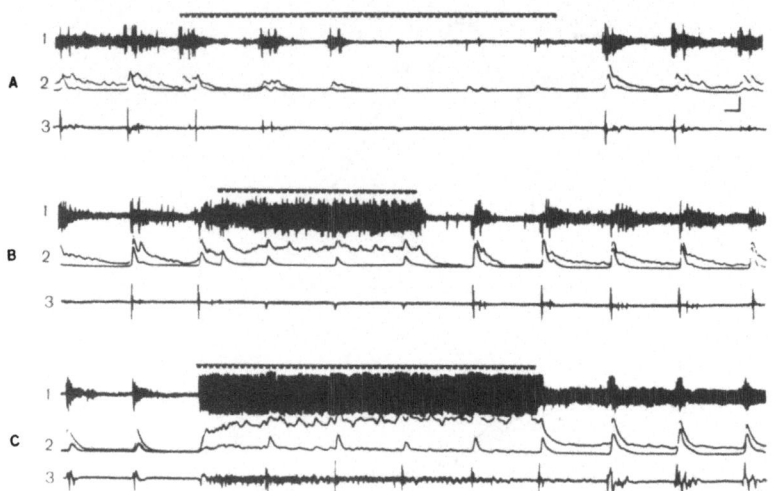

Fig. 47. Effect of ES of the giant-cell nucleus on EA in the left
          and right semitendinous muscles at different stages of
          local TT poisoning of the left lumbosacral segments of the
          spinal cord. Every fragment represents: 1: electric ac-
          tivity in the left semitendinous muscle, and 3: electric
          activity in the right semitendinous muscle; the middle
          curves (2) are recordings of integral EA in both muscles.
          Electric activity was evoked by regular single ES of the
          gastrocnemial nerve. The duration of ES of the giant-cell
          nucleus is marked by a line with black dots. A,B,C: first,
          third and seventh day, respectively, after injecting TT
          into the left femoral muscle. A: the beginning of the
          disturbance of the inhibitory processes and the formation
          of GPEE; spontaneous activity of polysynaptic origin was
          recorded; it was still suppressed when the giant-cell
          nucleus was stimulated. B: neither evoked nor spontaneous
          EA was suppressed when the giant-cell nucleus was stimu-
          lated; instead, EA sharply intensified; EA was suppressed
          on the opposite side. C: the same as in B,1, but the
          stimulating effect was more pronounced when the giant-cell
          nucleus was stimulated; the processes of the disturbance of
          inhibition and the formation of GPEE begin to manifest
          themselves on the opposite side (3). Long discharges after
          single ES of the nerve are recorded on the right part of
          curve 1 on all the fragments (activation of GPEE). Ampli-
          tude: 200 µV; time: 0.2 s.

activity, as was observed in the above-mentioned cases. A similar
effect was observed during the action of penicillin, which blocked
hypocampal EPSPs, unmasking prolonged EPSPs (Dingledine, Gjerstad,
1979).

The replacement of inhibitory effects by facilitating ones can play an important role in the functional structure of GPEE and in the establishment of the behavior of its neurons as a mechanism which enhances both the synaptic activation of the GPEE neurons and the summary production of excitation.

THROUGHPUT CAPACITY OF THE GPEE OUTPUT NEURONS

Apparently, excitation produced by GPEE will be pathogenically significant only when it overcomes control at the outlet of the structure of the central nervous system in which GPEE is formed.

The state of control at the output of the GPEE neuronal population was studied when GPEE had been formed in the ventral horn of the lumbosacral spinal segments (Kryzhanovsky, Dyakonova, 1964; Kryzhanovsky, 1968). Since the muscular Ia fibers directly terminate on the motoneurons (Lloyd, 1943; Kostyuk, 1959), the throughput of the motoneurons as the terminal link of the GPEE output was estimated on the basis of the reproducibility of the monosynaptic reflexes when the muscular Ia fibers were stimulated at different frequencies.

Investigations with cats have shown that the frequency of reproduction of the monosynaptic reflexes is virtually the same as that under normal conditions when only the inhibition of motoneurons is upset. However, monosynaptic reflexes were reproduced at a very high frequency, i.e. 100-200 per second, when they were evoked under the conditions of additional polysynaptic stimulation, which as has been shown, is sharply enhanced under the given conditions (Figure 48). Sometimes, the frequency of their reproduction was 300 and even 500 per second in experiments with rats. In this case, the source of polysynaptic stimulation was insignificant, since it could be produced by stimulating high-threshold afferents of either the same muscle nerve or the cutaneous nerve. What was important in this respect is that it had to be directed at the same motoneurons that produced action potentials, which constitute monosynaptic reflexes. A noteworthy fact is that polysynaptic reflexes grow when the frequency of stimulation increases while the force of stimulation remains the same. It has been shown (Figure 20 A) that an individual motoneuron can generate an action potential at a very high frequency (500 per second) under these conditions. Additional polysynaptic stimulation did not substantially enhance the reproducibility of the monosynaptic reflexes in experiments with healthy animals or with animals involving the side opposite to the side where TT was injected and where the inhibitory mechanisms were not disturbed. Thus, the throughput capacity of the GPEE's output is significantly enlarged. This effect is due to the enhanced synaptic activation of the neurons of the GPEE's output when inhibitory control is upset. Both mechanisms are synergic and cause the output neurons to be depolarized. The throughput of these neurons increases directly due to the depolarization of their membrane when GPEE is formed and activated.

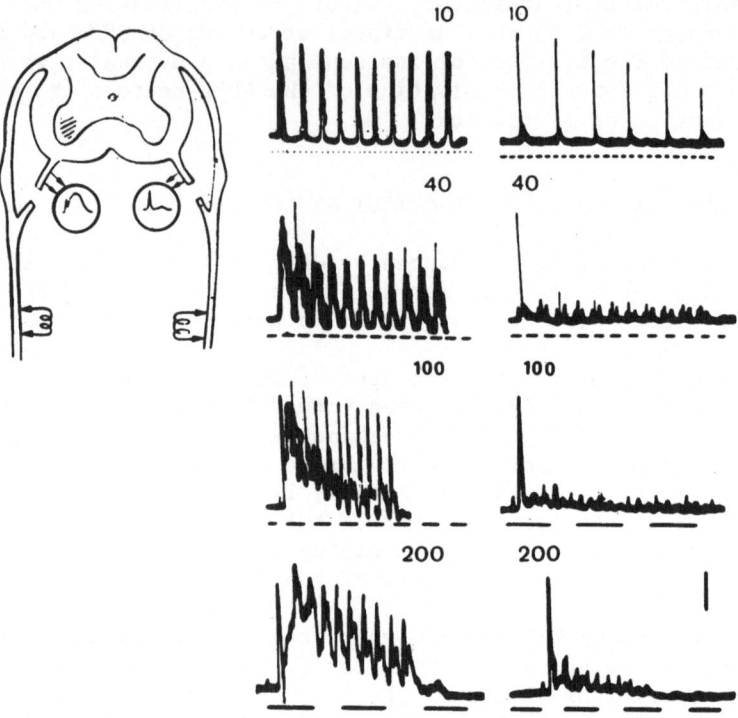

Fig. 48.   Reproduction of extensor monosynaptic reflexes during
           combined monosynaptic and polysynaptic stimulation after
           the local unilateral poisoning of the lumbosacral segments
           of the spinal cord with TT. The combined monosynaptic and
           ·polysynaptic responses were evoked by ES of the gastro-
           cnemial nerve with specially selected power. The illus-
           tration also shows how polysynaptic activity grew on the
           GPEE side as the frequency of stimulation increased.
           Amplitude: 1 mV; time: 20 ms. The scheme of the experiment
           is given on the left.

      An increase in the throughput capacity of GPEE's output neurons
is very important pathogenetically, since it allows impulses produced
by the GPEE neurons to go to other structures of the central nervous
system, to overcome the local control mechanisms in these structures,
and to involve them in the pathologic process.

SOME FEATURES OF GPEE's ACTIVITY

      A study has shown that there are two main types of GPEE ac-
tivity: tonic and phasic (or clonic) activities. The first type of
activity is recorded in the form of long-lasting, continuous dis-

charges, while the second type is recorded in the form of separate synchronized bursts (see Figure 4).

These types of activity can be observed in various CNS areas where GPEE is formed. A Comparison of the electrographic patterns of activity described by many authors in the case of cortical epilepsy with those which have been recorded when GPEE was in the spinal cord (Chapters 1, 2, 6) shows that they are fundamentally similar. There are single-typed mechanisms of seizure electrogenesis in various CNS areas.

It is important to ascertain the mechanisms of tonic and clonic activities. The question is whether both types of seizure activity are formed by the same structures of the central nervous system or whether they have different sources. Opinion is divided among neuropathologists and epileptologists in this respect. It has long been held that the tonic and clonic types of seizure activity are generated by different structures in epilepsy.

Both types of activity originate even on the spinal level when GPEE is in the propriospinal system as the spinal cord is cut low and the segments are deafferent (see Chapter 2). This fact makes it possible to more easily ascertain the source and mechanisms of the given types of seizure activity. It has been shown (Goff, 1973) on the basis of mathematical models, which include exciting and inhibitory neurons, that continuous harmonic and discontinuous undulations (which are similar to tonic and clinic epileptic activities) originate in such systems when interneuronal connections are potentiated in analogy with epileptic activity. These data suggest that both types of activity can originate as if during the activity of the same GPEE. It has been shown by taking the example of Leão's spreading depression that pulsation of a constant potential, which corresponds to a change in neuronal activity, can originate when the wave is delayed (Kuznetsova, Koroleva, 1977). An analysis of the reverberation mechanisms suggests that the activity of a circulatory nature should be discontinuous, like a 'volley' (Shuranova, 1977).

However, that question remains open, and many of its aspects should be specially studied. The cause of the transition of tonic activity to discontinuous activity is not quite clear: could it be due to the triggering of the inhibitory mechanisms or to the suppression of the neurons according to the cathodal depression type, or perhaps both types of inhibition participate in this process, just as when the epileptic focus is suppressed in the cerebral cortex (Okujava, 1969). In this respect, it should be noted that in generalized spinal myoclonia, clonic activity is exhibited at the last stages of the pathologic process. At first, only tonic activity originates (Chapters 1, 6). This fact is noteworthy also because the inhibitory mechanisms are greatly disturbed and the excitability of the GPEE neurons is enhanced at the late stages. GPEE induced by TT

in the spinal efferent output system produces tonic activity, while
clonic activity appears only at the late stages in some cases
(Chapter 2). This effect is probably due to the involvement of the
propriospinal interneurons of the given segments of the spinal cord
in the process. Only clonic activity originates in local spinal
myoclonia caused by penicillin (see Chapter 2). Therefore, it is
still note clear whether the same GPEE neurons produce tonic and
clonic activity or whether this is done by various neuronal pools
which are territorially sited in a given GPEE. These types of activity are affected differently by anticonvulsants. Investigations
have shown (Grafova, Danilova, 1981) that clonic activity, which
originates either independently or after tonic activity, is the first
to be suppressed, while tonic activity is very stable and remains
even when clonic activity disappears. This effect occurs during the
action of various anticonvulsants, e.g. diazepam, phenazepam, car-
bamezepin, diphenylhydantoin, etc. The effect is apparently due to
the differences in the reactivity of various groups of GPEE neurons
(or neurons which are in different functional states) that participate in the production of two different types of activity.

     Those effects are characteristic of the action of anticonvul-
sants of GPEE in the propriospinal interneuron system. The effects
of the same convulsants, e.g. diazepam, in certain doeses have some
peculiarities when GPEE is present in LGN (Rekhtman et al., 1979) and
in the cerebral cortex (Rekhtman et al., 1980b). The suppression of
ictal discharges and an increase in the frequency of interictal
spikes are a characteristic feature which was observed (Figure 49
A2). A similar effect has been noted also by other authors when
different anticonvulsants were used (Velasco et al., 1978; Bustamante
et al., 1981. Fasiello et al., 1981). Some authors interpret it as
the result of the opposite influences of anticonvulsants on various
types of epileptic activity (Bustamante et al., 1981). It has been
shown (Ayala et al., 1973) that the interictal spike recorded from
the cortical surface corresponds to the appearance of both PDS (by
intracellular recordings) and a high-frequency discharge against the
PDS background. Afterwards, hyperpolarization usually originates.
Its extent and duration differ in various neurons (Figures 36-39).
Hyperpolarization has been recorded in many neurons of the epileptic
focus and around it during the interictal spike (Prince, Wilder,
1967; Prince, 1968a; Dichter, Spencer, 1969b). Post-PDS hyperpolarization is regarded as a factor which limits the frequency of its
origin (Prince, Futamachi, 1970). When hyperpolarization occurs in
many neurons, the dimensions and power of GPEE decrease and epileptic
activity no longer spreads. During the ictal discharge, post-PDS
hyperpolarization is absent and a prolonged continuous depolarization
shift occurs; high-frequency discharges originate against the background of this prolonged depolarization shift (Ayala et al., 1973).
Anticonvulsants, particularly benzodiazepines, enhance the inhibitory
mechanisms as they activate the GABAergic apparatus (see Chapter 13)
and Na-K-ATPase in the epileptic focus (Samsonova et al., 1979;

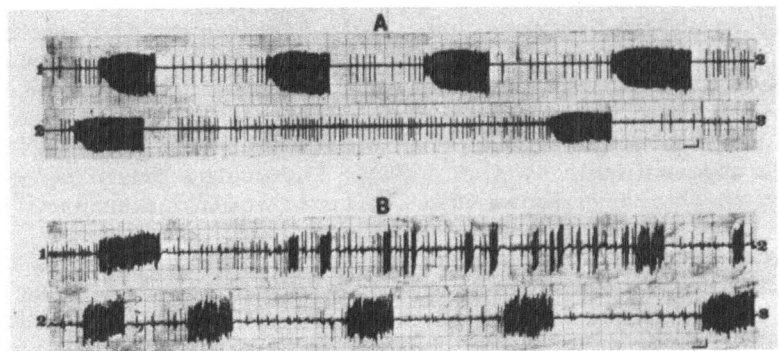

Fig. 49. Change in EA in the zone of the epileptic focus, formed by
          applying penicillin to the rats' cerebral cortex, after the
          intramuscular injection of diazepam. A: diazepam (0.01
          mg/kg) injected (↓) 25 min after applying penicillin. B:
          diazepam (0.01 mg/kg) injected 20 min before applying
          penicillin; the ECoG recording was made 10.4 min after
          applying penicillin. Amplitude: 400 µV; time: 10 s.

Rekhtman et al., 1980a). Consequently, the hyperpolarization of the
neuronal membrane is restored to a certain extent, as a result of
which the continuous depolarization shift is 'split up' and individual PDSs of different duration appear. A prolonged tonic ictal
discharge 'breaks up' into brief ictal discharges of different duration (Figure 49 B1) or individual interictal spikes of greater
frequency (Figure 49 A2). The enhanced frequency of interictal
spikes shows that PDS can be generated by high frequency, indicating
that the duration of restored hyperpolarization is reduced. The
extracortical mechanisms which suppress epileptic activity [the
activation of the structures of the antiepileptic system (see Chapter
13) and a change in the activity of the thalamocortical synchronizing
system, particularly the suppression of the activity of thalamic
nuclei, which apparently participate in the origin of ictal discharges in the penicillin foci (Noebels, Prince, 1975; Schwartzkroin,
Prince, 1975)] may become involved in those local processes, too.
Possibly, benzodiazepines can also directly intensify interictal
activity. For instance, the administration of diazepam before
penicillin is applied on the cortex reduces the duration of the
latent period and increases the frequency of the initial interictal
spikes (Rekhtman et al., 1980a) The dual effects of benzodiazepines,
just as other anticonvulsants and GABA preparations, on epileptic
activity have complex focal and extrafocal mechanisms; these effects
may be connected with the attenuation of inhibition due to substances
which act on the GABA receptor system, with the activation of endo-
geneous proconvulsants, etc.

In considering the peculiarities of the suppression of GPEE activity, it should be noted that clonic discharges and interictal spikes are not the same types of epileptic activity. Clonic discharges are a part of general ictal discharges. They originate as the level of the continuous depolarization shift characteristic of the tonic phase decreases and individual summary PDS of different magnitude appear (Ayala et al., 1973). Therefore, they are less resistant to anticonvulsants than the tonic ictal discharges.

Hence, the main types of manifestation of GPEE activity, namely, individual interictal spikes, clonic discharges and continuous tonic discharges, are an expression of the different extent of synchronization of the hyperactive neurons and their aggregates. The neuronal aggregates are the subunits of a general GPEE, i.e. they are likewise GPEEs whose dimensions can change in accordance with the current conditions. Tonic discharges may be maintained also by the alternating involvement of such subunits and individual neurons in a reaction. It has already been shown that the operating GPEE consists of not only neurons with more or less stable epileptic properties (the existence of PDS, etc.), but also neurons with varying membranous changes, and initially normal neurons (see the previous pages).

Some features of the operating GPEE at different stages of its formation should be mentioned.

It follows from the data in Chapter 2 that when GPEE is being formed, thresholds diminish and the characteristics of its activation change: relatively faint activity originates at the early stages in response to stimuli, and it may increase during a reaction. At the late stages, these stimuli or even far fainter stimuli produce an explosion of powerful high-amplitude and high-frequency activity, which immediately attains the maximum. During this period, the activation of GPEE either spontaneously or in response to stimulation is paroxysmal. In this case, both weak and strong stimuli cause a maximum reaction. This creates the impression that GPEE is activated according to the 'all-or-nothing' law. Figuratively, GPEE at this stage is a 'powder keg' which can explode at any moment by lighting it with a torch or a match. Such a type of reaction was observed when virtually all the GPEEs in various areas of the central nervous system were studied.

There is another noteworthy feature: the latent period of GPEE triggering by the same stimulus diminishes together with the above changes in the features of activity. The latent period is the minimum at the height of the process.

A comparison of the dynamics of the changes in GPEE's activity with that of the functional effects produced by it (e.g. the inhibition or facilitation of spinal reflexes during the activation of GPEE in bulbar nuclei) shows that the changes in these parameters are in complete accord with one another: the latent period of the effects produced and the duration of the attainment of the maximum by them are reduced as GPEE is being formed. The curve of the growth of the effect sharply rises at the height of the process (see Figures 40-43).

Several conclusions can be drawn from the peculiarities of GPEE's activity. Apparently, the latent period of the reaction of GPEE's neuronal pool is determined by the state of its inhibitory control. This fact is of definite interest also for neurophysiologists. It shows that the mechanisms which establish the latent period of a reaction of a neuronal pool and that of an individual neuron are not the same. Obviously, the neurons of GPEE become more capable of undergoing hypersynchronization as it is being formed. This capability conditions the latent period and causes both GPEE to wear away in an explosive way and the reaction maximum to rise sharply.

TRIGGER STIMULATION

There are two main stages of the formation of GPEE in accordance with the role which external stimuli play in activating it. This applies, in particular, to trigger stimulation. This question was studied at first in experiments involving generalized spinal myoclonia caused by producing GPEE in the system of propriospinal neurons (see Chapters 1, 6).

Figure 50 I shows that GPEE is activated by impulses which it directly receives via the regional afferent canals at the early stages of its formation. The nociceptive stimuli applied to the skin of the toes and the foot on the side where GPEE is located are especially effective. Such an effect is not produced by any afferentation coming from other receptive fields, including similar stimulations of the symmetrical fields of the opposite leg, since they do not activate GPEE (see also Figures 1-4) and do not cause general seizures. At the late stage of the process, general seizures are caused also by stimuli from other receptive fields (Figure 50 II). First of all, the field of effective nociceptive stimulation expands, covering not only the toes and the foot, but also the whole leg. General seizures can be induced also by applying nociceptive stimuli to the skin of the toes and the foot of the opposite leg and some other parts of the body. As for the initial zones, nociceptive stimuli from them cause more severe and more prolonged general seizures. However, not only nociceptive stimuli, but also light, tactile stimuli of these zones produce the same effect. At this stage, general seizures occur also spontaneously.

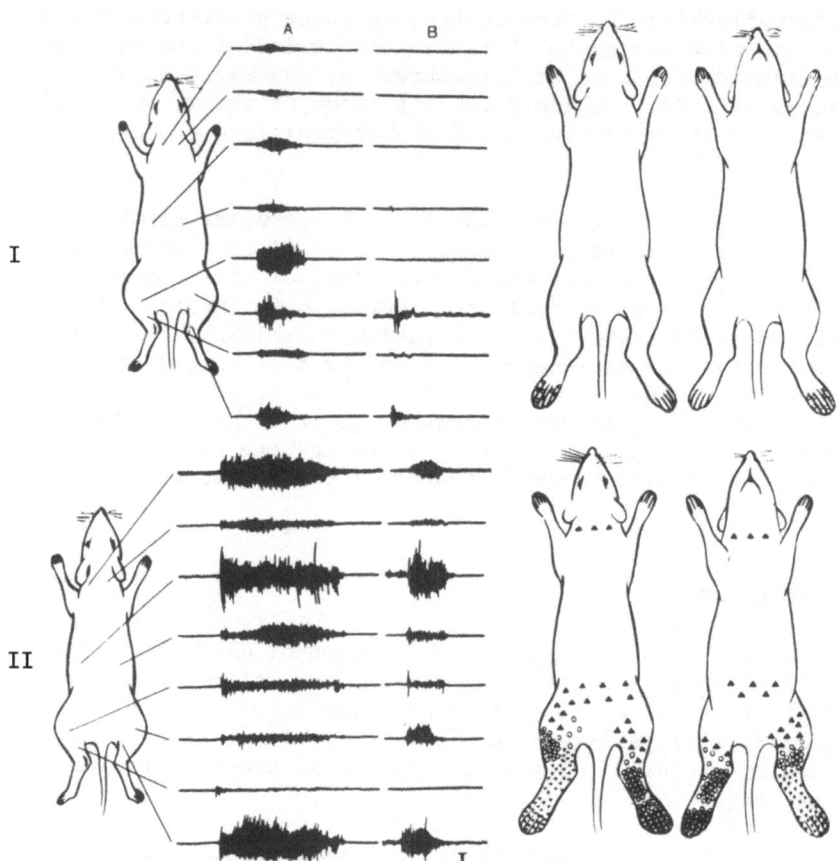

Fig. 50. Provocation of a general seizure by stimulating various
          receptor fields during the 'universal dispatch station'
          phenomenon. Receptor zones whose stimulation produces
          general seizure activity are shown. These zones expand,
          the seizure reaction becomes more intense, and the thresh-
          olds of its production decrease as the pathologic process
          develops. I,II: stages of the development of the syndrome.
          The following stimuli were applied: strong nociceptive
          stimulation (triangles), nociceptive stimulation of medium
          strength (white circles), and tactile stimulation
          (crosses). The closeness of the signs indicates the inten-
          sity of a seizure when a given receptor field is stimu-
          lated. Center: EA in various muscles evoked by the stimu-
          lation of the left (A) and right (B) hind legs. Left:
          schemes of EA recordings. GPEE was produced in the spinal
          lumbosacral segments on the left by injecting TT into the
          left gastrocnemial muscle. Electric activity was weakened
          in the left gastrocnemial muscle due to the blocking action
          of TT on neuromuscular synapses.

Those features of the production of general seizures by external stimuli can be satisfactorily explained with respect to the generator mechanisms of the given phenomenon. The expansion of the receptive zones of stimuli, which can produce seizures, at the late stage shows that the dimensions of GPEE have increased (see Chapter 2). As a result of this growth, the new zones also become the direct afferent fields of GPEE. The fact that nociceptive stimuli from the skin of the toes and the foot of also the opposite leg are effective at the given stage suggests that GPEE in these segments originates on the opposite side, too.

At the early stage of the formation of GPEE, external stimulation not only acts as a trigger, but also sustains the generator's activity: the duration of its work largely depends on the duration of afferent stimulation. At the late stages, such stimulation is more important as a trigger mechanism: it is enough to briefly and virtually instantly squeeze the skin of the toes or the foot in order to cause severe general seizures. An important fact is that stimuli of different modalities from diverse receptive fields, including audiogenic (claps), vibrational (jolting the table) and light (exposures to light), are effective and can cause seizures at the late stage of the process.

It will be shown that these main forms of GPEE's relationships to trigger stimulation are observed during neuropathologic syndromes caused by the activity of GPEE in various areas of the central nervous system. They are clearly exhibited, for instance, during the pathologically enhanced scratch reflex (Chapter 4). Photogenic epilepsy (Chapter 9) caused by the formation of GPEE in LGN is induced by very intensive light stimuli at the initial stage of the process. At the late stages, faint light stimuli and stimuli of other modalities become provocative, and seizures can occur also spontaneously. Clear-cut relationships between the activation of GPEE and trigger stimulation are exhibited during the central pain syndromes (Chapter 7).

In speaking about the neurophysiologic mechanism of the origin of trigger zones, it can be assumed that the receptor zones connected with the generators' most excitable neurons are the trigger ones. These are probably the zones from which impulses go to the generator's trigger neurons.

Those two main stages of a change in GPEE's reactivity to trigger stimulation give an idea about the peculiarities of the provocation of seizures during many neuropathologic syndromes. Modal features of a given system and strength relations remain at the initial stage, when GPEE is activated only by stimulation which is characteristic of it and which is quite strong. At the late stages, stimulation loses its modal specifics and its biological feature, becoming merely a trigger stimulus.

The 'trigger stimulation' concept covers a wide range of diverse stimuli. Stimuli of the organism's external environments and internal media, verbal stimuli, associative images, motivation, and so forth, can act as specific trigger stimulation which, by its modality, can correspond to both the system in which GPEE originated and the neuropathologic syndrome produced by it.

Obviously, the specific activation of GPEE at the early stages of its formation (when modality and the strength of stimulation retain their significance) reflects the regularities of the determinant's activity.

SOME TERMINOLOGICAL POINTS

The term 'focus of excitation' has long been used in literature as an ordinary term. However, it is descriptive and is more probably a figure of speech than a neurophysiological term. The term which we use, 'generator of excitation' (GE), has a neurophysiological meaning. As we have seen, GEs in the central nervous system are neuronal populations with a definite functional organization. The neurophysiological and neurochemical structure of GE, the features of its activity, its properties, power, and other parameters can be established.

The concepts of the role of GE in the activity of the nervous structures are being gradually used in descriptions of investigations and in electrophysiological literature. It is now not unusual to speak about the generation of action potentials or of prolonged bursts, firing, special generator mechanisms in a soma, dendrites or an axon, and the factors and conditions of these mechanisms' inactivation. Until recently, the concepts of the generator mechanisms of electrogenesis covered mainly the cellular processes. Currently, cellular pacemakers are the subject of special consideration (see Carpenter, 1982). We apply the term 'generator of excitation' especially to the activity of the neuronal populations which produce and maintain summary electric activity. Of course, GEs have their own mechanisms at every level of the structural and functional organization of the nervous system. They can also be very intricate constellations of neural formations of various areas of the central nervous system.

Physiological pacemakers in various areas of the central nervous system and in some organs are also GEs. The physiological term 'pacemaker' usually means a preformed functional structure which operates in a definite mode and produces a definite pattern of activity (e.g. pacemakers of various types of rhythmic cerebral activity, the pacemaker of physiological tremor, classical cardiac and gastrointestinal pacemakers of rhythmic and periodic activities, etc.). In our sense, GE originates as a new dynamic functional

formation. It is formed under definite conditions and its nature
varies, depending on several factors and the peculiarities of the
reaction induced by it.

The term 'pacemaker' somewhat resembles the term 'determinant'.
However, the latter has a broader meaning. A physiological pacemaker
is an expression of the phylogenetically and ontogenetically consolidated activity of specialized structures. Often, it is represented
as an electrophysiological mechanism of rhythmic activity (see
Carpenter, 1982). The determinant has a systemic meaning: it is a
new functional structure which establishes the behavior of a physiological or pathologic system. Under physiological conditions, the
determinant is formed due to the integrative activity of CNS as the
operant part of a program apparatus of a reaction that meets the
requirements at a given moment. It disappears together with the
whole functional system when the latter attains its programmed result
(see Chapter 4). Under pathologic conditions, the determinant can
originate under the influence of diverse pathogenic agents in various
functionally important parts of the physiological system, turning the
latter into a pathologic system. It has been shown that the determinant is a system-forming factor: it not only establishes the
system's behavior, but also forms it.

The 'generator of pathologically enhanced excitation' (GPEE) is
a neuropathophysiological term. It is a functional structure which
originates under pathologic conditions. As we have seen, neuronal
populations constituting GPEE produce powerful, prolonged and uncontrollable excitation. Such GPEEs are able to self-sustain hyperactivity and can preserve themselves for a relatively long time
regardless of the outcome of the reaction which they induce. These
properties of GPEE make it possible to explain the peculiarities of
the exhibition and occurrence of the neuropathologic syndromes which
are based on GPEEs.

At present, there is a clear-cut tendency to use the generator
mechanisms for explaining the nature of the pathologic states and
abnormal reactions (Loeser, Melzack, 1978; Calvin, 1975, 1980; Hrbek,
1980). In such cases, it is not always pertinent and correct to use
the term 'epileptic focus'. The word 'focus' is descriptive, while a
disadvantage of the word 'epileptic' is that not all the neurons
which constitute GPEE are epileptic ones in the strict sense of the
word. Many of them are simply hyperactive to a certain extent and
produce EPR. The advantage of the term 'GPEE' over the term 'Epileptic focus' is that it is free of syndromic and nosologic specifics. GPEE is a universal neurophysiological category. GPEEs can be
formed in virtually all the areas of the central nervous system,
differing from one another in their structural, functional and neurochemical peculiarities. It will be shown in Part 2 that the production of GPEEs in certain areas of the central nervous system may
engender definite syndromes which are specific of systems that are

drawn into the pathologic process (e.g. parkinsonism, choreiform
hyperkinesis, vestibulopath, central pain syndromes, pathologically
prolonged sleep, psychosis-like states, etc.). If the term 'epileptic focus' was used instead of the term 'GPEE' in such cases, all
these syndromes as well as similar clinical syndromes should be
called 'epilepsies', which should also include the syndromes of
regulatory diseases that are caused by the origin of GPEE in the
mechanisms which regulate vegetative functions. Such extrapolation
would leave out syndromic and nosologic specifics.

SUMMARY

     The formation of GPEE is followed by both the disturbance of the
initial functional distinctions in neurons which constitute the
operant part of GPEE and the appearance of new properties in them.
Neurons with relatively stable and maximally defined epileptic properties become free of local and integrative inhibitory control,
produce intense impulse action, and draw other neurons into the
pathologic process, engendering epileptic properties in them and
synchronizing their activity and imposing a new pattern of activity
on them. They act as trigger neurons (neurons of group I) and can be
regarded as primary microgenerators in GPEE. They constitute the
initial critical mass of epileptic neurons, which form and activate
GPEE. There must be a definite critical mass of all the neurons of
GPEE, i.e. GPEE must have definite power if it is to become pathogenetically significant, that is to say, if it is to pathologically
activate the appropriate system. The power of GPEE depends on the
number of the operant neurons of GPEE and the ability of each of them
to generate action potentials with an increased frequency and for a
long time. When all other conditions are equal, the power of GPEE
corresponds to its dimensions. It increases with the development of
the pathologic process due to the involvement of new neurons into the
process under the influence of factors which originate in the neuronal population of GPEE itself and to the continual action of the
pathogenic agent.

     The insufficiency of inhibitory control, which is an important
prerequisite of the origin of GPEE and its activity, can be either
primary (when the inhibitory mechanisms are directly upset under the
influence of pathogenic factors) or secondary (when the exciting
neurons are initially depolarized). In the latter case, inhibitory
connections remain, but they are functionally insufficient. Apparently, both factors are combined under natural conditions.

     Positive interneuronal connections in GPEE and external positive
connections which converge on the GPEE neurons are important in
maintaining and increasing the activity of GPEE. The replacement of
inhibitory effects by excitatory ones in the populations of the
GPEE's operant neurons is an important pathogenetic factor: the

distortion of the functional effects and an increase in the strength
of the influence produced by GPEE as well as the weakening of integrative inhibitory control are connected with that replacement. An
increase in the throughput of GPEE's output neurons due to both the
insufficiency of the inhibitory mechanisms and intense synaptic
activity allows the flow of impulses produced by GPEE to pass more
easily and go to either the subsequent areas of the central nervous
system or the effector structures at the periphery.

Modally specific stimuli are trigger ones for GPEE at the early
stages of its formation, when the inhibitory mechanisms are still
relatively effective and neuronal excitability is not great. At the
late stages, GPEE can be activated by various stimuli, and they
develop maximum activity almost instantly. Such paroxysms are realized, as it were, by the 'all-or-nothing' law. Trigger stimulation
is probably such afferentation which goes to the GPEE's trigger
neurons (neurons of group I).

The term 'GPEE' has a strict neuropathophysiological meaning,
unlike the term 'epileptic focus', which is more probably a figure of
speech and is descriptive. Its advantage over the latter term is
that it is free of syndromic and nosologic specifics. GPEE is a
general neuropathologic category which can be used for all the syndromes characterized by the hyperactivity of systems. It is inex-
pedient to use the term 'epileptic focus' for all the syndromes,
since this would leave out syndromic and nosologic specifics.

4
Hyperactive determinant structures and pathologic systems

To understand the nature and specifics of the pathologic system, its properties should be compared with the basic characteristics of the physiological system, especially with the properties of the physiological system as a functional organization which maintains and realizes the central nervous system's integrative activity (Sherrington, 1906; Ukhtomsky, 1923-1927; Pavlov, 1923, 1927; Gellhorn, 1943; Bertalanffy, 1962; Berstein, 1966; Anokhin, 1971, 1975; Szentágothai, Arbib, 1974).

PHYSIOLOGICAL SYSTEM

The physiological system's activity is of adaptive significance, i.e. it is intended to achieve a result which is biologically useful. Its program corresponds to the requirements at a given moment, the action of the environmental agents and to the experience gained either during previous encounters with these agents or in a similar situation. The disharmony between a result which is programmed and a result which is attained acts as a stimulus for the system's further activation. This stimulus is operative until the programmed result is attained. Then, the stimulus disappears together with the functional system as a dynamic organization. Feedback information from the performing organs and from formations which produce the intermediate links, and also afferentation from other sense organs that tells about the course of a reaction are important in establishing conformity between the result which is programmed and the result which is attained.

All that can be vividly illustrated by taking the classical example of the scratch reflex as a systemic reaction. When a tactile

100

stimulus is applied to the anterior part of the body, the animal begins to scratch the skin of that part with its ipsilateral hind leg. The extent and peculiarities of the scratch movements correspond to the stimulant's action. The reflex ceases when the stimulus is eliminated by this reaction, and definite signals which inform about the attained result go to the central nervous system. Hence, the functional system, which originates as a dynamic organization for realizing a given reaction, disappears when a programmed result is attained. There is no longer any need for it. The disappearance of such a system is the outcome of not only the cessation of stimuli which activate the system (passive process), but also its active inhibition due to other systems' influence, and general integrative control by the whole brain (active process). The elimination of the functional system as a dynamic organization is just as indispensable and expedient as the system's origination. Otherwise, new systems cannot be formed and, consequently, they cannot operate.

The determinant structure, which originates as a functional formation in accordance with the requirements at a given moment, is normally the working part of a programming apparatus. It forms the pattern of a functional message, which reflects the specifics of the acting stimuli and corresponds to both the requirements at a given moment and the experience gained in encounters with these stimuli or similar ones. The formation of the functional message is under multifactorial intersystemic and intersystemic control, making it physiologically adequate. This property of the functional message allows its structures being formed to act as a physiological determinant.

The physiological determinant establishes, through the functional message, the pattern of the activity of the system's parts and the behavior of the system itself. It can do this because the program message not only reflects the properties of the acting stimulus, but also corresponds to the system's functional abilities. Thus, the functional message acts as the operant of a controlling and organizing mechanism. During its realization, the system becomes consolidated and perfected, and its parts become more and more functionally complementary with respect to one another and help achieve a result.

PATHOLOGIC SYSTEM

Phenomena which are contrary by their nature and biological significance can be found in pathologic systems, particularly in those which originate under the influence of hyperactive determinant structures based on GPEE. The activity of such systems is of no adaptive significance, and its result may be biologically unfavorable for the organism and may act as a direct pathogenic factor. This feature is decisive for the pathologic system. It can be vividly shown by taking the example of the scratch reflex which is patho-

logically enhanced (Figure 51) and which originates when GPEE is
produced in the scratch reflex system in the brachial parts of the
spinal cord (Lutsenko, Kryzhanovsky, 1973; Kryzhanovsky, Lutsenko,
1975). When the reflex acts as the foreleg is stimulated on the side
of GPEE, the animal begins to intensely scratch the leg with its
ipsilateral hind leg. Gradually, the thresholds of this reaction
decrease, scratch movements become more frequent, more intensive and
longer, and can be provoked even by slight stimuli at the height of
the process. At this stage, these movements occur also spontaneously, without any special stimuli, i.e. they are automatic reactions. Induced or spontaneous seizures become longer and consist
of movements which are repeated many times. The animal begins to
scratch the site so intensively that it can tear the tissue. However, it cannot stop these movements, and it continues to 'scratch'
the site during the next spontaneous or induced fit. At the late
stages, the animal may use both hind legs to scratch the site, and
this may have the nature of a generalized spasm.

     That form of pathology clearly shows the properties of the
hyperactive pathologic system. To begin with, it is characterized by
the disadaptive nature of the result of the animal's activity, which

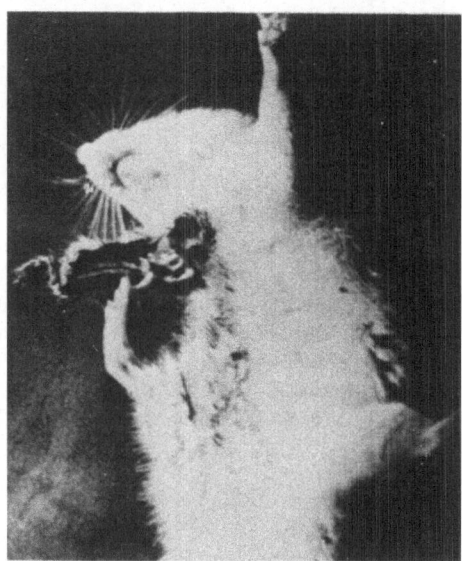

Fig. 51. Clinical picture of a pathologically enhanced scratch
          reflex caused by the production of GPEE in the right
          brachial segments of the spinal cord. The rats continu-
          ously and intensively scratched the right foreleg with the
          right hind leg. The tissue of the foreleg was damaged, and
          the bones were revealed.

is biologically unfavorable for the organism at the height of the
process. This activity does not correspond to stimulation, the
organism's experience in encountering similar stimulation, and to the
requirement at a given moment. Such abnormal activity can originate
also spontaneously, i.e. without any external stimulation. It
reflects the properties of not stimulation, but the hyperactive
determinant structure as GPEE which produces the given effect.

Hence, environmental stimulation loses its biological significance when the pathologic system originates: it begins to act as a
trigger that merely 'starts' GPEE, which activates the system. When
GPEE is activated, it no longer needs trigger stimulation, but develops its own activity independently. At the late stages, it can be
activated spontaneously, without the action of any external irritation; it can also be activated by random stimuli which do not pertain
to a given system. Therefore, this type of activity cannot be called
a reflex in the strict sense of the word: there is no longer any
reflex as such: there is only the pathologic form of activity which
is determined by the endogenous generator that acts as a hyperactive
determinant structure.

It follows from the above that the pathologic system has other
features. Its activity is not corrected during a reaction, and the
system does not disappear when the 'result' is attained: scratching
continues even when the 'reflex' is realized. In this case, the
animal tears the leg tissue and even the pain which then originates
cannot stop the paroxysm. Pathologic behavior stops not because the
'result' is attained, but because GPEE no longer functions for some
reason.

All those features of the hyperactive pathologic system and the
peculiarities of its activity are engendered by the activity of one
part of the system, namely, the hyperactive determinant structure.
The other parts, which are not influenced by the determinant, remain
normal at a given stage. When they are parts of the other physiological systems, the adequately perform their function and produce a
normal physiological reaction. This can be seen from animals with
the pathologically enhanced scratch reflex, the central efferent part
of which remains unchanged (Chapter 1, Figure 12), and other neuropathologic syndromes which will be described further, i.e. generalized spinal myoclonia, photogenic epilepsy, pain syndromes, etc.
When there is no paroxysm, the locomotion of such animals remains
more or less normal: they move, satisfy their physiological requirements, etc. This means that the respective locomotive centers normally perform the physiological function, and only during a fit do
they become a part of the pathologic system, obeying the determinant
structure.

The pathologic system is freed from integrative control also due
to the determinant's hyperactivity. In such a system, feedback

connections from the intermediate units and the effector organs are preserved, and afferentation from other sense organs, which inform about the result, is realized. However, this signalization is hardly effective. It is not functionally significant in controlling a reaction due to both the insufficiency of the inhibitory mechanisms and the hyperactivity of such a determinant structure as GPEE (Figure 52). Negative feedbacks, which normally control a reaction through the inhibitory mechanisms, are virtually ineffective at the height of a process, although they are apparently activated more than is normally the case. Feedback connections of other parts of a given system are also hardly effective, since the flow of GPEE-induced impulse activity that passes through these structures is very intense and can scarcely be corrected. Moreover, the impulse flow going to GPEE may not suppress, and may even enhance it activity (the phenomenon of the replacement of inhibitory effects by exciting ones; Chapter 3).

It follows that a hyperactive determinant structure does not produce a modally new biological system. When the stimuli produced by GPEE enter other, unchanged structures of a given system, the engender merely intensified activity which is characteristic of these structures. The example of the scratch reflex shows that the reflex did not lose its functional modality as it became pathologic under the influence of the hyperactive determinant. It has been shown in Chapter 1 that the normally existing inhibitory and facilitating effects on the lumbar level are enhanced when GPEE is in the brachial segments of the spinal cord. When GPEE is produced in the lateral vestibular nucleus, spinal monosynaptic reflex facilitation characteristic of the nucleus is augmented, and when GPEE is produced in the giant-cell nucleus, spinal reflex inhibition is enhanced (Chapters 2, 3). In all the cases, the phenomenon consists in hyperbolization and in the pathologic intensification of natural physiological effects, which are characteristic of a given system.

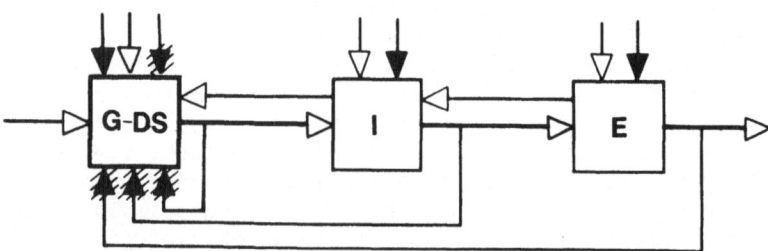

Fig. 52. Diagram of the pathologic system induced by the hyperactive determinant structure. G-DS: generator-determinant structure. I: intermediate links of the system. E: effector links of the system. White triangles stand for the stimulating links, black triangles, for the inhibitory links, and crossed black triangles, for the disturbed inhibitory links.

PATHOLOGIC SYSTEM'S RESISTANCE

There are many opportunities for structural-functional and plastic reorganizations in the central nervous system. Nevertheless, if the pathologic systems which originate in the central nervous system are not eliminated or can remain there, it means that there are mechanisms which sustain pathologic systems. Among these mechanisms, the activity of the determinant structure is most noteworthy. It has been shown by taking the example of pathologic systems in the spinal cord and epileptic complexes in the cerebral cortex (Chapter 1) that a pathologic system exists as long as there is a determinant; a pathologic system breaks up and disappears when the determinant is eliminated. It will be shown later that the neuropathologic syndromes disappear when the determinant structures are suppressed (Parts 2 and 3). When the brain's homeostatic mechanisms cannot suppress the hyperactive determinant structure, they are incapable of eliminating the pathologic system, although they can reduce it and limit its activity and development.

The determinant structure's background activity seems to be a probable neurophysiological mechanism which sustains and consolidates the pathologic system. GPEE in the giant-cell nucleus has already been taken as an example to illustrate enhanced tonic background activity at the height of a process (Chapter 2). Interictal activity in the epileptic focus of the cerebral cortex (Chapter 2) is of similar importance. The pathogenic significance of enhanced background activity is vividly exhibited when GPEE is sited in the efferent output system of the spinal cord (Chapter 2). Such activity sustains muscular hypertonicity, which later becomes muscular rigidity.

When excitation is generated during the period between fits in the form of continuous synchronous impulse flow or discharges, not only is the pathologic system sustained, but also the latter becomes ready to realize a pathologically enhanced message produced by GPEE in provocation or spontaneous activation.

However, it is not only the determinant structure which makes the pathologic system stable. An important fact is that the appearance of other foci under the influence of the determinant focus makes the latter as well as the system which originates more stable. The origin of even a 'mirror' focus makes the determinant focus more stable. When there are other foci, the determinant structure is protected, as it were, by other parts of the system and is stabilized under their influence. THus, the positive links between the determinant structure and other parts of the pathologic system consolidate the system. This phenomenon is clearly exhibited in investigations involving the suppression of the activity of the hyperactive complex in the cerebral cortex when some cerebral structures are electrically stimulated. It will be shown in Chapter 13 that epileptic activity is effectively suppressed in one focus (even if it is produced by sufficiently strong epileptogenous influence) when the caudal reticular pontis nucleus is stimulated. However, if the foci of subepileptic activity, which are produced in other cortical zones, form an epileptic complex under the influence of the determinant focus, the ES of that nucleus is ineffective, i.e. it cannot suppress the determinant and the complex as a whole (Figure 106 C). THe complex gradually breaks up as the nucleus is continuously stimulated for a long time (Figure 106 F). Hence, it is the system that becomes resistant, which is due to the interaction of its parts.

Thus, the pathologic system acquires new mechanisms of resistance that are characteristic only of it during its formation and activity.

All that is in accord with the fact that the period during which the epileptic complex continues to operate after the determinant structure is eliminated depends on the duration of the action of the whole complex as a system: the longer does such a system work, the longer is it preserved after the determinant is eliminated. This also applies to the role of the connectors in consolidating artificial functional systems in the brain (e.g. the effect of ethimisole) (Smirnov, Borodkin, 1975, 1979; Smirnov, 1976).

The pathologic system is probably fixed by fundamentally the same mechanisms as those which fix physiological systems. An interesting fact in this respect is that GPEE remains active longer when sick animals are treated with the $ACTH_{4-7}$ peptide, which can consolidate long-term memory (see Chapter 14; De Wied et al., 1975; Ashmarin et al., 1978). The fixation of a given system's engram probably results in the pathologic system's gradual consolidation. It has been shown by taking the phenomenon of the long-term artificial stable functional connections as an example (Smirnov, 1976; Smirnov, Borodkin, 1975, 1979) that various artificial psychic states produced by electrically stimulating the cerebral structures can remain for a long time after stimulation. Such maintenance of the pathologic syndromes (i.e. syndromes of hyperphagia and hyperbolized alimentary motivation) is observed when the lateral hypothalamus is no longer electrically stimulated (see Chapter 12). The peculiarities of the functional structure of GPEE on which the hyperactive determinant is based, just as the specifics of the pathologic system being produced by the determinant, should promote the realization of the neurophysiological mechanisms, which are regarded as memory mechanisms, namely, synaptic changes in neurons that facilitate the conduction of stimuli (Eccles, 1953; Kostyuk, 1960), stable changes in the neuronal membrane (Matthies et al., 1978; Brazier, 1979; Kruglikov, 1981), changes in perisynaptic glias (Roitbak, 1973, 1979), etc. The prolonged circulation of stimuli in the GPEE neuronal population and in the connections of the pathologic system, being caused by the peculiarities of their structural and functional organization, can promote and consolidate those changes (Chapter 3).

A peculiar state originates when the cerebral physiological systems and the respective mechanisms of homeostasis cannot suppress the pathologic system, but restrict its activity and development. In this state, interconnected intrinsic pathologic mechanisms and mechanisms which compensate and stabilize a process are exhibited. Besides, this state is an expression of a pathologic state, adaptation to a pathologic state, and adaptation to the environment under the given conditions. It is described as a 'stable pathologic stage' (Bekhtereva, 1974-1980; Bekhtereva et al., 1978).

DISORGANIZATION AND DISINTEGRATION OF SYSTEMIC ACTIVITY

The formation of a hyperactive determinant structure and the pathologic system is coupled with the disorganization and disintegration of intersystemic and intrasystemic physiological relationships.

The exposition in the preceding sections shows that the disorganization of both integrative nervous activity and intersystemic relationships, being caused by the pathologic system, may be effected by several mechanisms. The involvement of new cerebral structures in the pathologic process due to the distribution of pathologically enhanced excitation from a hyperactive determinant structure to other areas of the central nervous system is important in this respect. The formation of secondary hyperactive structure, i.e. generators of excitation, which can act as independent determinants that are capable of producing new pathologic systems, is a more profound and, pathogenically, a more stable expression of that process. Multifocal epilepsy is a vivid example of the process. The pathologic process spreads and new structures become involved also because GPEE grows (see Chapter 3). The so-called Jacksonian march is a clear illustration of the clinical expression of such a process.

Besides the pathologic system's expansive development, the coupled inhibition of other physiological systems' activity, being caused by the pathologic system, is important pathogenically. This mechanism becomes more and more important as the pathologic system develops and its activity grows. It is exhibited especially during paroxysms of the pathologic system's activity. However, it is also manifested during the period between paroxysms and even at the latent stage of the pathologic process. The pathologic system's activity keeps the mechanisms of homeostasis, particularly inhibitory control, in constant stress. Consequently, a large amount of energy and plastic material are consumed and the structures which effect that control deteriorate more rapidly. The pathologic system's development, which engenders new pathologic systems, ultimately leads to the disorganization of the brain's integrative activity.

The formation of an epileptic activity complex in the cerebral cortex or a pathologic system in other regions of the central nervous system under the determinant's influence is in itself a peculiar form of integration. This integration is effected by neurophysiological mechanisms, being of a pathologic nature. It is characterized by the rigidity of intrasystemic relationships produced by the determinant. This rigidity is more pronounced (i.e. the system is modulated less by several influences), the more active is the determinant structure and the more intensive is the influence produced by it.

The disintegration and disorganization of physiological relationships, which originate during the pathologic system's formation and activity, begin to occur already at the level of nuclei and neuronal populations. This has been shown when GPEE's functional organization was analyzed (Chapters 2, 3). GPEE is a structure with unusual integration, which is pathologic as regards both its nature and GPEE's activity. The hypersynchronization of the GPEE neurons' activity is an expression of such pathologic functional integration.

Those forms of functional integration of the pathologic system and GPEE originate only at the initial stage of the pathologic process, when, as we have seen (Chapter 3), the specific trigger and strength relations between stimulation and GPEE's activity remain. They are upset at the late stages, being replaced by the phenomena of the system's breakdown and the appearance of independent activity by its parts. The pathologic process then becomes more diffused.

All that can be illustrated by taking the example of a relatively simple system, i.e. the lateral geniculate nucleus (LGN) - ipsilateral visual cortex (VC), in which these links are connected monosynaptically and GPEE is produced in lGN (Figure 53) (Kryzhanovsky et al., 1976c; 1978e). The evoked potentials in both LGN and ipsilateral VC increase during the early period of the formation of GPEE. In this case, the value of EP in lGN corresponds to that in VC. Later, the values diversify: EP decreases in LGN and increase in VC, i.e. the generation of potentials does not correspond to the message from LGN, and the VC structures act as secondary GPEE. In LGN, a slowly rising wave with small low-amplitude discharges originates instead of the high-amplitude discharge which was hypersynchronized earlier. This indicates that the neuronal groups which generate the discharges are functionally dispersed.

Consequently, the pathologic changes in the central nervous system, are, functionally, hyperbolized forms of the main types of normal functional connections in the nervous system, i.e. firmly programmed and stochastic activities. Hyperbolized firmly programmed activity as an expression of the pathologic state is the result of the hyperactivity of the determinant structure, which does not allow the system's behavior to be corrected. The pathologic stochastic form of activity is the outcome of the further disturbance of inhibi-

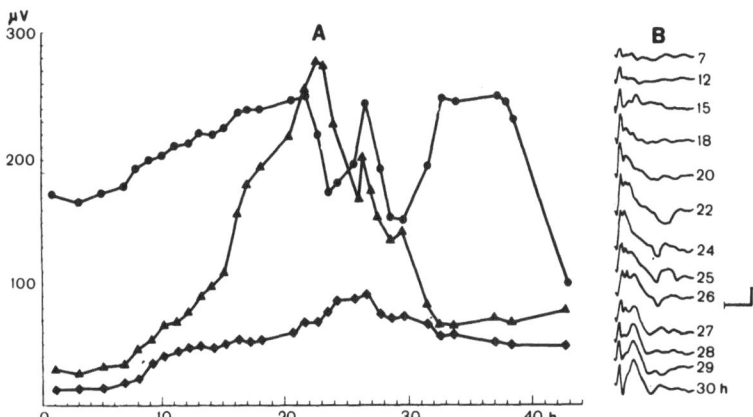

Fig. 53. Dynamics of the changes in visual EPs during different
periods of the development of photogenic epilepsy. A:
change in the amplitude of primary EPs in LGN (triangles)
and the visual cortex (circles) on the side of GPEE formation and in the opposite LGN (circles). B: dynamics of a
change in the averaged EPs in LGN during different periods
of GPEE formation after the injection of TT before the
first epileptic seizure. The time (hours) is indicated on
the right. Amplitude: 100 µV; time: 250 ms.

tory control, the system's disintegration, and the disinhibition and
auonomization of its components.

    At the late stages of a chronic pathologic process, secondary
pathologic changes promote the disintegration of the physiological
systems and the disorganization of nervous activity. These changes
have the nature of standard, nonspecific mechanisms. They occur
during various forms of CNS pathology and act as new pathogenic
factors. They include dystrophic neuronal changes, disturbances of
metabolism and microcirculation, hypoxia, intensive lipid peroxid-
ation, autoimmune processes, and morphological changes in the nervous
system.

PATHOLOGIC SYSTEMS AND NEUROPATHOLOGIC SYNDROMES

    It has been shown that the pathologic system which originates
under the influence of a hyperactive determinant structure on the
basis of a damaged physiological system retains the modal sign of the
latter. But this functional sign is hyperbolized so greatly that a
pathologic state originates. This state has its clearly defined
characteristic clinical features, i.e. a corresponding neuropathologic syndrome, which is consequently a clinical expression of the
behavior of the pathologic system and the result of its activity.
Its peculiarities depend not on the activity of the determinant

structure, but on the system in which this structure originated, while the characteristics of the syndrome's manifestation and its occurrence depend on the determinant.

It has been shown in Chapter 1 that the realization of the determinant's influences depends on many conditions under which its activity is carried on, including the state of the structures which receive the determinant's influence and the brain's general state. All these factors considerably affect the pathologic process's development and manifestation. Nevertheless, the specifics of a hyperactive determinant structure's activity are important in the syndrome's clinical expression.

A comparison of the peculiarities of GPEE's activity that were considered in Chapter 3 with the occurrence of the neuropathologic syndromes, particularly the nature of paroxysms, shows that the peculiarities are exhibited also in the syndromes' clinical expression. The growth of GPEE's power, the long production of excitation, the rapid attainment of both maximum impulse flow and the appropriate effect, the reduction of the latent period, the triggering of GPEE as if according to the 'all-or-nothing' law, its facilitated provoked and spontaneous activities, and other peculiarities of GPEE's operation are clinically expressed at the height of a pathologic process in the exhibition of syndromes and in the characteristics of paroxysms. Indeed, the above-mentioned (Chapters 1, 2, 4) forms of pathology show that paroxysms become more and more severe and prolonged at the height of a process. They originate under the action of not only strong stimuli, but also weak stimuli and are exhibited with the greatest intensity. The latent period of the production of paroxysms diminishes, and spontaneous ones occur more and more frequently.

It will be shown (Chapters 6-12) that such peculiarities are characteristic of also other syndromes, which include those whose pathogenic structure and clinical expression are very complicated and which are based on hyperactive systems. The peculiarities of GPEE's operation correspond to the peculiarities of a paroxysm also with respect to trigger stimulation: at the initial stages of GPEE's formation, the modality of the stimuli that activate GPEE and induce paroxysms corresponds to the system in which GPEE originated. At the late stages, when GPEE is activated by stimuli of different modality, paroxysms are provoked by stimuli which are just as different. Two main forms of activity of GPEE with an output to the motor system, i.e. tonic and phasic forms, determine two main forms of seizure, which is either tonic or clonic, while separate discharges engender slight jerks. The prolonged after-discharges or GPEE's virtually continuous activity result in either prolonged paroxysms or a subacute pathologic state, whose extent depends on the degree of GPEE's activity.

A hyperactive determinant structure may indirectly participate in the determination of the features of neuropathologic syndromes.

This participation is expressed in the involvement of new parts of a given system or parts of other systems in the pathologic process due to the growing intensity of the functional message formed by a determinant structure as a GPEE. When new parts are involved, a new, intricate set of syndromes or new components of a syndrome originate, and when other systems are involved, additional syndromes originate. Such a complication of the syndrome can be seen by taking the example of the above-mentioned pathologic scratch reflex, which is realized not by one, but by both hind legs at the late stages, and then it becomes a peculiar general seizure: the animal bends, and the hind legs simultaneously effect rapid convulsive motions, as if scratching.

Consequently, the latent period of the origin of the neuropathologic syndrome, which is based on a hyperactive system, ultimately depends on the period of GPEE'S formation, the origination of a hyperactive determinant structure, and the induced pathologic system.

SUMMARY

The pathologic system originates from the physiological system under the influence of a hyperactive determinant structure. An important aspect of the pathologic system is that the result of its activity, unlike that of the physiological system's activity, is of no adaptive significance. Moreover, such a result may be biologically unfavorable for the organism, i.e. it can be a pathogenic factor. This property of the pathologic system is connected with the fact that the system escapes the brain's integrative control due to the activity of a hyperactive determinant structure, which is also out of integrative, intrasystemic and intersystemic control. The pathologic system's induced activity reflects the properties of not the acting stimulant, but GPEE, which constitutes the neuropathophysiological basis of the hyperactive determinant structure. Such a system can also be activated spontaneously due to GPEE's spontaneous activation. A hyperactive pathologic system does not change the functional significance of the physiological system on the basis of which it originated, but merely pathologically enhances it.

A hyperactive determinant structure makes the pathologic system's functional organization and activity especially rigid. This rigidity levels out other systems' modulating influence. The extent of this rigidity depends on both the intensity of the functional message produced by the determinant structure and the characteristics of the determinant.

Unlike the physiological system's activity, which is corrected due to feedbacks as the programmed result is being achieved and which ends after it is achieved, the pathologic system's activity is not corrected by either an intermediate or a final 'result' in spite of the presence of feedbacks and other afferentation. Feedbacks from the system's intermediate and terminal links can even be functionally

intensified, but they are ineffective, since their influences cannot
be realized both in the determinant structure region due to the
insufficiency of the inhibitory mechanisms and at the intermediate
stages owing to the excessively strong flow of impulses. The pathologic system does not disappear when the 'result' is achieved. It
continues to act even if its 'result' is pathogenic as regards the
organism. The pathologic system's activity ceases only when the
determinant structure stops its activity.

All the main features of the pathologic system depend on the
determinant structure's activity. If the other links of the system
are not influenced by th4 determinant structure (e.g. if the determinant structure is not active for some reason at a given moment),
they can remain normal and, becoming parts of other physiological
systems, can adequately perform their functions.

The pathologic system is resistant. A hyperactive determinant
is the main factor which establishes the pathologic system's resistance. At the same time, pathologic structures of secondary origin
play an important role in this process. These structures constitute
the pathologic system and enhance the stability of both the determinant and the whole system. As the pathologic system develops, it
acquires new mechanisms and becomes more resistant.

The appropriate neuropathologic syndrome is the clinical expression of both the pathologic system's behavior and the result of
its activity. Its specifics depend not on a hyperactive determinant,
but on a system in which the structure originated. The determinant
establishes the peculiarities of the course and clinical expression
of a neuropathologic syndrome. The nature of a paroxysm depends on
the characteristics of the action of GPEE as the hyperactive determinant's neuropathophysiological mechanism.

The formation of a hyperactive determinant structure and the
activity of the pathologic system are connected with the physiological system's disintegration and disorganization. The pathologic
system can develop due to both the involvement of new cerebral structures in the process under the determinant's influence and the growth
of GPEE's power. This development entails the suppression of the
activity of antagonistic and, later, other physiological systems as
well as protective and compensatory mechanisms. Consequently, the
central nervous system may become disorganized.

At the late and especially last stages of a pathologic process,
the pathologic system itself may disintegrate, its parts may become
autonomous, and new pathologic structures may be formed.

Functionally, the nervous system's pathologic state is a hyperbolized form of the main types of functional connections in the
nervous system, i.e. rigidly programmed and stochastic ones.

Part II
Modeling neuropathologic syndromes

5
Methodological approaches

The syndromes of the disturbances of the activity of the nervous system and particularly the activity of the brain cannot be completely reproduced experimentally in animals not only because of the anisomorphosism of the structural and functional organization of the brain of man and animals, but also because of the specific social environment which is of particular significance to man. This environment plays a great role not only in the origin of various forms of the nervous system's pathologic state, but also in the establishment of the characteristics of their manifestation, course and elimination. From the methodological point of view, however, pathogenetically adequate model equivalents of the pathology of the human nervous system can be created. This is a matter of not reproducing nosologic forms and diseases which are specific of man, but creating models of syndromes which may occur in the case of various affections of the central nervous system. That is the topic of this part of the book.

The neurochemical and neurophysiological mechanisms of the activity of the homologous physiological systems and regions of the central nervous system of man and higher animals are fundamentally the same. These mechanisms are universal and reflect the general regularities of the structural and functional organization of the nervous system in man and higher animals. Therefore, it can be assumed that the neurobiological basis, i.e. the neurochemical and neurophysiological mechanisms of the disturbances of nervous activity in man and higher animals, is also of a single type when comparable forms of pathology are involved. Numerous investigations and especially recent researches have confirmed this view. However, the clinical manifestations of the pathogenetically same forms of affections of the central nervous system are different in man and animals.

115

They also differ in animals of diverse species. Ecological factors and the species behavioral peculiarities are fundamentally significant in this respect. In this sense, it is clear that 'cat psychosis', just as human psychosis, cannot be reproduced in rats. Only the model of 'cat psychosis' can be reproduced in rats if this is necessary.

An experimental model of a syndrome can be produced after tackling two main tasks: (1) the revelation of the neurobiological basis, i.e. the neurochemical and neurophysiological mechanisms of a given form of a pathologic stage, and (2) the establishment of the adequacy of the clinical signs of the experimental model as the symptom equivalents of the form of a pathologic state that is being modeled.

The creation of a pathogenetically adequate model is connected with a study of the neurobiological mechanisms of the modeled syndrome. The production of such a model directly verifies the basic concepts of the mechanisms of the syndrome being studied. Broadly, it is a criterion of the correctness and potentialities of the theory on the basis of which the working principles of modeling are formed. The common features of the neurobiological mechanisms of a model and the form of pathology that is being modeled makes it possible to work out therapeutic principles on the basis of the model and to use the principles clinically. This is how the biological methods of treating nervous and mental diseases in man had originated. Biological psychiatry owes its origin largely to the investigations involving experimental modeling and experimental therapy.

Moral and ethical questions are an important aspect of the modeling problem. Pharmacological agents which are capable of changing or even suppressing the manifestation of pathologic syndromes cannot be used, at least at the initial stages of study, owing to the need to thoroughly study animal behavior and determine the clinical syndrome equivalents by the model. This particularly applies to the models of the central pain syndromes, in the case of which an animal's pain reactions during its free behavior are important tests.

However, agents which alleviate the animal's suffering should be used whenever possible. The elaboration of experimental therapy itself is important in this respect. We tried to produce the models of the pain syndromes of central origin under the most merciful conditions, ending a process after it has been manifested to the necessary extent, and have adhered to the recommendations made by the Committee for Research and Ethical Issues of the International Association for the Study of Pain (IASP, 1980).

The above-mentioned general concept of the role of the determinant structures in the activity of the nervous system and the theory of the generator mechanisms of the neruopathologic syndromes were the theoretical basis on which the experimental models of various neuro-

pathologic syndromes were produced. A single methodological principle of the establishment of experimental models was worked out on the
basis of those theoretical concepts. Its essence is that a hyperactive structure is created in the functionally significant part of the
physiological system, i.e. in a definite part of the central nervous
system; owing to its functional peculiarities, it acts as a determinant of the given system; under its influence, this system becomes
hyperactive, escapes from integrative control and turns into a pathologic system (see Chapter 4). The appropriate neuropathologic syndrome is the clinical expression of the activity of such a pathologic
system. Either the neuronal population with an insufficiency of the
inhibitory mechanisms that forms GPEE or the disinhibited hyperactive
apparatus of transmitter secretion can be the operant part of the
determinant structure.

To activate the structures of the central nervous system under
chronic conditions, use is frequently made of electric stimulation
(ES) with implanted electrodes. This method, being widely employed
in experiments and clinical practice for diagnostic and therapeutic
purposes, has greatly helped in studying the brain, the function of
its structures and systemic reactions. However, it has several
disadvantages, because a rather large amount of various neural elements (neurons, fibers, synaptic structures) gets into the field of
the stimulating electrode. These elements belong to different physiological and neurochemical systems, have different functions and
diverse thresholds of excitability, and constantly change under the
influence of various factors. Therefore, the involvement of certain
elements and, consequently, the systems which are connected with them
in a reaction depends on ES conditions and on many influences which
change the reactivity of the neural structures. A reaction can
change not only quantitatively, but also qualitatively, and a dia-
metrically opposite effect may even be produced when ES parameters
change.

Two methods were used in the investigations given further to
produce GPEE: (1) the primary disturbance of the inhibitory mechanisms in a neuronal population, and (2) the primary depolarization of
neurons. In the former case, inhibitory control is primarily insufficient and neurons are out of inhibition; these neurons form
GPEE. In the latter case, the insufficiency of the inhibitory mechanisms is secondary and can be of relative significance at the early
stages of the processes, i.e. the inhibitory mechanisms are not
sufficiently effective for hyperactivated neurons (see Chapter 4).

Tetanus toxin (TT), strychnine, picrotoxin and penicillin were
used mainly in the first method, and ouabain, potassium chloride, and
ES in the second method.

The disturbance of the inhibitory mechanisms is the main effect
of TT. The drug suppresses various forms of inhibition in the spinal

cord when the IA and IB afferents, the II and III muscle afferents
and the skin afferents are activated; it also suppresses Renshaw's
recurrent inhibition (Brooks et al., 1957). It upsets the inhibition
of not only motoneurons, but also interneurons, as a result of which
polysynaptic reflexes substantially increase and become synchronized
(brooks et al., 1957; Kryzhanovsky, Dyakonova, 1964; Kryzhanovsky,
1966, 1967, 1973). It also suppresses the inhibition of gamma-
neurons (Takano, Kano, 1973) and that of the sympathetic neurons in
the spinal cord (Pear, Wellhöner, 1973). Tetanus toxin disturbs
descending inhibition (Curtis, 1959; Kryzhanovsky, Sheikhon, 1973)
and the inhibition of the basket cells in the cerebelum (Curtis et
al., 1973), and also reduces not only postsynaptic, but also presynaptic inhibition (Sverdlov, 1965, 1969; Curtis et al., 1973). Inhibition if disturbed under the influence of TT due to the blockade
of the spontaneous and evoked secretion of the inhibitory transmitters by the presynaptic terminal, and not to the disruption of
synthesis. TT blocks the release of glycine and GABA (Curtis, de
Groat, 1968; Guschin et al., 1970; Curtis et al., 1973; Curtis,
Johnston, 1974; Collingbridge et al., 1979; Davies, Tongroach, 1979;
Aliev, Kryzhanovsky, 1979; Bigalke et al., 1981; Kryzhanovsky et al.,
1982b) as well as dopamine (Collingridge et al., 1980; Kryzhanovsky,
Aliev, 1980; Collingridge, Davies, 1982). In small doses, TT inhibits Na release from the synaptosomes (Habermann, 1981), while in
large doses, it intensifies this process (Kryzhanovsky et al., 1980g;
Habermann, 1981). TT interferes with not only inhibitory synapses,
but also exciting synapses, as has been seen in experiments on the
crayfish muscle (Kano, Ishikawa, 1972), on the neuromuscular junction
of homoiotherms (Kryzhanovsky, 1967; Duchen, Tonge, 1973;
Kryzhanovsky et al., 1974c; 1981d; Pozdnyakov et al., 1981) and
poikilotherms (Mellanby, Thomson, 1972). It blocks the release of
acetylcholine, as a result of which large clusters of vesicles are
formed in the terminal of the neuromuscular synapse so that the
terminal is packed with vesicles (Kryzhanovsky et al., 1974c). Such
a large number of vesicles has been observed also in the presynaptic
terminals of the spinal cord during tetanus intoxication
(Kryzhanovsky, 1973). The blockade of the cholinergic synapses of
the iris under the influence of TT has been described (Ambach et al.,
1948).

Thus, TT can act on various types of synapses with different
transmitters, upsetting their conduction. It follows that there is a
universal mechanism of TT action on the secretion of transmitters
(Kryzhanovsky, 1973, 1981a). However, inhibitory synapses, particularly glycine and GABAergic synapses, which are the first to be
damaged under natural conditions as the minimum amount of TT acts,
are the most sensitive to TT. The content of K^+ in synaptosomes
decreases under the influence of TT (Lutsenko et al., 1982). The
data obtained in experiments involving both synaptosomes with the use
of a lipophilic cation, i.e. tetraphenylphosphonium (Ramos et al.,
1979), and neuronal culture (Dimpfel, 1979) show that TT may produce

depolarizing action. However, experiments in vivo have shown that
the electric properties of the membranes of the spinal motoneurons do
not substantially change even when the inhibition of these neurons is
greatly disturbed (Sverdlov, 1969) (see also Chapter 2). The depolarizing action of TT should be specially discussed (Kryzhanovsky,
1981a; Wellhöner, 1981). If this effect is produced under natural
conditions when TT acts on the central nervous system of animals, it
can play an additional role in the mechanisms that hyperactivate the
neurons and reduce the thresholds of the sensitivity of neurons which
form GPEE. An important fact is that the postsynaptic membrane,
i.e. the membrane of the neurons which constitute GPEE, is not
blocked by TT, and the postsynaptic receptors of the inhibitory
mediators are not bound by TT (Curtis, de Groat, 1968; Guschin et
al., 1970; Kano, Ishikawa, 1972). TT itself does not cause a detect-
able change in the activity of Na, K-ATPase. However, the activity
of this enzyme is changed in the nervous tissue of the affected
segments of the spinal cord (the GPEE region). The mechanisms of TT
action have been considered in greater detail elsewhere
(Kryzhanovsky, 1981a, Wellhöner, 1981).

Owing to the above-mentioned features of TT action, it can be
used for modeling the endogenous formation of GPEE. This peculiarity
as well as the origin of the local effects when TT is precisely
injected in a micro amount into the selected regions of the central
nervous system, the dependence of the intensity of the pathophysiological effects on the amount of TT being injected, the long-term
duration (many days and even weeks) of the clinical symptoms, and the
possibility of observing the animals under free behavior conditions
have made TT the main agent in both producing GPEE and modeling
neuropathologic syndromes in our researches. We have indicated
(Kryzhanovsky, 1966, 1973) that TT is a very useful and special tool
in this respect and also in other neurophysiological researches.
This view is now being supported by other authors (Mellomby, Green,
1981; Wellhöner, 1981).

To produce GPEE on the basis of neuronal disinhibition, use was
made of strychnine, which disturbs postsynaptic inhibition by changing the sensitivity of the glycine receptors on the postsynaptic
membrane (Wermann et al., 1968; Curtis et al., 1969; Curtis et al.,
1971; Curtis, Johnston, 1974), picrotoxin, which disturbs GABA inhibition (Curtis et al., 1969; Gallindo, 1969) and penicillin, which
disturbs GABA inhibition (Curtis et al., 1972; Davidoff, 1972;
Roberts et al., 1976), reduces the inhibitory driving force (Deisz et
al., 1979), and causes the depolarization of the neuronal membrane
(Futamachi, Prince, 1975; Hocher et al., 1976; Heyer et al., 1981,
1982).

To produce GPEE by directly hyperactivating the neurons, use was
made of such depolarizing agents as K^+ (KCl) and ouabain, which
suppresses the activity of Na, K-ATPase (Thomas, 1972). Moreover,

the method of electrically stimulating the brain structures for a short or long time by paired microelectrodes was used. This method makes it possible to compare the functional effects of GPEE and the effects of the direct activation of the same structures of the central nervous system. To form GPEE, the method of chronic electric stimulation of the kindling type (Delgado, Sevillano, 1961; Goddard, 1969; Goddard et al., 1969; see Wada, 1976) was used.

The study of the GPEE's effects produced by various methods has revealed the most general regularities of the pathologic process caused by GPEEs of a different nature, on the one hand, and has made it possible to ascertain the specifics of their neurochemical organization and activity needed for creating the models of neuropathologic syndromes as well as their rational pathogenetic therapy, on the other.

Besides microinjections, the 'agar depot' method was used to introduce substances into the spinal cord tissue. By this method, agar plates with substances were put on a certain place of the spinal cord surface. Owing to diffusion from the 'agar depot', substances slowly passed into the nervous tissue, as a result of which GPEE and the corresponding syndrome developed. The rate, intensity and duration of the syndrome depended on the concentration of substances in the agar place and its dimensions.

SUMMARY

It is impossible to completely reproduce the syndromes of nervous disturbances, particularly the human brain's pathologic states, in animal experiments. Methodologically, however, pathogenetically adequate equivalents of human nervous pathology can be produced by modeling certain syndromes, and not by reproducing human diseases. This problem can be solved by ascertaining the neurobiological basis, i.e. the neurochemical and neurophysiological mechanisms, of the pathologic states being modeled and by establishing the adequacy of the clinical features of an experimental model as symptom equivalents of a given pathologic state. Owing to the common nature of the neurobiological basis of a model and the pathologic state being reproduced, therapeutic methods elaborated by means of experimental models can be used in clinics. At the same time, the neurobiological mechanisms of apathologic state under consideration can be established by producing a model and elaborating the respective therapeutic methods.

Ethically, neuropathologic syndromes should be so modeled that the animal's suffering is maximally relieved. Whenever possible, drugs which alleviate suffering should be used. The elaboration of experimental therapy is important in this respect.

The above concept of the role of determinant structures in nervous activity and the theory of generator mechanisms of neuro-pathologic syndromes were the starting point in experimentally re-producing neuropathologic syndromes in our investigations. They were the basis of a unitary methodological principle of modeling neuropathologic syndromes, i.e. forming hyperactive structures which act as determinants and which transform physiological systems into pathologic ones, the latter being clinically expressed in the appro-priate neuropathologic syndromes. Generators of pathologically enhanced excitation (GPEEs), the augmented, uncontrollable secretion of transmitters, and the supersensitivity (an increase in the number) of receptors on the postsynaptic membrane can be the operant mechan-isms of pathologic determinants.

Either of the following two methods was employed to produce GPEEs: the primary disturbance of inhibitory mechanisms in a certain neuronal population or primary neuronal depolarization. For this purpose, use was made of drugs, including tetanus toxin, strychnine, picrotoxin, penicillin, KCl, and ouabain. Some of them produce a mixed effect (the disinhibition and depolarization of neurons). Owing to certain peculiarities, tetanus toxin became the main agent in producing GPEE and modeling neuropathologic syndromes.

6
Generalized spinal myoclonia

The syndrome of generalized spinal myoclonia, which has several characteristic features, originates when GPEE is produced in the anterior horns of the lumbosacral segments of the spinal cord by TT (see Chapter 1) at the late stages of a pathologic process. Clinically, it is expressed in severe general convulsions involving virtually all the muscles, including the respiratory muscles, when a seizure is provoked or when it originates spontaneously. As a result of these convulsions, respiration is inhibited, the hind legs stretch out, the tail rises, the back straightens out, the head is thrown back, the jaws clamp tightly, the eye slits narrow down, and the eyes close. In this phase, the spasm is tonic. The more severe is the process and the later do convulsions originate, the longer and more intensive is the tonic phase. During the seizure period, a long powerful EA burst can be recorded in all the muscles. This burst consists in high-amplitude and high-frequency discharges (Figure 54), and lasts as long as a general tonic spasm.

At the end of the seizure, the tonic phase passes into a clonic one, i.e. at first, general tonic muscular tension decreases, respiration recommences, the extensor hypertonicity of the hind legs, back, and neck attenuates, and so forth. That is followed by a series of contractions of the same muscles which are of short duration and have a clonic character. These contractions are also synchronized and originate simultaneously in all the muscles. Their intensity and frequency depend on the general severity and stage of a pathologic process. When the process is not severe, muscular contractions can be brief and faint and can be expressed in jerks.

Synchronized discharges with a high frequency and a different amplitude of potentials are recorded in the muscles during their

122

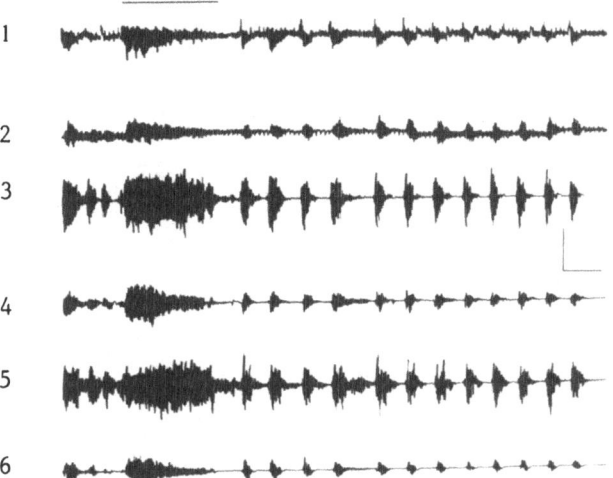

Fig. 54. Evoked burst of tonic seizure activity and spontaneous
clonic seizure activity. The burst of EA was provoked by
stimulation (squeezing the foot; indicated by a horizontal
line) on the side where GPEE had been formed in the lumbosacral segments (L_5-S_1) after microinjecting TT into their
ventral horns. Recordings of EA: 1,2: left and right neck
muscles; 3,4: left and right back muscles; 5,6: posterior
group of left and right femoral muscles. Amplitude: 500
μV; time: 1 s.

contractions. The initial discharges usually have the highest amplitude. Afterwards, the intensity of the discharges diminishes and
becomes rather small at the end of a fit (Figure 54). These group
discharges originate simultaneously in all the muscles and correspond
to one another with respect to their duration. They are separated by
intervals when electric activity may be completely absent. These
intervals can also be synchronized in all the muscles.

Synchronized EA bursts in all the muscles and clonic muscular
contraction which corresponds to them can originate also spontaneously. Such a phenomenon is characteristic of the late stages, when the
process becomes especially severe. In this case, brief group discharges of certain frequency (which can change in the course of a
seizure) occur in the muscles.

We termed that phenomenon of generalized synchronized discharges
in all the muscles 'strychine-like tetanus' (Kryzhanovsky, 1966-1973;
Kryzhanovsky, Lutsenko, 1975). This term was given to it because a
similar phenomenon was observed earlier by Bremer (1941, 1953) when
small laboratory animals were poisoned with strychnine. In the case

of strychnine poisoning, individual synchronized group bursts occur more regularly and their frequency is more steady.

However, the main difference is in the interpretation of the mechanisms of the given phenomenon. Bremer believed that the phenomenon was based on the diffusive change in the excitability of the spinal cord throughout its whole length due to general strychnine poisoning. In this respect, he believed that the leading role was played by the spinal motoneurons and their electrotonic excitation, which involved the whole column of motoneurons at once. Hence, according to Bremer, this phenomenon is realized well in animals with a rather short spinal cord.

All the above-mentioned data and particularly the data on the mechanisms of the phenomenon show that Bremer's assumption cannot be accepted. Investigations have revealed that this phenomenon is engendered by the activity of GPEE in the system of the propriospinal neurons (see Chapter 2). This is confirmed by the fact that when only GPEE is suppressed by glycine without influencing the whole spinal cord, the entire syndrome will disappear together with its tonic and clonic seizure phases and characteristic clinical and electrographic manifestations (see Chapters 1 and 13). We succeeded in reproducing this phenomenon in animals of diverse species with a different length of the spinal cord (Kryzhanovsky, 1966). Although it is still not quire clear how tonic and clonic activity originates in the same source (see Chapter 3), the source is obviously GPEE and we know its localization and origin.

The above-mentioned phenomenon reproduced experimentally in animals in its 'pure' form is apparently not observed clinically in man. However, account should be taken of the fact that such a phenomenon can originate when GPEE is formed in the system of propriospinal diffuse connections (see Chapter 2). Some pathologic forms which are either equivalent or similar to that phenomenon may be observed in clinical practice after either injuries to the spinal cord or local hypoxia, especially when much time passes. Therefore, it seemed expedient to describe this phenomenon as a syndrome and to define it as generalized spinal myoclonia or 'spinal epilepsy' with regard to the above-mentioned peculiarities. The syndrome can be considered an equivalent of focal epilepsy with secondary generalization. The involvement of supraspinal brainstem motoneurons is not contradictory in this respect, as the given spinal form of myoclonia can originate also when the spinal cord is cut at a high level (Figure 55). Under these conditions, generalized seizures are exhibited even better. The supraspinal parts of the central nervous system apparently inhibit the realization of the epileptic syndrome in the spinal cord.

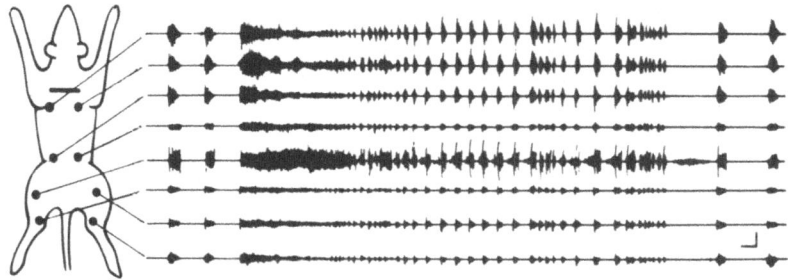

Fig. 55. Tonic and clonic seizure activities during generalized
          spinal myoclonia under the conditions of spinal cord tran-
          section. Stimulation (squeezing the foot on the side of
          GPEE is indicated by a horizontal line. Left: recordings
          of EA in various muscles. Section of the spinal cord at
          the Th_5 level 24 hours before testing (48 h after injecting
          TT into the left hind leg muscles). Amplitude: 1 mV; time:
          1 s.

SUMMARY

    The production of GPEE in the lumbosacral segments of the spinal
cord in the system of propriospinal neurons entails the syndrome of
generalized myoclonia, which is expressed in the origination of a
general seizure accompanied by bursts of enhanced EA in all the
muscles, when GPEE is activated from definite receptive fields
(trigger zones). Muscles with supraspinal (bulbar) innervation are
involved in the seizures. At the height of the syndrome, general
seizures originate just as spontaneously, and they have tonic and
clonic phases. When GPEE is suppressed, the syndrome disappears
during the action of the inhibiting agents. When the spinal cord is
cut at a high level, the syndrome (seizures involving the distal part
of the body and hind legs) may be exhibited more intensively. The
above-mentioned characteristics make it possible to term the syndrome
'spinal epilepsy'. The syndrome of 'spinal epilepsy' can be used in
model neurophysiological investigations of hyperactive systems, the
pharmacological screening of epileptic preparations, and the study of
some types of hyperkinesis.

7
Vestibulopathy

Vestibulopathy was produced by creating GPEE in the system of vestibular nuclei (Kryzhanovsky et al., 1976c). The functional structure of the vestibular complex is such that it may become the basis of the formation of GPEE of vestibular neurons under pathologic conditions when the inhibitory mechanisms are upset. The distinguishing features of the vestibular neurons includes their constant background impulse activity (De Vito et al., 1956; Gorgiladze, 1966; Wilson et al., 1966; Ito et al., 1969), which may be connected with the presence of proper multisynaptic pathways in the nucleus (Precht, Shimazu, 1965; Shimazu, Precht, 1965). These pathways are believed to be connected with the multipeak EPSP which originate in the secondary vestibular neurons when the vestibular nerve is stimulated (Ito et al., 1969b) and with the tonic vestibular neurons which originate after the monosynaptic response of the prolonged secondary discharges. Multisynaptic excitation in vestibular nuclei can be produced by intranuclear interneurons via the feedback system (Brodal et al., 1962). Besides the intranuclear mechanisms, the vestibular complex has many activating connections (Brodel et al., 1962). At the same time, vestibular neurons can apparently independently generate spontaneous impulses, as these impulses were recorded under the conditions of deep nembutal anesthesia, when the reticular formation and other areas of the central nervous system were 'switched off', and under the conditions of curarization, the removal of the cerebellum and the section of the vestibular nerves (Gorgiladze, 1966). The spontaneous impulse activity of the secondary neurons may be possibly caused by definite properties of their membrane (Bruggencate et al., 1972). The population of the secondary vestibular neurons is heterogeneous; more than five types of neurons have been ascertained in it (Shimazu, Precht, 1966). The neurons of type I are divided into tonic and phasic ones (Precht, Shimazu, 1965; Shimazu, Precht,

1965). Tonic neurons possess the properties of spontaneous impulse
activity and a low threshold of their reaction to acceleration, and
they slowly change the frequency of the impulse during acceleration.
Kinetic neurons do not possess the properties of spontaneous impulse
activity; they have a high excitation threshold as regards acceleration, and intermittently change the frequency of the impulse flow.

The connection between vestibular nuclei of both sides is of
great functional significance. It has been shown that contralateral
vestibular stimulation causes the excitation of the neurons of type
II and the coupled inhibition of the neurons of type I (Shimazu,
Precht, 1966). Neurons of type II can be inhibited by the influence
of many areas of the central nervous system (Markham et al., 1966;
Ito et al., 1969). Contralateral inhibition is sensitive to strychnine, picrotoxin and bicuculin, but it is completely removed only
when these agents are used simultaneously (Precht et al., 1973). The
commissural inhibition system consists of at least two types of
inhibitory processes, whose probably mediators are glycine and GABA
(Obata, Highstain, 1970; Felpel, 1972; Fukuda et al., 1972). The
constant dynamic equilibration of the activity of the nuclei of both
sides and their reciprocal relationships, which are of great importance to physiological reactions associated with the position of the
body, the head, and so forth, become especially significant in pathology when a function fails or the nuclei of one side are hyperactive.

It has been shown (Chapter 3) that GPEE originates when the
inhibitory processes are disturbed after TT is injected into Deiters'
nucleus. The activity of this GPEE was exhibited in an acute experiment with a motionless animal, when the extensor monosynaptic reflexes were facilitated. In chronic experiments in which animals
behave freely, GPEE produced in Deiters' nucleus causes a specific
rotation syndrome, i.e. vestibulopathy.

The pathologic process develops in the following way: the first
signs of the rotation syndrome appear a few hours after TT is injected in a microdose into LVN in rats. Only tic-like motions of the
head towards the injection of TT (ipsilateral motion) and upwards are
observed at the initial stages of vestibulopathy. Afterwards, the
ipsilateral front leg extends, the head is intensively thrown upwards, and the body turns around lengthwise contralaterally to the
site of GPEE (Figure 56).

At the initial stages of vestibulopathy, rotations could be
triggered by specific stimulation (when the animal moved its head
abruptly). Rotation occurred slowly, and the animal turned around
once or twice during one fit. Afterwards, the rotation syndrome
became more pronounced: paroxysms either originated spontaneously or
were provoked by unspecific sensory stimulation (sound, tactile
stimulation, light, etc.); they occurred more quickly, the animal
making 5-7 rotations during one paroxysm.

Fig. 56. Rotations after the production of GPEE in the right lateral
          vestibular nucleus. The sketches were made from a motion
          picture.

     In the interval between paroxysms, the animals showed no signs
of pathology (especially at the early stages of vestibulopathy); they
normally oriented themselves in space and moved freely. At the last
stage of vestibulopathy, any motion by the animals caused them to
make a series of rotations.

     Those rotations could be caused in healthy animals by electric-
ally stimulating one of the LVNs. In the case of single stimulation,
the animal moved its head towards the stimulation site and then
upwards. Such motion was observed at the early stages of vestibulopathy after TT was injected into LVN. When LVN was stimulated by a
series of impulses, the animals made rotations similar to those
observed at the height of the syndrome after TT was injected into LVN
(Figure 57). In this case, the number of rotations was determined by
the number of impulses in a series, while the rat was determined by
the frequency and force of ES. When LVN was stimulated with a low
frequency (once every second), there was slight displacement along
the rotation path due to every impulse. Hence, partial rotation was
observed.

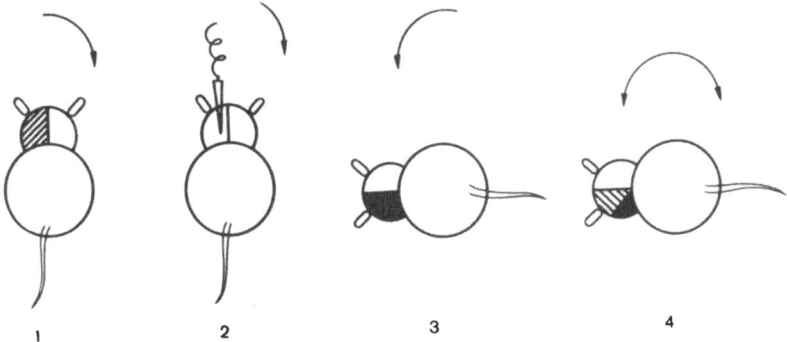

Fig. 57. Direction of rotations in the case of various interventions
          in the vestibular nuclei. 1: production of GPEE by TT in
          the left lateral vestibular nucleus; 2: ES of the left
          lateral vestibular nucleus; 3: destruction of the left
          lateral vestibular nucleus; 4: destruction of the left
          lateral vestibular nucleus in which GPEE was produced. The
          arrows indicate the direction in which the animal moves
          around the longitudinal axis.

    The effect of the unilateral vestibular 'switch-off' is ex-
hibited immediately after the nuclei are destroyed by electrocoagu-
lation or barotrauma. The functional failure of vestibular nuclei
differs from their hyperactivation in that the animal's posture is
changed and it lies on the affected side with its head and body
turned in the same direction. When there are no external restrict-
ions, the animal continuously rolls over ipsilaterally. These movements greatly differ by their nature from the rotations which are
made when GPEE is being formed in the same nucleus (Figure 57 3).
The animal rolls over in the following manner: the front part of the
body turns by 180° with respect to the hind part, and then the body
turns around lengthwise in the ipsilateral direction. When the
animal is rolling over, its legs move in a manner which helps it to
do so. The animal may roll over continuously. However, it stops
when there are obstacles which press against it or when it assumes a
pose in which it can remain for some time.

    That effect was observed in healthy animals after coagulation
caused by direct current (5 µA for ten seconds). In animals with
vestibulopathy caused by the creation of GPEE in LVN, like in intact
animals, the destruction of LVN by direct current (5 µA for ten
seconds) produced the 'switch-off' effect (the disturbance of both
the posture and the rolling movement in the ipsilateral direction).
Although the vestibular function failed in these animals, it intensified again soon or a little after coagulation: the ipsilateral
rolling movements occurred together with the contralateral movements
(Figure 57 4). The rotation syndrome caused by the intensification

of the activity of vestibular nuclei had completely disappeared only
when the vestibular apparatus was coagulated by a current of 5 μA for
60-80 seconds. Then, the vestibular complex was extensively damaged
and the GPEE in it was apparently completely eliminated.

Thus, the paroxysmal disturbance of the body's equilibrium is
the main manifestation of the activity of GPEE produced in the system
of vestibular nuclei. In this respect, vestibular stimulation is
needed to provoke rotation at the early stages of vestibulopathy. In
other words, GPEE is activated at this stage by stimulation of specific modality. At the late stages, paroxysms either are provoked by
different stimuli or originate spontaneously. Thus GPEE either may
be activated by various stimuli or become 'self-excited' at this
stage. The paroxysmal nature of the origin of rotation shows that
GPEE works out in an explosive way.

The apparatus which enhances the incoming signal can first of
all be the neuronal system which is capable of ensuring both the
formation of GPEE in LVN and its activity.

The difference between GPEE's activity and the suppression of
the vestibular complex's nuclear activity is noteworthy. In the
first case, motions occur paroxysmally and are phased, but during the
interval between fits, the animal shows no pathologic signs. However, the disturbance of the posture is tonic when the functions of
the vestibular nuclei are suppressed. Magnus (1924) has shown that
rolling around lengthwise is an animal's locomotive act when equilibrium is tonically disturbed. Secondary kinetic vestibular neurons
may possibly play the main role in organizing motion when LVN is
stimulated or GPEE is created in it. But when LVN is destroyed,
tonic vestibular neurons play the main role in disturbing the posture. These neurons determine he balance between the activities of
the vestibular nuclei of different sides.

Experiments involving the coagulation of LVN in which GPEE was
at first produced show that there are two somewhat independent vestibular systems (tonic and kinetic ones). It has been shown that the
posture of animals with vestibulopathy was disturbed by coagulation;
at the same time, they roll in the direction of coagulation and make
rotations in the contralateral direction due to the activity of GPEE
in the part of the vestibular nuclei that remains after coagulation.
GPEE stopped functioning only when LVN in which it was produced had
been greatly destroyed. Thus, a small amount of kinetic vestibular
neurons which can constitute GPEE is enough to enhance the descending
synchronous message that causes rotation. In other words, GPEE which
consists of a small amount of neurons can be functionally significant, i.e. it can cause a pathologic syndrome.

SUMMARY

   When GPEE is created in a lateral vestibular nucleus (LVN), a
characteristic rotation syndrome is produced as the animal turns
around lengthwise in the contralateral direction (to the GPEE side).
The syndrome occurs paroxysmally, and the rate of rotations, their
number during a fit and the frequency of the latter depend on the
GPEE's power and the peculiarities of its activity. A similar effect
is produced when LVN is electrically stimulated. The syndrome is
temporarily suppressed when the nucleus which contains GPEE is partially coagulated by an electric current. This effect is produced as
the nucleus function (rolling over ipsilaterally) fails. The syndrome disappears when nucleus coagulation is extensive, i.e. when
GPEE is completely eliminated. The syndrome can last even when a
small part of GPEE remains after coagulation.

8
Photogenic epilepsy

It has been shown in Chapters 2 and 3 that the functional relationships between the nucleus neurons are disorganized and the
neurons form GPEE in LGN when the inhibitory mechanisms are disturbed. A characteristic neuropathologic syndrome of photogenic
epilepsy is engendered by the origin of GPEE in LGN (Kryzhanovsky et
al., 1976b).

Photogenic Epilepsy in Cats

Spontaneous seizures or fits caused by light stimulation (either
a single flash or rhythmic flashes) appeared in animals on the first
day after injecting TT into LGN. Other stimuli did not cause any
seizures. In most cases, characteristic prodromes were observed
before generalized seizures originated: at first, the animal made
tracing movements with its eyes in the direction opposite to the GPEE
site, and then the head and the body began to move; the animal (cat)
fell on its side (where GPEE was localized) and a seizure began. At
the beginning of the seizure, the muscles of the contralateral front
and hind legs and the contralateral half of the body were involved in
the spasm, and then the muscles of the whole ipsilateral side were
drawn into this process. During the seizure, the animal's pupils
were enlarged, and abundant salivation and urination were observed.

The clonic-tonic phases usually alternated during a seizure.
High-amplitude hypersynchronized convulsive discharges were recorded
in LGN in various areas of the brain and in the muscles during a
seizure (Figure 58). The fit often ended with a muffled cry as a
result of the spasm of the breath muscles. After the seizure, the
animal lay prostrate motionless. However, it could stand up and move

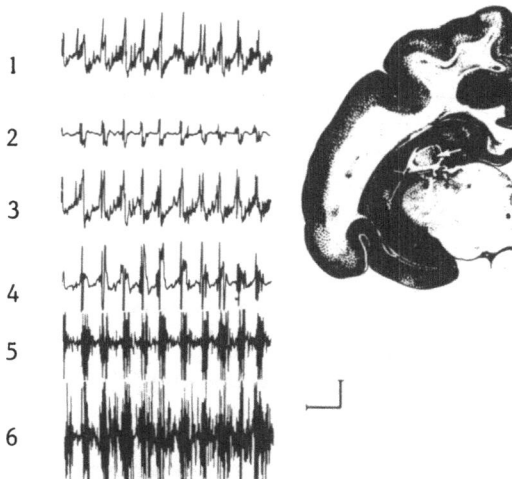

Fig. 58. Electric activity in the cerebral structures and muscles
          during the clonic phase of a spasmodic fit in a cat.
          Recordings: 1: ipsilateral visual cortex (VC); 2: contra-
          lateral VC; 3: ipsilateral sensorimotor cortex; 4: contra-
          lateral sensorimotor cortex; 5,6: electromyogram of the
          flexor (5) and extensor (6) of the femur. The arrow on the
          brain section shows the TT injection site in LGN. Ampli-
          tude: 500 μV; time: 1 s.

rather soon afterwards. At the last stage of the syndrome, single
seizures could be replaced by the epileptic status, resulting in
death.

Photogenic Epilepsy in Rats

The whole period between the administration of TT and the
animals' death could be nominally divided into three phases: the
latent, preseizure and seizure stages. Behavioral and EEG abnormali-
ties were absent in the latent phase. In this phase, the animals
adequately reacted to various external stimuli. For instance, they
became rapidly accustomed to light stimulation, normally satisfied
their physiological requirements, and could fall asleep when there
were no stimuli.

A change in the potentials in LGN and the visual cortex (VC) on
the side of TT administration was the first sign of the preseizure
phase (Figure 59). The amplitude of the primary response components
grew. A late complex of slow negative waves of the potential in VC
was recorded in some animals. Figure 59 B shows that the ordinary
components of the evoked potential are followed by the replacement of

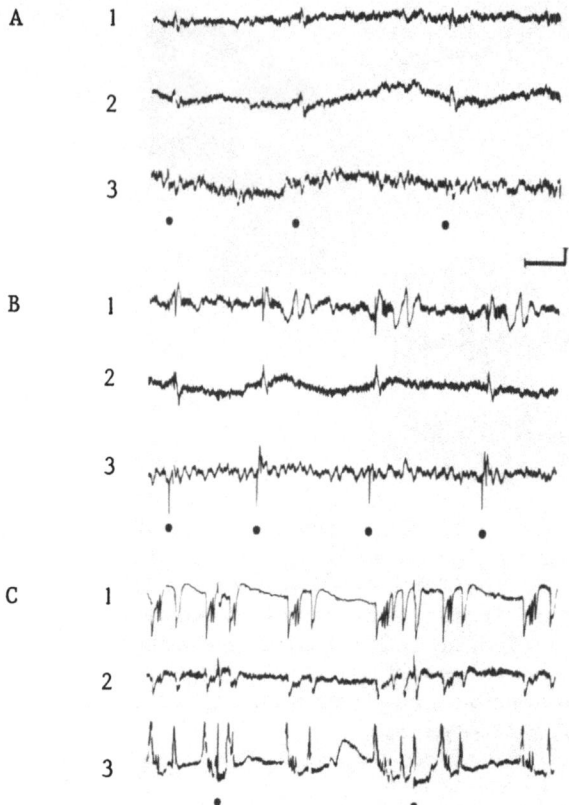

Fig. 59. Light EP in different phases of photogenic epilepsy. A:
          3 h after the injection of TT into the right LGN; B: 9 h
          after injecting TT into it; C: 21 h after injecting TT into
          it. Recordings: 1: right VC; 2: left VC; 3: right LGN.
          Light flashes are indicated by dots. Amplitude: 200 μV;
          time: 1 s.

the late response by several low-amplitude high-frequency spikes with
the after discharge in the form of two slow high-amplitude waves.
Such changes have not been recorded in contralateral VC.

     The presizure phase was characterized by elevated motor ac-
tivity: the rats did not react to food and water, slept more rarely,
and disoriented search reactions periodically occurred. Somewhat

after the beginning of the preseizure phase, single light stimuli
evoked not only local exhalted reactive potentials, but also motor
responses, i.e. starts and retroverse and lateral jumps. At the end
of the preseizure phase, the EEG showed paroxysmal spontaneous generalized potentials. This phase ended with the appearance of the
first epileptiform EEG paroxysm or a convulsive fit.

Just as the duration of all the phases, the time of the appearance of the first seizure depended on the TT dose. The interval between the first and second fits lasted 2-4 hours, and then the intervals between the seizures progressively diminished. In most cases, the initial seizures were provoked during a light test by a single flash or a rhythmic series of light stimuli (Figure 60). Rhythmic light stimulation (Figure 60 A) was followed by the appearance of convulsive activity (Figure 60 B). This process began in LGN, where GPEE was localized; that is where hypersynchronized discharges of a growing amplitude originated (Figure 60 trace 5). Ipsilateral VC was simultaneously drawn into this process, and then

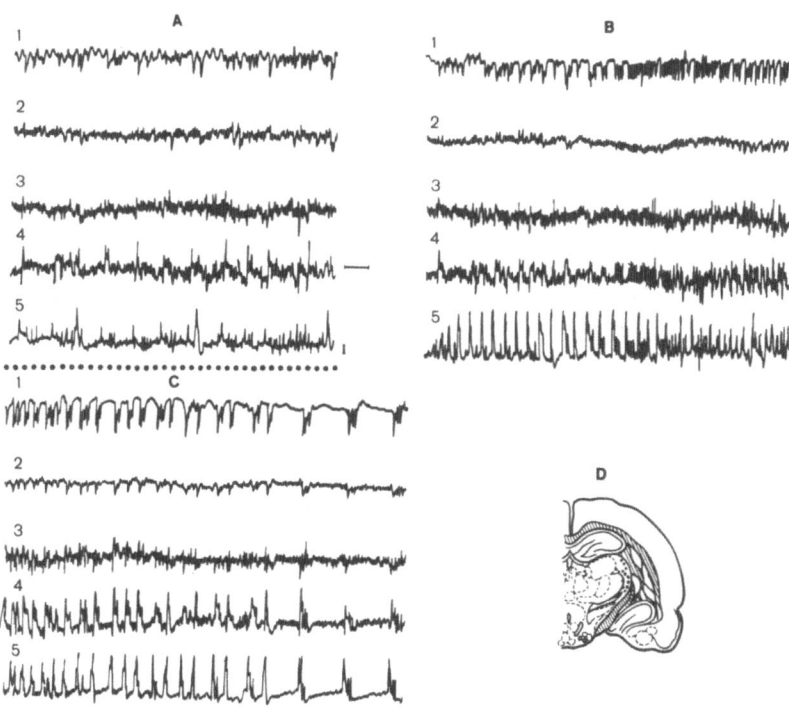

Fig. 60.   Formation of a seizure provoked by rhythmic light stimulation. A: EEG during light flashes (indicated by dots) with a frequency of 3 Hz; B: beginning of a seizure after light stimulation was ceased; C: formation of the clonic phase of the seizure. A,B,C: successive fragments of the recording; D: points of TT injection that produces photogenic epilepsy (indicated by dots) on the scheme of rat brain section. Recordings: 1: right VC; 2: left VC; 3: sensorimotor cortex on the right; 4: sensorimotor cortex on the left; 5: right LGN. 21 h after injecting TT into the right LGN. Amplitude: 200 μV; time: 1 s.

seizure activity appeared in other cerebral structures. The short-
term tonic phase of the seizure (see Figure 60 end of B and the
beginning of C) was replaced by the clonic phase (end of C). During
the period of chaotic locomotive movements after the clonic phase of
the seizure, the rats' reaction to light was restored and they could
jump in response to every light flash.

The evoked potentials of this period are given in Figure 59 C.
Such a potential resembled spontaneous sharp peak waves. Its amplitude was very large, and it was accompanied by after discharges.
Spontaneous peak potentials with afterdischarges have been observed,
too. Similar seizure activity in VC was also recorded. In the
seizure phase, photogenic compulsive reactions became more intensive
and occurred in response to every light stimulus. Generalized synchronous responses in various areas of the cortex and the subcortical
structures corresponded to the compulsive motor reactions on the EEG
recordings. Spontaneous generalized potentials began to occur regularly, their amplitude increased, and they could originate independently of the focal afterdischarges evoked by light stimulation. A
noteworthy fact is that seizure activity in the form of multiple
high-frequency synchronized discharges with a growing amplitude
originated in the system of both LGN and VC on the side of GPEE
(Figure 61). Seizures in the form of massive two-sided myoclonias

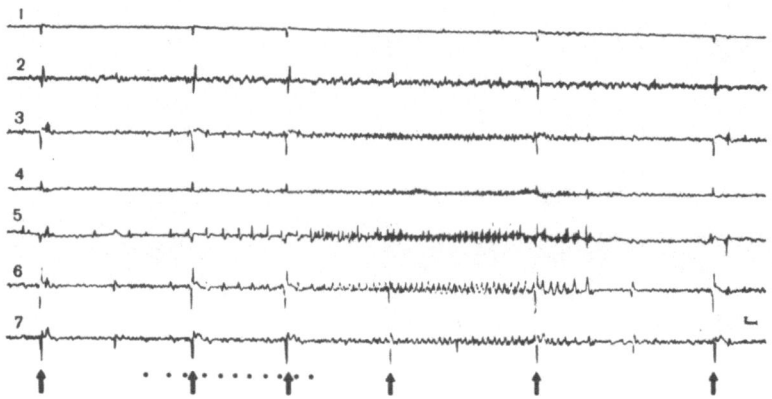

Fig. 61. Formation of the focal after discharge in VC against the
          background of spontaneous generalized potentials. The
          seizure after discharge develops after a series of light
          flashes (indicated by dots). Spontaneous generalized
          potentials are indicated by arrows under the recordings.
          Recordings: 1: left basal amygdaloid nucleus; 2: left
          dorsal hippocampus; 3: right indefinite thalamic zone; 4:
          sensorimotor cortex; 5: right VC; 6: right LGN; 7: left
          LGN. Amplitude: 200 µV; time: 1 s. The potentials were
          recorded 17 h after injecting TT into the right LGN.

and clonic-tonic paroxysms were the most typical form of the manifestation of this phase. Such seizures began with furious chaotic
running and high jumping, being followed by a new fit of chaotic
locomotive activity, i.e. rhythmic jumps to which high-amplitide
generalized peak waves corresponded on the EEG. At the end of this
phase, epileptic seizures increased and developed into an epileptic
status, and then the rats died.

Epileptic paroxysms in rats usually began with high-amplitude
generalized potentials which ensued directly in response to light
stimulation or somewhat after it. An example of the formation of
such an epileptic discharge is shown in Figure 62. During the initial period of a paroxysm, overwhelming activity was recorded in LGN
on the side of TT administration, and then other structures were
drawn into the process and the afterdischarge could abruptly cease.
However, the relatively short-term burst of generalized activity was
not followed by postseizure depression; the light flash produced a
clear generalized response on the recording. Later, a new after-
discharge developed with activity at first prevailing in the contralateral LGN and then passing over to a prolonged convulsive fit.

Besides the generalized epileptic discharges accompanied by
clonic-tonic convulsions, there were convulsiveless epileptiform EEG,
i.e. paroxysms and short-term focal discharges. Such focal epileptic
discharges either originated in response to light stimulation or
developed spontaneously. An example of a spontaneous focal discharge
in LGN on the GPEE side is given in Figure 63.

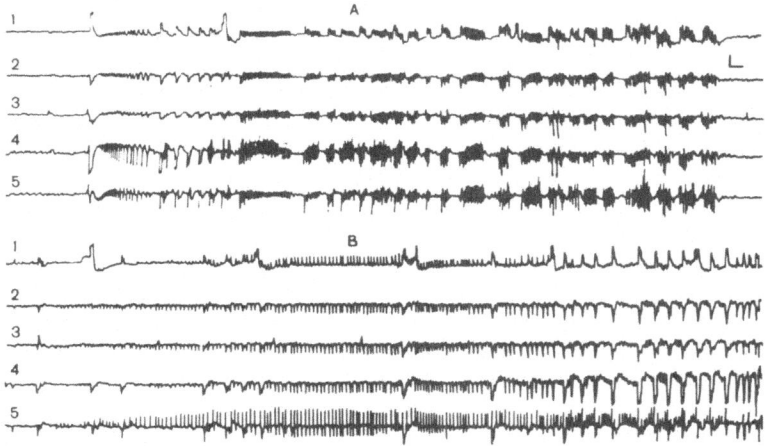

Fig. 62. Epileptic activity in the cerebral structures during a
          seizure. A,B: successive fragments of recordings. Re-
          cordings: 1: left ventral hippocampus; 2: medial raphe
          nucleus; 3: right VC; 4: right LGN; 5: left LGN. Ampli-
          tude: 500 µV; time: 1 s. The recording was made 18 h after
          injecting TT into the right LGN.

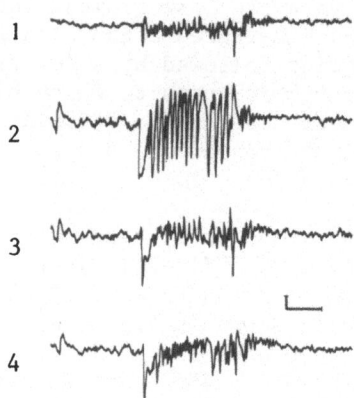

Fig. 63. Spontaneous focal epileptic discharge in LGN on the side of
          GPEE and in other cerebral areas. Recordings: 1: VC; 2:
          right LGN; 3: left LGN; dorsal raphe nucleus. Amplitude:
          200 µV; time: 1 s. The recording was made 16 h 25 min
          after injecting TT into the right LGN.

     Focal epileptic discharges occurred also in other structures,
such as the sensorimotor cortex, the visual cortex, and on the side
of GPEE in LGN. Such potentials were not engendered by spreading TT
to these areas, since afterdischarges did not originate in the control experiments when TT was directly injected into them. The spatially limited focal patterns of the hypersynchronized rhythmic activity of that type often evolved into a generalized afterdischarge
involving other regions of the brain and were accompanied by general
seizure activity. Such generalization is represented in Figure 64,
which shows the gradual involvement of various cerebral areas in the
synchronized generalization of the epileptic spikes. This process
was accompanied by changes in rhythm and amplitude and could be
abruptly replaced by depression. Figure 64 F shows the development
of a short-term afterdischarge from the hippocampus during postseizure depression.

     Experiments involving the coagulation of LGN during different
periods after TT injection were carried out in order to ascertain the
pathogenic role of GPEE in the formation of the syndrome of photogenic epilepsy and to establish the stages of the process.

     Investigations have shown that the limited coagulation of LGN in
the part where TT was injected (region of the top of the micropip-
ette), being produced during the whole latent phase of photogenic
epilepsy, had prevented the syndrome from developing in more than 70
per cent of the animals. Similar coagulation during the preseizure
phase curtailed the syndrome's further development in most rats.
Coagulation became less effective as it was produced later and later

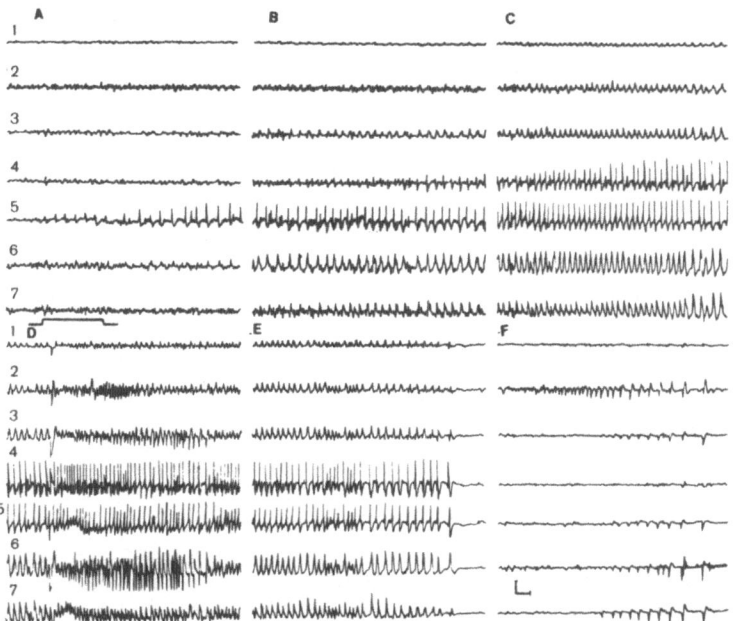

Fig. 64. Secondary generalization of the focal epileptic after
          discharge in VC on the side of GPEE in LGN after rhythmic
          light stimulation. The six successive fragments of record-
          ings (A-F) shows the gradual involvement of various cere-
          bral structures in the epileptic process, which begins
          locally in VC on the side of GPEE in LGN after rhythmic
          light stimulation. Recordings: 1: basal amygdaloid
          nucleus; 2: left ventral hippocampus; 3: indefinite
          thalamic zone; 4: sensorimotor cortex; 5: right VC; 6:
          right LGN; 7: left LGN. Stimulation with a frequency of
          3 Hz is indicated (the horizontal line under the recording
          turns upwards when light stimulation occurs). Amplitude:
          500 μV; time: 1 s. The recording was made 19 h after
          injecting TT into the right LGN.

during the preseizure phase. Such coagulation during the initial
hours of the seizure phase virtually did not influence the dynamics
of the subsequent occurrences. In this case, most animals died, just
as the control ones, during the same periods in the epileptic status.
In such rats, epileptic fits could be provoked by light stimulation
(Figure 65). The seizures were accompanied by a characteristic
prolonged high-amplitude discharge in the undamaged part of LGN.

Unlike the coagulation of the limited part of the nucleus, the
complete coagulation of LGN during different periods and even at the
initial stage of the seizure phase eliminated signs of photogenic
epilepsy. In most cases, seizures did not occur again after such

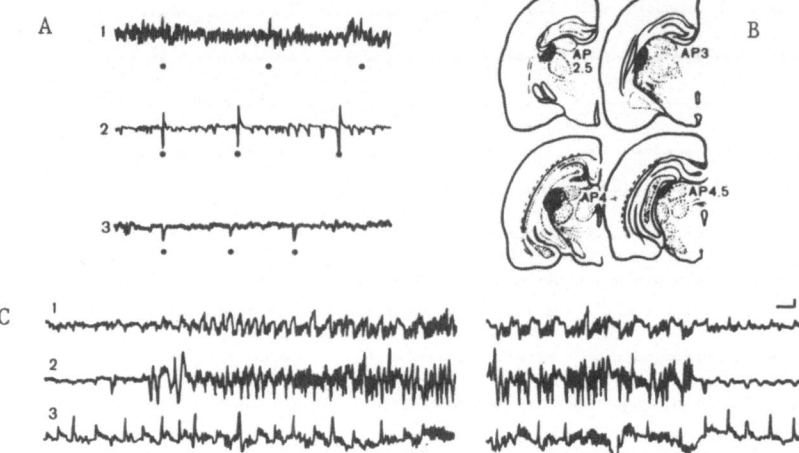

Fig. 65. Photogenically induced electroencephalogram: epileptic
paroxysm under the conditions of the partial coagulation of
LGN with GPEE. A: EP in LGN (light flashes are indicated
by dots): 1: latent phase (2 h after injecting TT into
LGN); 2: seizure phase (22 h after TT injection) before the
coagulation of LGN; 3: seizure phase (23 h after TT injection and one hour after the limited coagulation of the TT
injection site in LGN). B: coagulation area (indicated by
a black spot on the sections of the rat brain). C: the
beginning and the end of a seizure under the conditions of
the partial coagulation of the right LGN 22 h after injecting TT there. Recordings: 1: right VC; 2: right LGN; 3:
left LGN. The interval between two fragments of recordings
is 12 s. Amplitude: 200 µV; time: 1 s.

coagulation. However, such coagulation of LGN at the late stages
usually did not improve the animals' state, but merely precipitated
their death.

The given data can be interpreted in the following way: the
limited coagulation of LGN in the area of TT administration during
the latent period eliminated not only GPEE which was still not large,
but also the conditions of its further formation owing to the distribution of TT, which became bound with the coagulated cerebral
tissue. However, such interference became less effective as GPEE's
dimensions increased due to the spreading of TT throughout the nucleus. When GPEE involved a large amount of the nucleus elements,
such coagulation of the tissue of LGN only partially destroyed GPEE.
There was enough of it still left to cause seizures provoked by light
stimulation and to allow photogenic epilepsy to develop. It is known
that even when 5 per cent of the fibers of the visual tract are

preserved, the evoked potentials which are virtually unchanged in
comparison with the normal level can be produced in VC (Frommer et
al., 1968). Unlike the limited coagulation of LGN, its complete
coagulation caused GPEE's full destruction and was effective even
during the initial stage of the seizure phase: it not only eliminated
the syndromes of photogenic epilepsy, but also prevented the syndrome
from developing further. Thus, the pathologic process at this stage
depends on GPEE, which acts as a significant pathogenetic factor.
The ineffectiveness of the elimination of GPEE at the late stages of
the syndrome attests to the generalization of the pathologic process
and the secondary epileptical involvement of other cerebral structures, i.e. the origination of secondary GPEE.

Thus, the formation of GPEE in LGN is a significant initial link
in the development of the syndrome of pathogenic epilepsy. At the
initial stages of the process, GPEE in GN is the only source of
epileptic activity. However, other sources induced by primary GPEE
originate at the late stages of the process.

Two aspects are noteworthy when the mechanisms of photogenic
epilepsy are being analyzed. Both of them pertain to the general
problem of the brain's epileptogenesis.

Enhanced activity in the specific sensory system may not be
enough for engendering a generalized epileptic reaction without
additional factors which augment the brain's general excitability
(Hunter, 1960). Therefore, GPEE in LGN is pathogenically significant
as regards the syndrome of photogenic epilepsy not only in that it is
a hyperactive determinant structure which induces and synchronizes
epileptic activity of other cerebral structures, but also in that the
enhancement of general excitability and the origination of the
brain's seizure activity are connected with GPEE's activity. This
aspect is thus a manifestation of the determination of type III (see
Chapter 1), i.e. the unspecific intensification of activity under
the influence of the hyperactive structure. The following hypothesis
can be set forth for the mechanisms of that process: the LGN neurons
are normally released from inhibition under the influence of the
mesencephalic reticular formation (MRF), which suppresses the inhibitory neurons in LGN (Fukuda, Iwama, 1970). In other words, stimulation from MRF can disinhibit the neurons of LGN and, consequently,
engender GPEE in it. This effect should evoke an inverse reaction,
namely, the suppression of MRf for establishing functional homeostasis. Besides the actuation of negative feedbacks, such a reaction can be realized by activating the 'synchronizing' structures
which are antagonistic with respect to MRF and which are directly
connected with LGN, e.g. the limbic structures (McLean, 1966;
Bergman et al., 1970; Zambrzhitsky 1972) or the unspecific thalamic
structures (Sheibel, Sheibel, 1958). Normally, the stronger the
excitation of the LGN neurons, the greater is their subsequent inhibition due to the actuation of similar servomechanisms. However,

this mechanism cannot be realized when inhibition is disturbed in LGN
(just as when GPEE is formed) (see Chapter 3), while the 'synchronizing' structures, which facilitate cerebral epileptization, are activated (Kryzhanovsky et al., 1980b). The variations of the levels
of unspecific cerebral activity and especially the periods during
which the activity of the 'desynchronizing' parts of MRF is suppressed are an important factor which enhances the animals' predisposition to form seizure states (Kruglikov et al., 1980; Halazs,
1972; Testa, Gloor, 1974; Nanobashvili et al., 1975; Wagner et al.,
1975; Woodruff, 1975; Kryzhanovsky et al., 1980b). Under these
conditions, a strong volley of excitation from LGN can evoke electrographic and clinical epileptic signs.

This model of epilepsy is interesting also as regards the still
unresolved problem of the genesis of so-called primarily generalized
or genuine epilepsy. One of the characteristic features of the given
model of photogenic epilepsy is that it can be exhibited as primarily
generalized epilepsy even at the early stages of the process. This
feature can become more pronounced at the height of the process, when
recordings of the potentials in the deep cerebral structures (including LGN, where GPEE is localized) may often not reveal any differences in the characteristics of the electric activity of these structures. The impression may be that the whole brain is in the primary
epileptic seizure, and when not all the circumstances are known, its
primary, determinant factor cannot be specified. After rest, local
epileptic activity which causes a general seizure is often recorded
not in LGN, where GPEE is localized, but in other structures. Such
foci can 'migrate'. However, factual material and the way in which
an experiment is set show that this syndrome is based on the primary
local origin of GPEE in LGN; moreover, the appropriate methods of
analysis have revealed the functional structure (see Chapter 1) and
activity of GPEE at different stages of the process. It will be
shown (see Chapter 13) that masked GPEE, which is not revealed
against the background of high seizure activity, can be exposed under
the action of benzodiazepines, particularly when its effect attenuates. Characteristic interictal activity in LGN with GPEE is exhibited under these conditions. Such interictal activity can be seen
earlier, when cerebral seizure excitability is at a high level, since
every discharge causes a generalized reaction. These facts and other
data (Kryzhanovsky et al., 1979a) corroborate the concepts that there
are many latent epileptic foci which are difficult to diagnose as
regards numerous forms of so-called primarily generalized epilepsies
(Saradzhishvili, Geladze, 1976). Hence, the role of the hyperactive
determinant in the genesis of the pathologic forms of the central
nervous system, which seem to be primarily generalized forms when
they are viewed in the usual manner, can be discussed positively and
more broadly.

SUMMARY

     The syndrome of photogenic epilepsy is produced when GPEE is
created in LGN. At the height of its development, the syndrome is
clinically exhibited in the form of generalized tonic-clonic seizures
which result in <u>status epilepticus</u> and death. Seizures either can be
provoked by light stimulation or can originate spontaneously. At the
early stages, the reaction to light stimulation is expressed in
intense jerks and retroverse and lateral jumps whenever an electric
flash occurs. A spontaneous seizure is preceded by a state which can
be regarded as the visual aura. Clinically, the process can be
nominally divided into the latent, preseizure and seizure phases.
The first sign of the preseizure phase is an increase in the amplitude and a change in the configuration of the evoked potential in LGN
and the visual cortex (VC). Paroxysmal generalized potentials originate on the EEG at the end of this phase. After rhythmic light
stimulation, seizure activity originates in LGN, where GPEE is localized. Hypersynchronized discharges originate then in other cerebral
structures. Besides the generalized epileptic discharges accompanied
by clonic-tonic seizures, there are seizureless epileptiform EEG
paroxysms and short-term focal discharges. At the late stages,
seizure discharges can originate at first in various cerebral structures. The determinant focus in LGN is clearly exhibited when general seizure activity is suppressed by benzodiazepines at this stage.
At the latent stage, the limited coagulation of LGN, where GPEE is
localized, prevents the syndrome from developing in most animals.
The complete coagulation of the nucleus at the beginning of the
seizure phase causes the syndrome to disappear, but such coagulation
is ineffective at the late stages of the phase.

9
Pain syndromes of central origin

SPINAL PAIN SYNDROME

The spinal dorsal horns are a very complex integrative mechanism with somatotopic organization that serves not only the transmission of nociceptive information, but also its primary processing. The presence of populations of functionally different neurons with multiple connections that produce peculiarities of the structural and functional organization of the dorsal horns allow powerful GPEE to be formed in them.

Pain Syndrome Caused by Tetanus Toxin

The pain syndrome caused by GPEE which is formed in the spinal dorsal horns under the influence of TT has the following distinguishing features (Kryzhanovsky, Grafova, 1972; Kryzhanovsky et al., 1973a; Kryzhanovsky et al., 1974a; Kryzhanovsky, 1976, 1979, 1980a): the animals become excited and aggressive and attack one another. The time it takes for these phenomena to occur after applying TT to or microinjecting it into the lumbosacral segments depends on the dose. Then, the animals begin to lick out hair in one region of their hind leg or another on the side of TT action. When relatively weak nociceptive stimuli are applied to such a region, the animals scream and begin to intensively lick the region. Such a reaction does not occur when similar stimuli are applied to the symmetrical region of the opposite extremity. The area of enhanced sensitivity, whose stimulation produces the given reaction, is usually localized on the outer surface of the hind leg. However, it may originate on the posterior and anterior surfaces of the femur, the crus, the foot, the paws and the tail (Figure 66). Its localization depends on the TT

microinjection site. As time passes, the rats more and more intensively lick and bite the skin of the given region, as a result of
which hair is removed and the skin may become exposed (Figure 67 I).
During this period, hyperalgesia in that region becomes especially
high and the slightest stimulation, even when it is tactile, causes a
severe reaction: while groaning and screaming, the animal begins to
furiously gnaw the leg and even bites the tissue. This reaction has
a pronounced after effect: it continues for a long time even when
stimulation is stopped. At this stage, other stimuli, i.e. stimuli
from other areas of the body, rapping, tapping the table, and so
forth, causes a general reaction and can also produce that fit.
Eventually, the region of hyperalgesia becomes larger, the animal's
reaction becomes more furious, and unprovoked spontaneous pain
paroxysms occur more frequently. If the process continues, self-
amputation may occur.

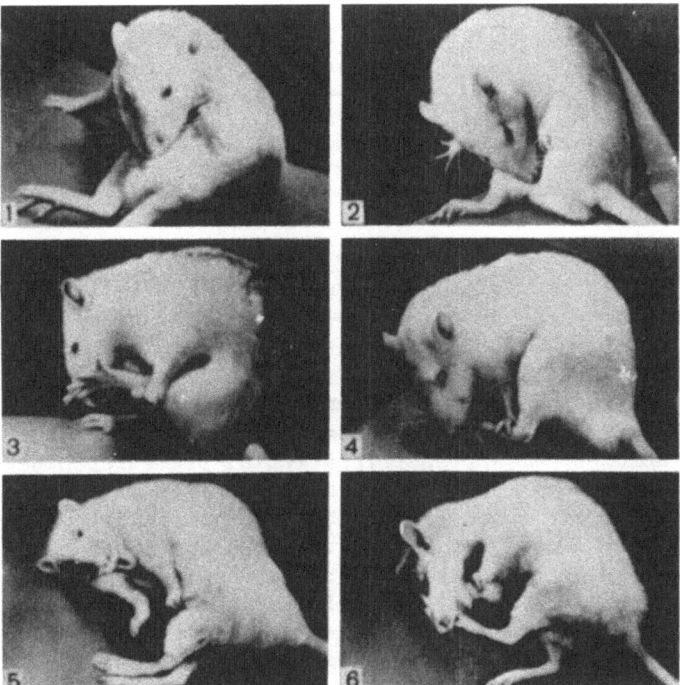

Fig. 66. Pain syndrome in a rat caused by GPEE formation in the
          dorsal horns of the spinal lumbosacral segments. Success-
          ive (1-6) clinical pictures of the development of the
          syndrome and the manifestation of pain fits from the first
          signs (licking and scratching) to the maximum. GPEE was
          produced by applying an agar plate containing TT on the
          dorsal surface of the lumbosacral segments on the left.

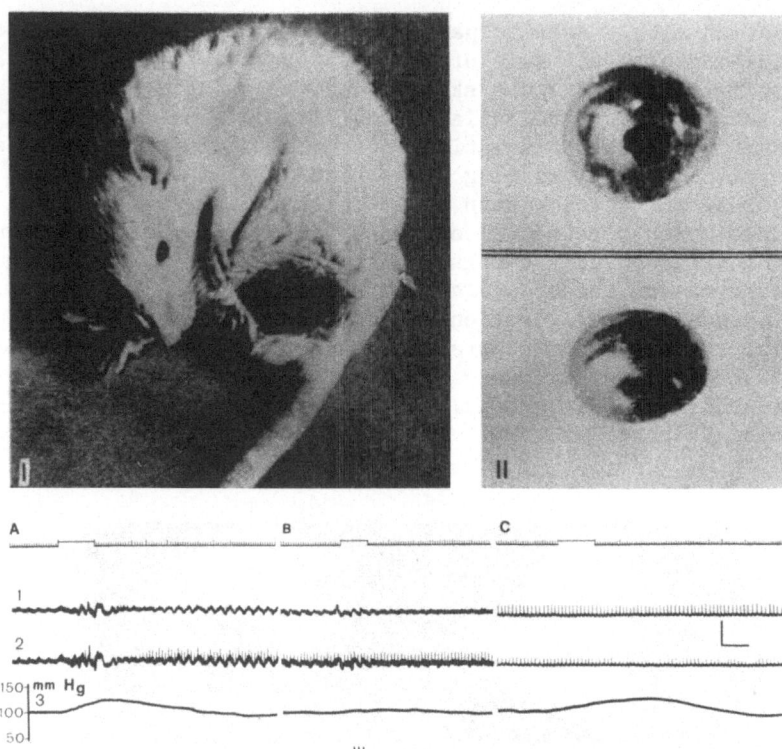

Fig. 67. Manifestation of a pain syndrome in a rat caused by GPEE
          formation in the dorsal horns of the spinal lumbosacral
          segments. I: origination of the pain projection zone
          (licking region), which is also the trigger zone; II:
          expansion of the pupil when the trigger zone was stimulated
          (top fragment); III: vegetative components of the pain
          syndrome. A: increase in arterial pressure when the
          trigger zone was stimulated; B: absence of reactions when
          the symmetrical region of the other hind leg was stimu-
          lated; C: reaction of arterial pressure when the trigger
          region was stimulated under the conditions of complete
          curarization and artificial respiration. Top to bottom:
          time and stimulation; electric activity of the intercostal
          muscles on left (1) and right (2); arterial pressure (3).
          Amplitude: 500 μV; time: 1 s.

     That behavioral reaction is accompanied by a set of vegetative
symptoms: when a stimulus is applied to a hyperalgesic area, the
pupil of the eye expands, the eyeball protrudes, the eyeslit widens
(Figure 67 II), respiration is inhibited and disturbed, the arterial
pressure grows (Figure 67 III). This elevation of arterial pressure
is not connected with muscular contraction and the animal's motor

reaction, since it occurs also in the case of complete curarization
and artificial respiration (Figure 67 III, c). The vegetative
symptoms which have been mentioned are a characteristic part of the
animal's reaction to nociceptive stimulation. They occur also in a
decerebrated animal. Therefore, Sherrington (1906) called them
pseudoaffective reactions. Such reactions are an objective indication of pain.

A study of the catecholamine balance in the development of the
pain syndrome (Kassil et al., 1972) has shown that the content of
catecholamines in the myocardial tissue and the hypothalamus increases and their content in the norepinephrine glands varies at the
initial stage of the process. The content of adrenaline, noradrenaline and ι-dopa in the norepinephrine glands, noradrenaline and
adrenaline in the heart, and adrenaline, noradrenaline, ι-dopa and
normetanephrine in the hypothalamus greatly decreases at the late
stages of the process. The level of adrenaline and noradrenaline is
especially low in the norepinephrine glands and the heart (Figure
68). Such a sharp diminution of the catecholamine content of the
norepinephrine glands and the tissues shows that the sympathoadrenal
system is exhausted. The administration of ι-dopa to animals at the
height of the pain syndrome increased the content of catecholamines
in the tissues and the hypothalamus, but did not greatly influence
the syndrome's development (Matlina et al., 1973).

An analogous pain syndrome was likewise produced in cats (Figure
69). At first, the cats carefully licked a definite region of the
hind leg on the side where GPEE was formed. They became excited,

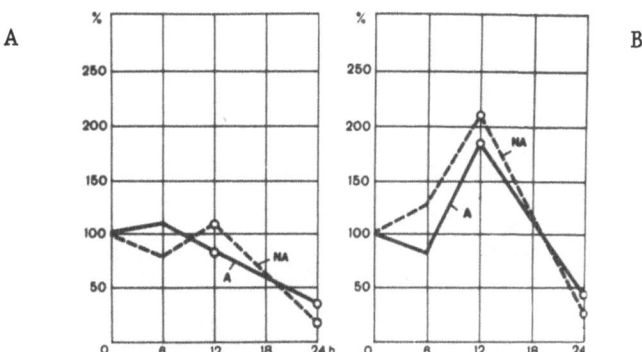

Fig. 68. Content of adrenaline (A) and noradrenaline (NA) in the
          rat's adrenal glands and the heart during the development
          of a pain syndrome of spinal origin. Ordinate: A and NA
          contents (% of the control level). Abscissa: time (h)
          after the injection of TT into the dorsal horns of the
          lumbosacral segment. A: adrenal glands; B: heart.

aggressive and did not allow anyone to touch that region. If someone managed to touch it, they screamed, attacking the experimenter's hand or the region being stimulated. As time passed, the animals licked the region more and more frequently and intensively until the hair on it was lost and the skin was revealed. When the slightest stimuli (even a weak blow of air) were applied to the region, the cats screamed and reacted vehemently, pounced on the region, or ran away to another place. At the height of the syndrome's development, the licking region could become so large as to involve the hind leg, the sacrum, etc. During a fit, the cats ran around the cage, furiously pounced on the objects which were in their way, gnawing and tearing them, jumped on the cage walls, etc.

Thus, the production of GPEE in the dorsal horns in a cat causes a severe pain syndrome which has the same characteristics as the pain syndrome in a rat. All these features correspond to the accepted criteria of pain in an experiment. It should be noted that such a pain syndrome was observed for the first time by Meyer and Ransom (1903) in cats and dogs when TT was injected into the dorsal horns. They believed that this form of a pathologic state was a particular expression of tetanus and called it tetanus dolorosus. It follows that this pain syndrome is not connected with the species specificity of the structural and functional organization of the central nervous system, since it may be produced in animals of different species.

That pain syndrome has a spinal origin. This is evident from the localization of GPEE in the dorsal horns, the disappearance of the syndrome when GPEE is suppressed (see further), and the fact that the section of the spinal cord above the level of localization of

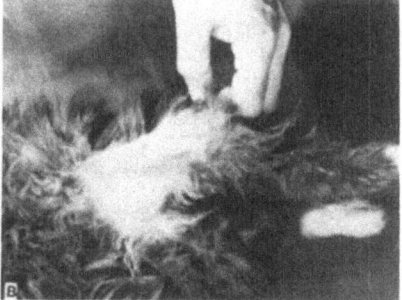

Fig. 69.   Pain syndrome of spinal origin in a cat caused by GPEE in the dorsal horns of the spinal lumbosacral segments. The illustration shows the stages of the development of the pain syndrome (A,B), the licked-out zone, which is hairless, at first on the outer femoral surface, and then on the sacrum. GPEE was produced by microinjection TT into the left dorsal horns of the lumbosacral segments.

GPEE causes the syndrome to disappear: the animals become calm and no longer pay any attention to the former hyperalgesic region which they previously intensively licked and gnawed.

The syndrome described above apparently occurs together with a severe itch. When GPEE was sited in the lumbosacral segments of the spinal cord, the region of enhanced sensitivity was confined to the hind leg. Therefore, the ordinary scratch reflex by the hind legs did not occur. Instead, the animal scratched out the hair with its teeth, just as when catching bugs. This may be regarded as an itch sign. In special experiments, we produced GPEE in the thoracic dorsal horns. In such cases, the region of enhanced sensitivity appeared on the side or the back. An extremely intensive character- istic scratch reflex originated together with the pain reaction, which was accompanied by a scream. The rats either intensively scratched the side with their hind leg or rolled on their back, rubbing against it or against some objects in an effort to scratch their back. Intensive scratch movements with both hind legs often occurred. They passed over at first to local convulsions (frequent convulsive movements only with the hind legs), and then to general convulsions. In such cases, a very severe itch apparently occurred together with pain.

These itch symptoms coincide with those described by other authors (Sherrington, 1906; Mehes, 1938; Koenigstein, 1948; Feldberg, 1971). An extraordinary fact is that such irritation can cause general convulsions and even an animal's death, which occurs instant- ly. Such a rapid manifestation of an acute paroxysm with general seizures was not, as a rule, observed during a pain syndrome caused by the production of GPEE in the lumbosacral segments.

Pain Syndromes Caused by Other Agents

Strychnine, penicillin, ouabain and KCl were used, besides TT, to produce GPEE in the spinal dorsal horns.

Investigations have shown that the production of GPEE in the spinal dorsal horns by those agents causes a pain syndrome with all the above-mentioned features (Figure 70) (Kryzhanovsky, 1976, 1980a; Danilova et al., 1979; Grafova et al., 1979).

Some time after the application of the agar plate with one of those agents (see Chapter 5) to the dorsal surface of the lumbosacral segments of the spinal cord either to the left or to the right, the animals grew restless and began to lick out a region of their hind leg on the side where the plate was applied, furiously combing and gnawing out the hair of that region. They screamed and ran around the cage. When several rats were put in the cage, they attacked and furiously fought one another. During the intervals between fits, the

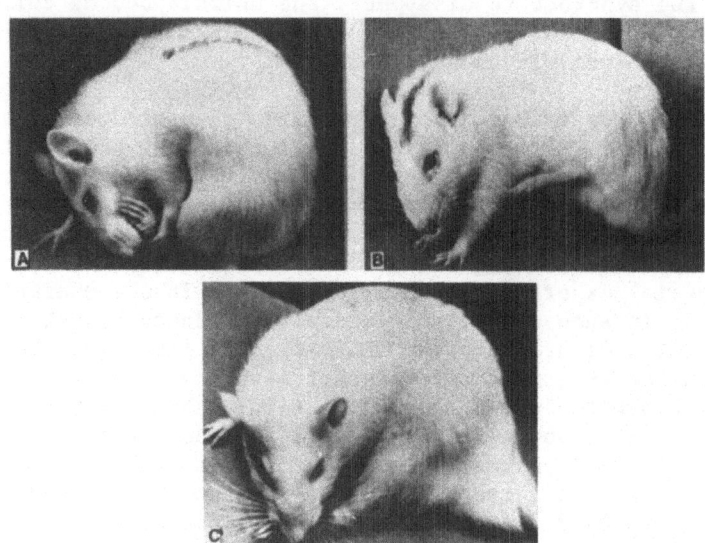

Fig. 70. Pain syndrome in rats caused by GPEE formation in the
          dorsal horns of the spinal lumbosacral segments after (A)
          KCl, (B) penicillin and (C) strychnine were applied to the
          left dorsal surface of the segments on the left.

rats carefully licked their leg, holding it bent in the air. Such a
posture was especially frequently observed in animals with a 'penicillin' GPEE. The animals frequently took a fixed position, fearing
to move. The latent period of the syndrome's origin, its duration
and intensity mainly depended on the concentration of an agent on the
agar plate, i.e. by the effective dose. The syndrome was exhibited
most acutely and intensively when penicillin was used even in comparatively small doses. The maximum duration of the syndrome was two
or three hours. The syndromes disappeared apparently due to the
diminution of the agent's concentration in agar. This is what distinguishes the syndromes caused by the above-mentioned substances
from the TT-induced syndrome, which develops as TT penetrates deep
into the spinal cord and is distributed in it.

     Thus, a clinically uniform characteristic pain syndrome of
spinal origin occurred regardless of the agent used to produce GPEE
in the spinal dorsal horns. It follows that the given pain syndrome
can be polyetiological: it can be caused by the action of various
agents, but its pathologic mechanisms are the same, i.e. GPEE
originates in the system of the spinal dorsal horns.

     The pattern observed by Black (1974) in the production of a
focus of epileptic activity in the spinal dorsal horns completely
coincides with the above-mentioned pattern of the spinal pain syn-

drome. Black also noted that animal behavior (pain reactions) was
the same in all cases, although the epileptic focus was produced by
different epileptogens. Characteristic features in this respect were
the presence of an incubation period, the occurrence of spontaneous
paroxysms of pain, the long duration of fits, hyperalgesia, the
possibility of inducing fits by slight tactile stimulation (even by
blowing) of definite dermatomes, the licking out of the appropriate
areas, the loss of hair, sparing the leg, etc.

Suppression of the Pain Syndrome

When the given pain syndrome is connected with the origin of
GPEE in the spinal dorsal horns, the suppression of GPEE's activity
can be expected to cause the disappearance of the syndrome. Investigations have confirmed this assumption (Kryzhanovsky, 1976, 1980;
Danilova et al., 1979). Experiments have shown that the spinal pain
syndrome disappears as glycine or GABA is injected into the affected
dorsal horns (Figure 71): the animal remains calm after anesthesia is
eliminated, spontaneous paroxysms do not occur, the application of
stimuli to the trigger zone (the licking regions), which caused a
vehement motor reaction and screams earlier, does not produce this
effect, and the animal reacts to such stimulation by withdrawing the
leg as usual. The effect lasts 30-40 minutes, i.e. when glycine is
effective, and then the syndrome reappears.

An important fact is that rather large doses of glycine should
be used and be injected into the dorsal horns of all the affected
segments in order to produce a full effect. It is not difficult to
carry out these and other procedures on the spinal cord of rats
because, unlike other animals, they have a compact lumbal enlargement
of the spinal cord. Experiments have shown that the pain syndrome
remains even if one segment is not treated by glycine. This fact
shows that GPEE in the dorsal horns is very powerful and rather
large; the syndrome which it produced lasts even when a small part of
GPEE is preserved. The suppression of GPEE when glycine is injected
into its site is not connected with any injury to the spinal cord.
It is due to the specific action of glycine. Such an effect was not
produced when an isosmotic solution of sodium chloride or neutral
amino acids with the same pH was injected into the dorsal horns: the
syndrome remained when the animal was no longer in a state of narcosis (Figure 71 B).

The inhibitory effects of glycine are connected with its ability
to hyperpolarize the postsynaptic membrane (Curtis et al., 1968;
Wermann et al., 1968; Gushchin et al., 1970; Curtis, Johnston, 1974).
SInce the inhibitory mechanisms are disturbed under the action of TT
because the release of the inhibitory transmitter is blocked (see
Chapter 5), the postsynaptic membrane remains intact and glycine can
thus become bound with the receptors, causing a specific effect of

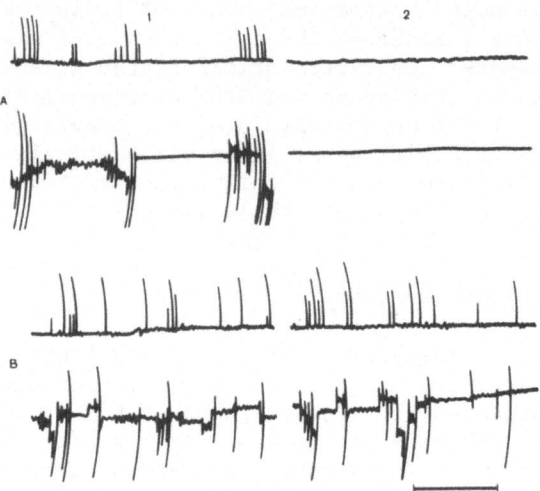

Fig. 71. Effect of glycine microinjection into the dorsal horns of
          the spinal lumbosacral segments in the region of GPEE
          formed by TT. The tope curves are phonograms, and the
          bottom curves are actograms recorded at the height of the
          pain syndrome in rats before (1) and 20 min after (2)
          microinjecting glycine (20% in $1-10^{-4}$ ml: A, 2) and an
          isosmotic sodium chloride solution (control; B, 2) (three
          injections) into the dorsal horns of the L_4-L_6 segments 2 h
          after injecting TT into the dorsal horns of these lumbo-
          sacral segments on the left. The horizontal line indicates
          5 min.

hyperpolarizing inhibition. That is the mechanism of the glycine
suppression of GPEE engendered by TT. It should be noted that GPEE
produced by various agents can be resistant to the appropriate inhibitory mediators. Therefore, the syndromes may remain under the
action of these mediators (for details, see Chapter 13).

    A similar effect of the suppression of GPEE and the whole pain
syndrome can be produced by applying GABA to the dorsal surface of
the spinal cord or microinjecting it into the dorsal horns in the
GPEE regions.

    Thus, investigations have shown that a characteristic pain
syndrome can be engendered by producing GPEE in the spinal dorsal
horns by one technique or another.

    The structural and functional characteristics of the neuron
systems of the spinal dorsal horn suggest that the neurons of Rexed's
V lamina, called 'wide dynamic range neurons' (Prince, Mayer, 1974),
are the operant part of the given GPEE. They constitute the bulk of

the lamina V neurons, and have inputs from small Aδ - and C-afferents
as well as from fast conducting afferents (Hillman, Wall, 1969;
Wagman, Price, 1969; Price, Wagman, 1970; Handwerker et al., 1975;
Dubner et al., 1976; Foreman et al., 1976; Price et al., 1976).
Small visceral afferents converge on them in the upper lumbosacral
and thoracic segments (Pomeranz et al., 1968; Kostyuk et al., 1968).
These neurons realize significant spatial and temporal summation
(Price et al., 1978b), being transmissive cells (T-neurons) and
forming the ascending pain pathways. They are also regarded as the
main neuronal substrate of the gate control theory (Melzack, Wall,
1965). The T-neurons are under multiple segmental and suprasegmental
inhibitory control and can be inhibited by various influences which
induce analgesia, i.e. ES of the dorsal columns (Foreman et al.,
1976) and the transcutaneous nerve (Handwerker et al., 1975), morphine (Mayer, Price, 1976), ES of the brainstem structures (Oliveras
et al., 1974; Mayer, Price, 1976) and ES of the contralateral nerve
(Wagman, Price, 1969). These neurons promote intensive activity
under the influence of noiceptive stimulation and generate prolonged
high-frequency discharges much longer and higher in comparison with
the afferent nociceptive impulses (Price, Mayer, 1975; Price et al.,
1976; Hayes et al., 1979). They possess some other properties which
are characteristic of GPEE activity. Hence, the populations of "wide
dynamic range neurons" can easily form GPEE when inhibitory control
is insufficient. Their ability to be activated by stimuli of different modality can explain the above-mentioned peculiarities of trigger
stimulation of GPEE. The selective stimulation of these neurons
evokes different kinds of pain sensations in man, ranging from acute
pain with appropriate localization to burning pain, depending on the
nature of stimulation (Mayer et al., 1975).

However, the marginal cells of lamina III apparently cannot be
excluded in the formation of GPEE in the spinal dorsal horn. These
neurons are regarded as specific nociceptive cells (Price, Dubner,
1977), because they respond only or almost only to nociceptive stimuli (Christensen, Perl, 1970; Willis et al., 1974; Kumazawa et al.,
1975; Price, Mayer, 1975; Cervero et al., 1976). They play an important role in subserving discriminative pain. These neurons are
under inhibitory influences from the cell of lamina II and the supraspinal structures (Cervero et al., 1976; Bishop, 1980; Kerr, 1980)
can can be inhibited during ES of the dorsal columns (Foreman et al.,
1976) and nerves (Cervero et al., 1976). It is noteworthy that the
somatotopic organization of the neuron has been found in the marginal
layer (Willis et al., 1974; Price et al., 1976).

It should be mentioned that the large myelin fibers (a$\alpha\beta$ of the
cutaneous nerves can also participate in the transmission of nociceptive information. Spatial summation is important in these mechanisms
(Willer et al., 1978, 1980; Albe-Fessard, 1979). Taking account of
the possibility of pronounced excitatory summation on the GPEE's
neurons (Chapter 3), it may be assumed that the large nerve fiber

apparatus participates in GPEE activity particularly as regards the role of wide dynamic range neurons in these processes.

The neurons of different nociceptive apparatus in spinal dorsal horns probably form their own GPEE, thus determining the character- istics of the given models of pain syndromes of spinal origin.

TRIGEMINAL NEURALGIA

GPEE was produced in nucleus caudalis of the trigeminal spinal tract in order to cause trigeminal neuralgia.

Nucleus caudalis plays a dual role in the activity of the trig- eminal sensory complex: it modulates the sensory inflow of different modalities in the rostral nuclei of the spinal tract (principal and oral nuclei) (Young, King, 1972; Denny-Brown, Yanagisawa, 1973; Sessle, Greenwood, 1973, 1976), and is important in the perception of oral-facial noxious stimuli. Being the continuation of the apex of the spinal dorsal horn, nucleus caudalis in its distal part has a laminar structure, while in the proximal part, lamina V becomes reticular and passes over to the bulbar reticular formation (Olszewski, 1950; Rexed, 1964; Gobel et al., 1977; Gobel, 1978). Trigeminal nucleus caudalis has the same laminae as the dorsal horns. The cells of these laminae perform the same function and have the same functional relationships as the appropriate cells of the dorsal stems. Wide dynamic range neurons are in subnucleus reticul- aris dorsalis. Nociceptive neurons are also in subnucleus reticul- aris ventralis (Mosso, Kruger, 1973; Price et al., 1976; Gobel et al., 1977; Yokota, Nishikawa, 1977; Gobel, 1978a, 1978b; Hayes et al., 1979; Yokota et al., 1979). Trigeminal nucleus caudalis has the same sensory inputs as the spinal dorsal horns, and the same nocicep- tive fibers terminate on its neurons. Nociceptive neurons ensure somatotopic organization as well as temporal and spatial summation, and they respond with prolonged discharges to repeated stimulation (Dubner et al., 1976; Price et al., 1976; Yokota, Nishikawa, 1977; Hayes et al., 1979; Yokota et al., 1979).

Those specifics of the structural and functional organization of nucleus caudalis and the great role it plays in the perception of oral-facial noxious stimuli suggested that GPEE can be produced in the nucleus by injecting TT into it, as a result of which the syn- drome of trigeminal neuralgia originates. Investigations have con- firmed this assumption (Kryzhanovsky et al., 1974b; Kryzhanovsky, 1976).

Some time after injecting TT (the duration of this period depend- ed on the TT dose), the animals began to scratch out definite regions of their face or head. In some cases, the scratching movements were preceded by washing movements with the forelegs. Regions of the face

or the head were scratched out by the respective hind leg only on the
side where TT was injected (Figure 72). After every scratching
movement, the rat carefully 'cleaned' the claws of the respective
leg, just as in the case of an ordinary scratch reflex (Figure 72 2).
The rats were calm during the intervals between scratching. Gradually, the fits became more frequent, longer and more intensive; they
occurred spontaneously and were 'silent'. Later, the fits were
furious reactions: the rats screamed and began to vehemently scratch
a given region on the head or the face. The animals grew restless
and aggressive. Paroxysms became more frequent and were easily
induced by stimulating the scratch region. They could be induced
even by applying slight tactile stimuli to the region. Their intensity and duration were virtually the same in all cases and did not
depend on the force of stimulation. The scratch region, which became
the trigger zone of easy provocation, grew larger. The skin of the
region was damaged due to scratching, and hair was lost (Figures
72 5,6). Sometimes, a large area of the face was damaged, and the
animals often tore the tissue in the region. The animals assumed a
very characteristic posture during the interval between paroxysms:
they sat with their forelegs pressed against their face and their
toes clenched, as if they were afraid to move (Figure 81 4). They
eye slit on the side of TT injection was narrow, and sometimes
lacrimation and intensive secretion from the nostrils were observed.
A refractory period set in after a paroxysm, and the application of
stimuli (just as other stimulation) to the trigger zone did not cause
a fit for some time.

    The intensity of the symptoms and the rate at which they grew
depended on the TT dose. The syndrome rapidly developed and the
symptoms were clearly expressed when relatively large doses were
used, while it developed slowly and could be observed for several
days when small doses were used. Apparently, tetanus intoxication
was the main cause of the animals' death. However, account should
also be taken of the disturbance of the catecholamine balance and, in
this respect, cardioplegia, as is the case in the spinal pain syndrome.

    The electromyogram of the head muscles has shown that a burst of
electric activity (EA) occurs in masseter, temporal and frontal
muscles during fits (Figure 73). Such bursts became longer and more
powerful as the process developed. In most animals, these bursts
were more pronounced on the affected side. In some cases, enhanced
background activity was recorded in the muscles, which was also more
pronounced on the TT injection side. These data coincide with the
results of the clinical electromyographic investigations, which
showed that electric activity was enhanced in the frontal and especially masseter muscles in the trigeminal neuralgia (Yerokhina,
1973).

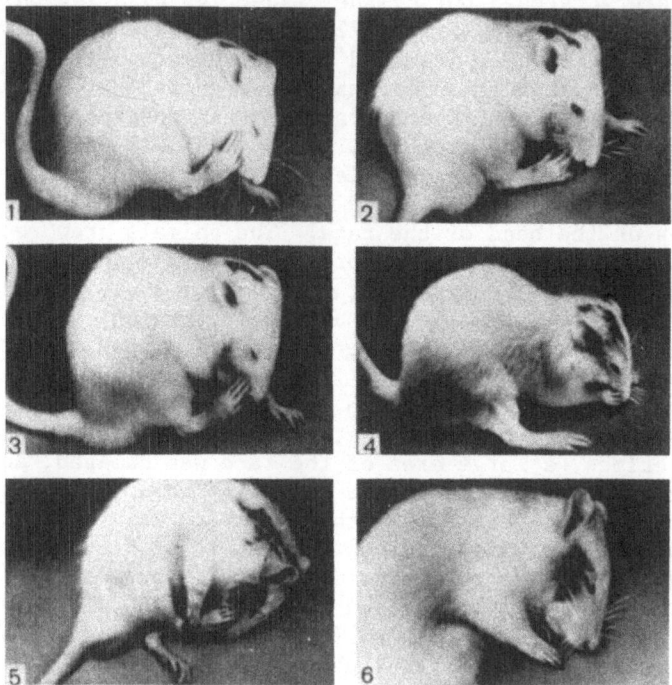

Fig. 72. Trigeminal neuralgia in a rat caused by GPEE in the
          trigeminal caudal nucleus. Clinical manifestation of the
          syndrome and the stages of its development (5-6) from the
          first signs, i.e. scratching (1), to its maximum develop-
          ment (5-6); the region of scratching is hairless and has
          areas of skin erosion (3-6). GPEE was produced by micro-
          injecting TT into the caudal nucleus of the spinal
          trigeminal tract.

        The injection of glycine into trigeminal <u>nucleus caudalis</u> in the
GPEE region had completely suppressed the syndrome during glycine
action (Figure 74): the paroxysms disappeared, the animals became
calm and stopped scratching or tearing a certain region, and the
application of stimuli to this region did not induce a fit. They
syndrome originated again when glycine no longer produced its effect.
In the control experiments, the administration of a saline did not
produce any effect, and the pain syndrome remained (Figure 74 B).

        A comparison of those data with the symptoms of trigeminal
neuralgia in man (Harris, 1940; Wilson, 1954; Kugelberg, Lindblom,
1959; Stockey, Ransonoff, 1959; Kerr, 1963; Krol, Fedorova, 1966;
Mikheev, Rubin, 1966; Bogolepov, Yerokhina, 1969; Lindsay, 1969;
Karlov et al., 1973; Smirnov, 1973, 1976; Yerokhina, 1973; Anderson,
Matthews, 1977) suggests that the above syndrome has characteristic

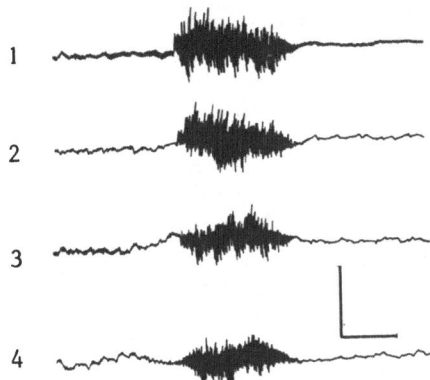

Fig. 73.   Burst of EA in the mastication muscle and head during a fit
           of the trigeminal pain syndrome. Recording of EA seven
           hours after injecting TT into the trigeminal caudal nucleus
           on the right. Recordings: 1,2: right and left mastication
           muscles; 3,4: right and left occipitofrontal muscles.
           Amplitude: 500 μV; time: 1 s.

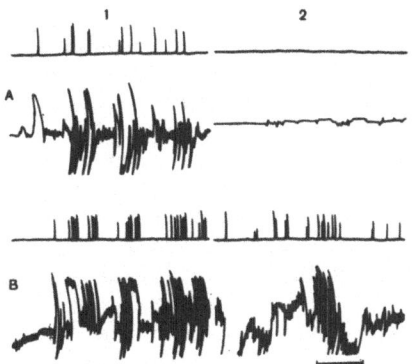

Fig. 74.   Effect of glycine injected into the trigeminal caudal
           nucleus with GPEE. Upper curves are phonograms and the
           lower ones are acrograms recorded in rats with a trigeminal
           pain syndrome before (1) and 20 min after (2) injecting
           glycine (20% in $4-10^{-4}$ ml) (A) and an isosmotic NaCl sol-
           ution (control; B) into the trigeminal caudal nucleus with
           GPEE; 72 h after injecting TT into the caudate nucleus.
           The horizontal line indicates 5 min.

features of trigeminal neuralgia. They include the paroxysmal nature
of pain origin, the presence of trigger zones, whose stimulation
causes a fit, bursts of electric activity in the masseter and head
muscles during pain fits, the existence of a refractory period after
a fit, during which another fit cannot be induced, some vegetative
components, etc.

Besides pain, itch may occur in the given syndrome. Itch was especially pronounced at the early stages of the pathologic process, being expressed in the scratch reflex. Trigeminal neuralgia in man can sometimes be accompanied by itch (Smirnov, 1973).

Those specifics of trigeminal neuralgia clearly show that the formation of GPEE in the caudal nucleus of the trigeminal nerve is the main pathologic mechanism of the given syndrome. This is evident also from the experiments involving the glycine suppression of GPEE: the inhibition of GPEE caused the whole syndrome to disappear during glycine action.

Such a syndrome of experimental trigeminal neuralgia was produced also by Black (1974) by injecting various epileptogens into trigeminal nucleus caudalis. The clinical pattern of the syndrome which he observed completely coincides with the clinical pattern of the above described syndrome. Epileptic activity has been recorded near the site of injection of those agents. A noteworthy fact is that this activity did not disappear when the nerve was peripherally blocked, but it was suppressed by dilantin and small doses of barbiturates.

Referring to our investigations, Japanese authors (Sakai et al., 1979) experimentally reproduced trigeminal neuralgia by microinjecting picrotoxin, penicillin, strychnine, brucine and D, L-homocystic acid into the medular part of nucleus caudalis. A noteworthy fact is that the clinical pattern of the pain syndrome (like the one described by us) was the same when various epileptogens were injected. The pain syndrome produced by picrotoxin could be suppressed by noradrenaline, dopamine, serotonin, GABA and glycine when they were applied to the picrotoxin action site (Sakai et al., 1981).

Trigeminal neuralgia can occur in the deafferentation of the trigeminal nuclei after rectogasserian rhizotomy (Anderson et al., 1971; Black, 1974) and the elimination of the pulp of all teeth on one side (Black, 1974). Under these conditions, epileptiform activity was recorded in nucleus caudalis (see also further).

The similarity of both the clinical picture and the characteristics of the occurrence of the experimental syndrome of trigeminal neuralgia, caused by the production of GPEE in nucleus caudalis of the trigeminal nerve, to those observed in trigeminal neuralgia in man suggest that some of its forms in man are connected with the formation of GPEE in the trigeminal system. However, this does not mean that the formation of GPEE in the sensory nuclei of the spinal tract of the trigeminal nerve is the only pathologic mechanism of trigeminal neuralgia. GPEE can apparently be formed in the appropriate areas of the central nervous system in the case of trigeminal neuralgia of central origin.

It is noteworthy that, as early as 1877, Trousseau called trigeminal neuralgia 'douleur epileptiforme', and that this syndrome can
be treated very effectively by using antiepileptic agents, particularly carbamezepine, diphenylhydantoin and clonazepam (Blom, 1963;
Cambier, Denen, 1971; Caccia, 1975).

PAIN SYNDROME OF THALAMIC ORIGIN

The pain syndrome of thalamic origin was reproduced by injecting
TT into a rat's gelatinous thalamic nucleus, which is related to
intralaminar thalamic nuclei (Pellegrino, Cushman, 1967). The syndrome had several clinical characteristics (Kryzhanovsky, 1976;
Kryzhanovsky, Igonkina, 1976).

A few hours after injecting TT, the animals became excited and
began to restlessly move around the cage and scratch out, with their
hind legs, regions of the skin on their back in the vicinity of the
scapula either on the right or on the left (Figure 75 1,2).
Paroxysms gradually became more pronounced: a sitting rat suddenly
screamed, jumped up and began to run around the cage; then, it
stopped just as suddenly and began to scratch the skin above the
scapulae; afterwards, it screamed again and began to run around the
cage, etc. Gradually, the fits occurred more frequently and became
longer and more intensive (Figure 76). The skin of the scratching
zone was damaged and eroded, and hairless regions appeared in it.
These regions became the trigger zones of easy provocation of fits,
and even their slight irritation caused a paroxysm. At the same
time, paroxysms occurred spontaneously. Gradually, they became more
frequent, longer and more intensive: the animals screamed, furiously
scratched the given regions, tried to lick them, and ran from the
place where they were sitting. In many cases, the scratching and
licking zones migrated. Such zones appeared on the body, the belly,
the back the perineum, the genital organs, and the legs (Figure
75 3-6). The animals became aggressive, vicious, furiously bit
objects which they came across, and screamed. Various stimuli, even
the slightest ones (slight tapping, light, blowing, touching, running
a brush across the hair, etc.), produced fits. Psychotic reactions
occurred in animals: the rat could keep running around the cage in
one direction, jump up high, jump on the cage walls, etc. If the
experimenter opened the cage, the animal could either jump out or
even jump on him.

Hence, a very severe pain syndrome occurs in rats when GPEE is
produced in the gelatinous nucleus of the thalamus. At the height of
they syndrome's development, fits which are provoked occur as if they
obey the 'all-or-nothing' law (regardless of the intensity of stimulation. The long duration of fits, the spontaneous origin of paroxysms, and some other signs clearly show that the syndrome has
generator mechanisms. The migration of the pain projection zones is

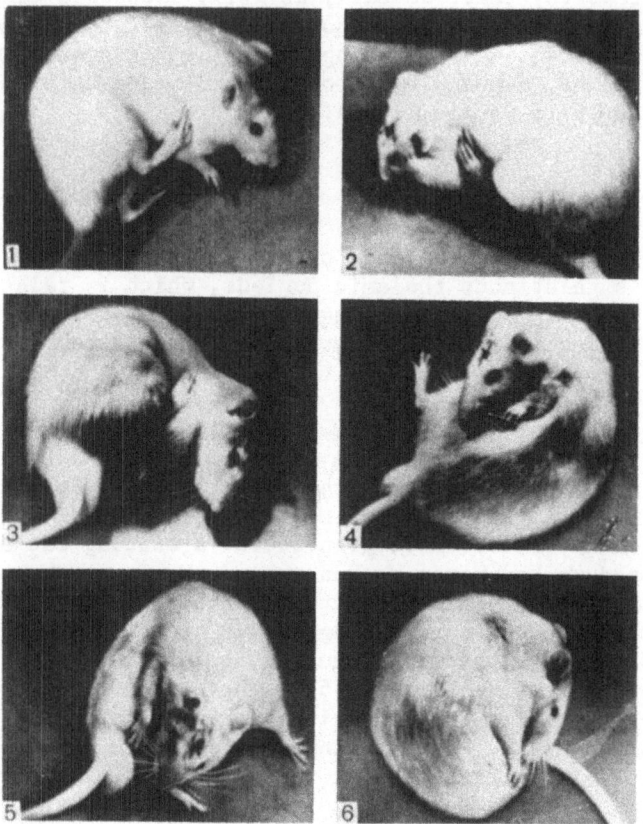

Fig. 75. Pain syndrome caused by injecting TT into the rat's
          gelatinous thalamic nucleus. 1-6: clinical picture of
          paroxysms and the migration of the zones of pain pro-
          jection.

a feature which distinguishes this syndrome from the above-mentioned
spinal and trigeminal syndromes. This distinguishing feature exists
apparently because GPEE is produced in the system of the unspecific
nuclei of the thalamus. This migration of the pain projection zones
is probably equivalent to diffuse pain in the case of some forms of
thalamic pains in man. That phenomenon may possibly be similar to
the one which occurs in some forms of cenesthopathy.

MODEL OF PHANTOM PAINS

     Phantom pains are a special form of the pain syndrome in the
case of which a patient feels pain the the nonexisting (amputated)
part of the body (e.g. fingers, leg, hand). This specific feature

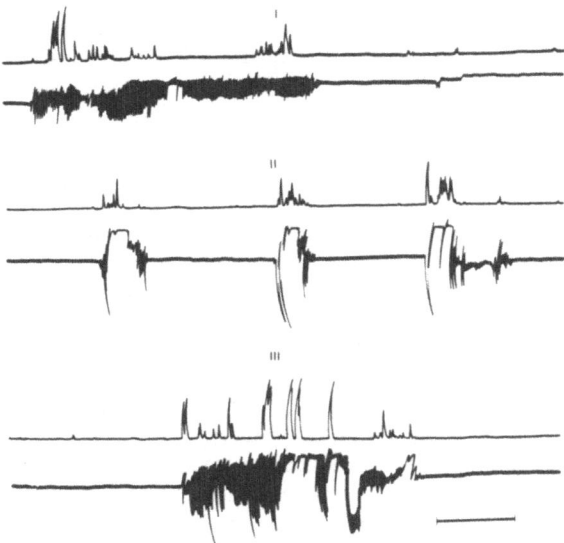

Fig. 76.   Characteristics of paroxysms in a pain syndrome of thalamic
           origin as shown by the phonogram (upper curves) and the
           actogram (lower curves).   I-III: successive recording
           fragments.   The horizontal line indicates 5 min.

of the syndrome suggests that the central generator mechanisms play a
definite role in its pathogenesis.   Taking account of all this,
experiments were carried out in modeling the phantom pain syndrome
(Kryzhanovsky et al., 1974a).   These experiments were mainly intended
to produce a characteristic pain syndrome by creating GPEE in the
dorsal horns of the lumbosacral segments of the spinal cord when
afferent stimulation from the periphery was cut off.   Afferentation
was switched off by either cutting the three main nerves of a leg
(the sciatic, femoral and obturator nerves) at a high level or cutting the posterior roots (L_2-S_2).   In a special series of experiments, the hind leg was amputated at a high level, modeling the
surgical or traumatic postoperative syndrome.   Deafferentation was
quite complete when the nerves and the posterior roots were cut:
sensitivity was absent in the fingers, the foot, the crus, and the
lower two-thirds of the femur.   Since a trauma can produce a certain
effect in the region where the afferents are cut, the operation was
carried out during different periods (from 3-5 days to 3 weeks)
before TT was microinjected into the dorsal horns.

The pain syndrome with all those characteristics originated when
GPEE was formed under the given conditions in the dorsal horns of the
cord on the side of deafferentation.   This syndrome differed from the
development of the syndrome under normal conditions in that the
latent period was longer, development was slower, and the symptoms
were less pronounced during the initial period.   However, all the

signs of a severe pain syndrome were exhibited later: when paroxysms occurred, the animals screamed, attacked their denervated or deafferent leg, and began to lick or bite definite regions (Figure 77 1,2). These regions corresponded to the site of the microinjection of TT into the spinal cord, i.e. to the localization of GPEE in the spinal segments. They corresponded to the identical regions in the control animals as GPEE was produced in the same spinal segments when afferentation was preserved. The behavior of animals with an amputated leg was interesting: when paroxysms occurred, the rats at first attacked the site where the leg was supposed to be and, as they 'missed' it, they began to lick the remaining regions on the uppermost part of the femur (Figure 77 3). At the height of the syndrome, fits became longer and more frequent. They either occurred spontaneously or could be provoked by sound and stimuli applied to various regions of the body besides the deafferent leg. The application of even the strongest stimuli to the licking region on the deafferent leg did not produce any reaction. At the late states of the process, there were cases of self-amputation of the leg. It did not matter when the nerve conductors were cut before TT was injected into the spinal cord. In all cases. the syndrome originated and occurred in the same way.

Pain projection was absent in some animals. Such animals ran around the cage, screamed, furiously attacked various objects in the cage, gnawed them, etc. These animals had all the signs of severe protopathic pain.

When the spinal cord was cut in the thoracic part (above the GPEE site), the pain syndrome always disappeared in animals with pain projection and animals with diffused, non-localized pain.

Several conclusions can be drawn from the above data. The central generator mechanisms are obviously important in the pathogenesis of phantom pains. When GPEE was produced in the spinal dorsal horns, a characteristic pain syndrome originated in all the three models used. Nevertheless, afferentation from the periphery was completely switched off.

It is believed that the phantom pains are the sad privilege of man and that they are connected with his psychic activity and are realized at the higher levels of the central nervous system. Apparently, this is not so. We know nothing about this syndrome in animals. Veterinarians may probably observe its equivalents. Investigations have shown that the phantom syndrome can occur in animals. As for the concept that the syndrome of phantom pains is realized at the higher levels of the central nervous system, two aspects should be taken into account. The sensation of pain is realized by cerebral cortical and subcortical structures, while the projection of pain occurs not only in these structures, but also at the level of the spinal cord.

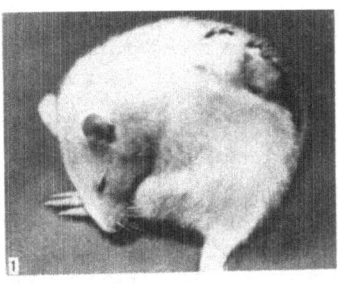
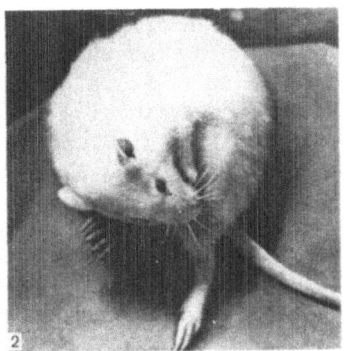
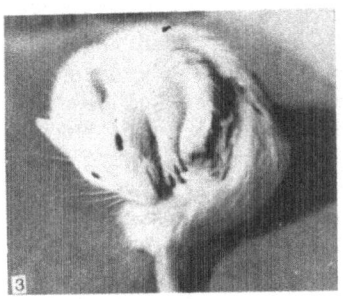

Fig. 77. Models of phantom pains of the hind leg. 1: pain syndrome
          in a rat with cut dorsal roots (L_2-S_2) on the left. TT (1
          DLM) injected into the dorsal horns (L_4-L_6) (three injec-
          tions) on the left three days after the roots were cut; 2:
          pain syndrome in a rat with cut nerves of the left hind
          leg; TT (1 DLM) was injected into the dorsal horns (L_4-L_6)
          on the left six days after the nerves were cut; 3: pain
          syndrome in a rat after the left hind leg was amputated; TT
          (1 DLM) was injected into the dorsal horn (L_6) on the left
          two weeks after amputation.

     Spinal mechanisms can play an important role in realizing the
phantom syndrome in man. This is evident from the data which show
that the coagulation of the gelatinous substance of the appropriate
segments of the spinal cord can eliminate the phantom pain in man
(Nashold et al., 1976). However, it should be taken into account
that a pathologic system (see Chapter 4) involving various areas of
the central nervous system can be formed and stabilized as the pathologic process develops.

     An interesting fact is that, in our experiments, local pain
projection was not observed in some animals. Instead, a syndrome of
severe generalized pain had originated. Apparently, it is a problem
of the relationships between epicritic and protopathic sensitivities
(head, Holmes, 1911; Head, 1920). Protopathic pain is especially
exhibited when epicritic sensitivity is switched off, as is the case
when the posterior roots and nerves are cut. When account is taken

of the severity and other peculiarities of the syndrome, protopathic pain occurs also under the conditions of preserved innervation as GPEE originates in the spinal dorsal horns, but it is marked by pain projection.

Phantom pains can originate not only when nerves are cut and a leg is amputated, but also when the spinal cord is completely or partially damaged, when it is completely cut across, and even after cordectomy (Davies, Martin, 1947; Freeman, Heimburger, 1947; Kuhn, 1950; Monroe, 1950; Bors, 1951; Li, Elvidge, 1951; Botterell et al., 1954; Melzack, Loeser, 1978). Since phantom pain can originate after cordectomy, the periphery is obviously not involved and noiceptive stimulation does not have anything to do with the syndrome's origin. In their studies of the phantom pains in the case of damages to the spinal cord, Melzack and Loeser (1978) have concluded that the syndrome is based on the origin of the 'pattern generating mechanism', being realized by pools of hyperactive neurons in the central nervous system. According to them the hyperactivation of such neurons is connected with the deafferentation of the neurons themselves, on the one hand, and the attenuation of the inhibitory descending influences, particularly influences coming from the formations of the brainstem that are normally maintained by tonic afferentation from the periphery, on the other. The significance of the central generator mechanisms in the phantom syndrome can be seen from the fact that the thresholds of pain sensitivity are higher in paraplegics with a phantom syndrome (Hazouri, Mueller, 1950). While not denying that the peripheral factors play a definite role in the origin of other central pain syndromes, Melzack and Loeser have concluded that the factors become either less significant or not significant at all in the maintenance of pathologic pain after the formation of the 'pattern generating mechanism'.

TRIGGER MECHANISMS

Trigger zones originate in many forms of pain syndromes, and their stimulation causes a pain fit. They are zones of easy provocation of a paroxysm, and sometimes they are the only receptive field whose stimulation can cause a fit. The presence of trigger zones is very characteristic of trigeminal neuralgia.

The question of the trigger mechanisms in the pain syndromes of central origin is especially noteworthy. There are trigger zones, whose stimulation causes a pain fit, in all the above forms of experimental pain syndromes. This regularity was rather clearly manifested in both trigeminal neuralgia and the pain syndrome of spinal origin.

In trigeminal neuralgia, such zones appeared in various parts of the face and head (the nose, cheeks, etc), but always on the side

where GPEE originated. In the pain syndrome of spinal origin, when
GPEE was produced in the dorsal horns of the lumbosacral segments,
the trigger zones could originate in various areas of the hind leg or
the tail, but always on the side where GPEE was formed. Hence, the
origin of trigger zones in some regions is determined by the localization of GPEE, which may be very limited and clear-cut localization
in <u>nucleus caudalis</u> of the trigeminal tract or in the segments of the
spinal cord, since the trigger zones may be small.

That has been completely confirmed in special investigations
involving a model of the spinal pain syndrome. Experiments have
shown that the licking zones, which are zones of hyperalgesia and at
the same time trigger zones, exactly corresponded to the TT injection
site, i.e. to the localization of GPEE, when TT was precisely injected in micro amounts into the dorsal horns of the lumbosacral or
sacral segments (Figure 78). A relatively large trigger zone corresponding to a large GPEE immediately originated when TT was injected
into two or three points of the dorsal horns. Gradually, the dimensions of GPEE increased and the hyperalgesia zone, which was also a
trigger zone of easy provocation of a pain fit, grew as TT spread in
the medullary substance.

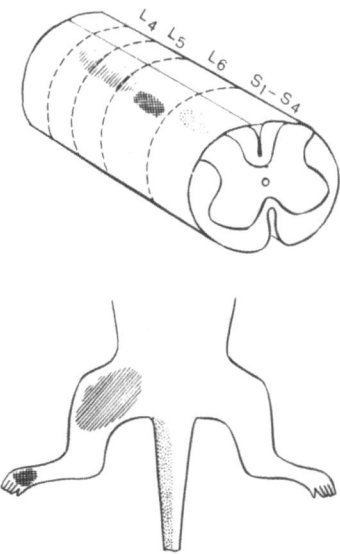

Fig. 78. Projection of hyperalgesic zones, or trigger zones, during
          the production of GPEE in the dorsal horns of different
          spinal lumbosacral segments. The pain projection zones on
          the hind leg and the tail, and the segments corresponding
          to them (into which TT was locally injected) have the same
          mark.

The peculiarities of the origin of trigger zones and their role in the provocation of pain fits can be clearly explained by the data on the dynamics of the formation and the functional organization of GPEE (see Chapter 3). It can be assumed that the trigger zones are zones from which afferentation goes to the 'leading' neurons of GPEE. These neurons are also trigger neurons, since they are the first to become excited (or are excited continuously), and they develop epileptiform activity, draw other neurons into the process of generating discharges, etc. (see Chapter 3). It is the connections between the trigger zones and the trigger neurons of GPEE that make it possible to easily activate GPEE even when local and very faint stimuli are applied to the trigger zones.

The migration of the trigger zones, being observed in the clinical picture of trigeminal neuralgia, should be regarded as the result of the changes in the excitability of the GPEE neurons, when the role of the 'leading' trigger neurons is played by various GPEE neurons. The incoincidence of the trigger zones with the pain projection zones in trigeminal neuralgia (Bogolepov, Yerokhina, 1966, 1969; Yerokhina, 1973; Calvin, 1979) can be attributed to the fact that the trigger neurons of GPEE, which ensure its start and which are activated from the trigger zones, and the neurons with which pain projection is connected, are different. In this respect, account should be taken of the complex somatotopic organization of the nuclei of the trigeminal nerve, which is connected with three branches, and also the fact that the realization of the phenomenon of pain projection depends on the higher cerebral structures.

The inactivation of the trigger neurons causes the trigger zones to disappear. This phenomenon is known in clinical practice. The reappearance of trigger zones before a fit shows that the enhanced excitability of the trigger neurons is restored.

A paradoxical phenomenon has been observed in the clinical picture of trigeminal neuralgia: strong stimuli applied to trigger zones can be ineffective, or they can even prevent a fit, while weak stimuli can provoke an attack (Kugelberg, Lindblom, 1959; Mikheev, Rubin, 1966; Yerokhina, 1973; Calvin, 1979). To explain this phenomenon, a hypothesis has been proposed to the effect that enhanced presynaptic inhibition is paradoxically converted into excitation (Calvin, 1979). However, it can also be explained by the fact that the wide dynamic range neurons, which constitute the operant basis of GPEE, are activated by impulses via large myelin fibers, i.e. by weak tactile and other stimuli. Strong nociceptive stimuli can activate presynaptic inhibition and cause the autoblockade of stimulus transmission. Another question which arises in this respect is whether hyperactivated neurons can be depressed by the influence of strong nociceptive stimulation.

The significance of the trigger zones in the pathogenesis of pain syndromes is apparently not confined to the easy provocation of fits. Our investigations involving the production of GPEE in the spinal dorsal horns and in <u>nucleus caudalis</u> of the trigeminal tract by injecting TT into them have shown that the pain syndromes of central origin are formed more rapidly when the zones are systematically stimulated. This means that afferent stimulation from the trigger zones promotes the formation of GPEE. It has already been shown by the phantom pain model that the formation of GPEE is held up and its power is reduced at the early stages of the process when afferentation from the periphery is switched off. In those cases, when afferentation from the periphery is important in triggering GPEE or maintaining its activity, a favorable therapeutic effect can be produced by eliminating stimulation from the periphery. Examples of this are trigeminal neuralgia treated by interventions into the nerve trunks, causalgia treated by neuromectomy, etc. It has been shown that chronic myofacial pain is based on a cyclic mechanism in which the trigger zones and the zones of reflected pains are switched on; in essence, this means that a complex multicontour generator is formed. The elimination of the trigger zones by procaine, which destroys such a multicontour cyclic generator, causes the syndrome to be eliminated (Travell, 1976).

PATHOGENESIS OF PAIN SYNDROMES IN THE LIGHT
OF THE GENERATOR MECHANISM THEORY

Phantom pain, multiform neuralgia and causalgia are the three main forms of pathologic pain, the explanation of whose mechanisms and peculiarities of manifestation are the touchstone of all the theories of pain. According to Melzack (1973), all the signs and symptoms of such pain syndromes as phantom pain, causalgia and neuralgia in patients suffering from pain are clearly exhibited in the above models of the central pain syndromes. The formation of GPEE in the appropriate areas of the nociceptive system was shown as the principal and general pathogenic mechanism of those syndromes. It is expedient to analyze some forms of pathologic pain more thoroughly in the light of the theory of the generator mechanisms and to consider some important data which have accumulated in clinical and experimental investigations.

RELATIONSHIP BETWEEN PERIPHERAL AND CENTRAL
MECHANISMS IN CAUSALGIA AND NEURALGIA

The neuroma is known to be a powerful source of pathologically intensive and constant impulses. This property of the neuroma is connected with the secondary changes and enhanced sensitivity of the terminals of the regenerating axons which form sprouting. They constitute a large receptive field with low excitability thresholds.

Ongoing activity in the Aδ and C fibers, whose source is the fiber sprouts in the neuroma, has been experimentally recorded (wall, Gutnik, 1974; Devor, Wall, 1976; Wall, 1979). However, such activity has not been detected in the Aβ fibers. Unlike normal fibers, which are not stimulated by noradrenaline, sprouting nerve endings are very sensitive to noradrenaline. Activity in nociceptive fibers sharply increases under the action of noradrenaline. Therefore, pain becomes more severe when the content of noradrenaline increases even slightly in the neuroma (as is the case, for instance, in stress situations). The efferent sympathetic fibers in the afferent nerve become more active in causalgia (Torebjörk, Hallin, 1979). Thus, regional positive pathologic feedback between the center and the periphery is formed in causalgia; there is a permanent nociceptive impulse flow from the neuroma, and sympathetic hyperactivity plays an important role in sustaining it (Leriche, 1939; Loh, Nathan, 1978; Bonica, 1979; Zimmermann, 1979). Sprouting nerve endings in the neuroma may probably be very sensitive also to other endogenous physiologically active substances which can either enter the neuroma or be secreted by its elements. The changed afferent fibers are known to generate many impulses, being influenced by mechanical, thermal and chemical effects (Wall, 1979). Fiber sprouts with a neuroma are spontaneously active and can generate impulses in response even to slight pressure (wall, Gutnick, 1974; Wall, 1979). Although the origin of ephaptic transmission in a neuroma is still debated, data show that there may be such a mechanism which activates the fibers at the late stages of nerve damage (Seltzer, Devor, 1979). It can be assumed that the number of proton receptors grows in the neuroma; they are activated when pH decreases, and it has been shown (Kristal, Pidoplichko, 1981) that they are connected with nociception. Thus, the neuroma with its sprouting fibers is, in essence, <u>peripheral GPEE</u>.

This constantly operating <u>primary</u> pathologic generator mechanism entails the formation of secondary GPEE in the central structures of the nociceptive system. In our case, the matter in question is the origin of GPEE according to the second type, i.e. due to the prolonged and intensive primary depolarization of neurons. The formation of secondary GPEE in the spinal dorsal horns is facilitated by the peculiarities of their structural and functional organization and by the properties of the nociceptive neurons and the specifics of their activation in nociceptive stimulation. The longer the existence of the neuroma and its activating influence on the nociceptive apparatus of the dorsal horns, the more likely is the origin of secondary central GPEE and its further development. At the late stages of the process, secondary GPEE can be stabilized and can become significant in itself. This explains the neurosurgically known fact that causalgia is not always eliminated when the neuroma is removed; also, there may be a relapse after temporary relief.

The formation of new GPEE is not confined to the region of the dorsal horns of the spinal cord: GPEE can also originate in the

thalamus under the influence of continuing nociceptive stimulation
coming from the neuroma and secondary GPEE in the dorsal horns.
Hyperactivity of nociceptive neurons in the spinal dorsal horns may
disappear at very late stages of the pain syndrome. Instead, hyperactivity of an epileptic character originates in the populations of
the thalamic neurons, thus not only sustaining, but also intensifying
the pain syndrome, which may be manifested by the same pain projections at the periphery (Albe-Fessard, Lombard, 1980, 1981).
Apparently, an operation on the dorsal horns intended to remove GPEE
or on the ascending spinothalamic tracts may be ineffective under
such conditions. The possibility that the pain syndrome lasts after
not only rhizotomy, but also even cordotomy has given rise to the
conclusion that the peripheral mechanisms play an insignificant role
in the pathogenesis of causalgia (Gerard, 1951; Melzack, 1971;
Sunderland, 1976; Loh, Nathan, 1978). At least, this is true of the
late stages of a disease. Considerable interest began to be taken in
the origin of an abnormal state of supraspinal structures, particularly the brainstem (Melzack, 1971).

Thus, the reasonings about the role of the peripheral and
central mechanisms in the pathogenesis of causalgia cannot be alternative ones. Such a conclusion is drawn also from an analysis of
clinical data (Bonica, 1977, 1979; Nashold, 1981; Parry, 1981; Tasker
et al., 1981). The significance of the peripheral and central components (accordingly, neuromas or damaged nerves of both peripheral
GPEE and secondary central GPEE in the nociceptive system) depends on
the conditions and stages of the development of the pathologic
process.

Such a conclusion on the role of the peripheral and central
components can be drawn also with respect to neuralgia. Neuralgia
has complex mechanisms in the case of a chronic trauma of a nerve or
the posterior roots, ischemia, or infectious toxic affections. The
damaged axons can exhibit their extent and the source of ongoing
activity, which increases with time (Wall et al., 1974; Calvin, 1979;
Torebjörk, Hallin, 1979; Wall, 1979). Such axons are very sensitive
to various influences, including mechanical ones. Therefore, even
slight brief pressure can cause prolonged (for tens of minutes)
impulses (Howe et al., 1977). When an axon is damaged, ectopic
impulse generators can originate in it (Calvin, 1979). These generators are more likely to appear in the region of the axon terminals.
The afterdischarge, which originates in the in-continuity lesion when
the impulses pass, can be either repetitive firing, which gradually
dies away, or a reflected impulse after the primary impulse passes
(Howe et al., 1977).

When the afferent nerve is damaged, spinal ganglia neurons can
also become the generator of impulses (Kirk, 1975). In the case of a
chronic trauma, degenerating and regenerating fibers possess the
properties of ongoing activity and respond to various stimuli by

bursts of impulses (Dyck, O'Brien, 1976; Wall, 1979). Thus, the
damaged afferents are the source of chronic nociceptive stimulation
in neuralgia. Exactly such stimulation enhances the sensitivity and
activity of nociceptive neurons (Mendell, 1966), i.e. it can be
optimum in the production of GPEE by the nociceptive neurons. These
conditions to a certain extent approximate those which are experimentally produced in kindling. Intensive but brief nociceptive
impulse activity, which causes acute pain, apparently cannot induce
the formation of GPEE in the spinal dorsal horns, since such nociceptive afferentation can cause an autoblockade by presynaptic and
postsynaptic inhibition of nociceptive neurons (Frankstein et al.,
1965; Franz, Iggo, 1968; Hongo et al., 1968; Zimmermann, 1968, 1979;
Burke et al., 1971; Jänig, Zimmermann, 1971; Kerr, 1976; Smolin,
1981).

The formation of secondary GPEE in the spinal cord under the
influence of chronic nociceptive stimulation can be promoted by
weakening inhibitory control at the afferent input of the spinal cord
as a result of the disturbance of conduction by large myelin fibers.
Clinical and experimental studies, which began with the classical
work by Gasser and Erlanger (1929) and, in this connection, with the
investigations carried out by Lewis and his associates (1931), have
shown that large myelin fibers are less resistant to damaging influences, particularly ischemia. Their conductance is upset, while that
of the C fibers is preserved (Van Hees, Gybels, 1972; Torebjörk,
Hallin, 1979). Consequently, the balance of afferent inputs of
different modality in the spinal cord is upset [Noordenbos (1959) has
termed this 'fiber dissociation'] and the tonic inhibitory influences
of large myelin fibers are eliminated, thus disturbing 'gate control'
(Melzack, Wall, 1965, 1970). This mechanism occurs in herpetic
neuralgia and tabes dorsalis (Noordenbos, 1959), trigeminal neuralgia
(Kerr, Miller, 1966), and some forms of chronic peripheral diseases
that are connected with ischemia of the nerve trunk (Gilliatt,
Wilson, 1953; Fullerton, 1963; Harding, Le Tann, 1977). In this
respect, the disturbances of gate control can be regarded as a factor
which exists in various forms of chronic diseases that are connected
with neural compression (tunnel syndromes, neoplasmic syndrome,
herniated disk syndrome, etc.). Neural compression engenders a set
of interconnected pathogenetic factors. Besides the disturbances of
the conductance of large fibers, the damaged fibers are very sensitive to ischemia, which also enhances fiber sensitivity to mechanical
influences. Apart from this local phenomenon, there are other
vicious circles which are connected with several peripheral and
central factors that maintain nociceptive stimulation, upset the
central inhibitory mechanisms and promote the formation of the self-
exciting neuron pools in the spinal cord (Bonica, 1977; Zimmermann,
1979). The immobility of the 'painful' leg as a result of sparing it
reduces afferent impulses via large fibers, thus also relaxing inhibitory gate control (Melzack, 1973, 1981).

The fact that an injury to large fibers does not always cause spontaneous pains (Nathan, Rudge, 1974; Dyck et al., 1976; Wall, 1978; Ochoa, Noordenbos, 1979, Thomas 1979) cannot be regarded as good grounds for denying the very idea that there is inhibitory control in the afferent input, including gate control (Wall, 1978; Thomas, 1979; Melzack, 1981), especially since, in such cases, hyperesthesia often occurs (although spontaneous pains are absent), and then even very faint non-nociceptive stimuli cause pain (Ochoa, Noordenbos, 1979). This shows that there are latent disturbances of inhibitory control and that the sensitivity of nociceptive neurons is enhanced.

For pains to originate in neuralgias, it seems that there must be two principal factors: the peripheral source of pains and the disturbance of central inhibitory control. The presence of only one of these factors is not enough for chronic pain to originate. These factors may include an increase in the sensitivity of the nociceptive neurons in chromatolysis due to retrograde degeneration or deafferentation (Wall, 1974). The combination of these factors is an optimum prerequisite for the formation of GPEE in the dorsal horns and nucleus caudalis of the trigeminal nerve. Under these conditions, GPEE is formed on the mixed-type basis, i.e. it is formed due to the constant depolarization of nociceptive neurons and their disinhibition.

The significance of the central component in the form of secondary GPEE can differ from case to case. Its significance is accentuated by therapeutic specifics; the surgical methods of treatment, while being intended to interrupt nociceptive stimulation going to the spinal cord, are far from always effective. However, an increasing amount of data show that pain can be cut by antiepileptics in some forms of neuralgia. In trigeminal neuralgia, which is characterized by paroxysmal pain, carbamazepine is one of the main agents used in medicamentous therapy. This agent is now being employed also in other forms of neuralgia. The electric stimulation of large afferents or the stimulation of the dorsal columns, just as the stimulation of the supraspinal antinociceptive structures (see Chapter 13), causes not only the modulation of sensory input and the restoration of inhibitory control, but also the suppression of GPEE which either is being formed or has already been formed. It should be noted that the plasticity of the nervous system plays a leading role in both the origin of GPEE and its elimination. This plasticity allows structural and functional reorganizations to occur in the central neural formations. Therefore, when the conditions are favorable, the elimination of the source of pathologic nociceptive stimulation or the steady modulation of sensory input is enough for eliminating GPEE which has already been formed.

DEAFFERENTATION PAIN SYNDROME

Numerous experimental studies have shown that a severe pain syndrome originates when the nerves or the posterior roots are cut; this syndrome can be a model of chronic pain in man (Loeser, Ward, 1967; Basbaum, 1974; Black, 1974; Dennis, Melzack, 1977; Ducrow, Taub, 1977; Albe-Fessard et al., 1979; Lombard et al., 1979a, 1979b; Wall, 1979; Wall et al., 1979; Albe-Fessard, Lombard, 1980, 1981; Yamaguchi, 1980; Levitt, Heybach, 1981; Levitt, Levitt, 1981; Wisenfeld, Lindblom, 1982). The syndrome occurs just as hyperalgesia with regional pain projection. The animals begin to scratch the respective areas of their leg. They do this so intensely that the tissue is damaged. The wounds may heal, but then they originate again. The animals bite their leg and sometimes bite away the tissue in the pain projection areas. In severe cases, self-amputation is observed. This occurs when deafferentation is quite extensive (when four roots are cut) and involves the brachial plexus. Animals with a deafferent leg become very sensitive and hyperactive to tactile stimuli, the light, etc.

That syndrome can be reproduced with a different extent of severity in animals of various species (mice, rats, cats, monkeys), showing that it develops according to a general regularity and that is mechanisms are universal. It is reproduced most easily in rats when the foreleg is deafferent, but it can also originate when the afferent pathways of the hind legs undergo intervention. We also observed the origin of a pain syndrome in some rats involving the self-amputation of the hind leg in the case of wide deafferentation (L_2-S_2).

Obviously, the clinical picture of that syndrome virtually coincides with the clinical symptoms of the above-mentioned pain syndrome, which originates when GPEE is produced in the dorsal horns of the spinal cord. In deafferentation, the pain syndrome can also be considered a model of phantom pain: it can be seen from common clinical characteristics and conditions of the origin of both syndromes. Hence, the basal pathogenic mechanism is the same for all the above pain syndromes, i.e. GPEE is formed in the appropriate areas of the nociceptive system.

Indeed, an electrophysiological analysis of the pain syndrome in deafferentation shows that pools of hyperactive neurons originate in the dorsal horns of the spinal cord, particularly in the V lamina. Such neurons generate abnormal bursts of impulses, and the stimulation of the adjacent intact roots by single electric stimuli causes prolonged discharges which continue for hundreds of milliseconds (Loeser, Ward, 1967). Neuronal hyperactivity appears a few days after the posterior roots are cut, and it gradually intensifies. Such a stage was seen during prolonged observation (six months). Abnormal activity is generated also in the cells of the dorsal horns

of the cervical segments (Lombard et al., 1981). It has been
observed also in the deafferent nucleus of the trigeminal nerve after
the roots were cut (Anderson et al., 1971) and when the pulp of all
teeth on one side was removed (Black, 1874). Neuronal hyperactivity
has been observed after partial deafferentation in the lateral cuneate nucleus (Kjerulf, Loeser, 1973; Kjerulf et al., 1973), the
gracile nucleus (Basbaum, Wall, 1969). Thus, this phenomenon is an
expression of a general regularity. Thalamic neurons (VP) become
abnormally active at very late stages of the pain syndrome after
deafferentation, and their population assumes the nature of 'quasi-
epileptic foci' (Lombard et al., 1979b).

All these data are in accordance with the possibility of suppressing the deafferentation pain syndrome by antiepileptics (Black,
1974; Ducrow, Taub, 1977, Albe-Fessard, 1981). Peripheral electric
stimulation, like the one used to alleviate pain, or contralateral
stimulation inhibits the syndrome (Albe-Fessard et al., 1979; Albe-
Fessard, Lombard, 1980, 1981). Experiments with monkeys have shown
that animal behavior, particularly biting the forelimb after rhizotomy (C_2-T_3), depends on the quickenss and extent of restoration of
the manipulating function of the extremity and on the state of the
motor function of the contralateral extremity.

Pain syndromes associated with deafferentation occur in man,
too. Even Livingston as early as in his day described a group of
chronic pain syndromes whose common pathogenetic feature was deafferentation; in accordance with his theory of the central mechanisms of pain (see further), deafferentation can cause central changes
which become an independent pain-producing mechanism. Loeser and his
associates (1968) described the hyperactivity of the cells of the
spinal dorsal horns when it was deafferent in patients. Recent
clinical studies give data of the years' long observations of hundreds of patients with a deafferentation pain syndrome, such as
causalgia or dysesthesia after various ruptures of the peripheral and
central afferent pathways (Nashold, 1981; Parry, 1981; Tasker et al.,
1981). After analyzing clinical examinations and the results of
therapeutic interventions, the authors drew the conclusion that the
hyperactivity of the central structures of the nociceptive system
plays an important role in the origin of that syndrome.

In the pain syndrome after deafferentation, GPEE is formed in
the nociceptive system apparently due to several factors. In deafferentation, we in essence come across denervation phenomena, and
account should be taken especially of the regularities and mechanisms
which form the basis of the denervation syndrome. The above-
mentioned disturbance of inhibitory control due to the elimination of
tonic afferentation, the disinhibition of the pool of nociceptive
neurons, particularly the neurons of the V lamina, and the enhancement of their sensitivity are important in this respect. Their
sensitivity grows due to not only deafferentation, but also the

transganglionary degeneration of the nociceptive neurons of the
dorsal horns of the spinal cord that originates when the sensitive
fibers are cut (Wall, 1974, 1979; Basbaum, Wall, 1976; Moradian,
Rustioni, 1977). Degeneration phenomena occur in the neurons of
spinal ganglia (Grant, Ardvisson, 1975; Moradian, Rustoni, 1977),
which consequently, can generate impulses (Kirk, 1975), and in the
neurons of trigeminal <u>nucleus caudalis</u> (Black, 1974; Westrum et al.,
1976; Gobel, Binck, 1977). The central terminals and the neurons of
the trigeminal nerve may degenerate even when the peripheral afferents are slightly damaged (as in the case of tooth extraction).
Primary afferent depolarization (PAD) and presynaptic inhibition are
eliminated, the dorsal horns are reorganized (Wall, 1981), and trophic transport from the neurons of spinal ganglia to the neurons of
the dorsal horn is disturbed (Knyihar, Csilik, 1976) when the primary
afferents degenerate in the dorsal horns. That disturbance may be
one of the causes of the degeneration of the neurons of the dorsal
horn. The phenomenon of the 'neuronal plasticity' or 'supersensitivity' of the partially deafferented neurons, i.e. neurons in which a
part of the inputs has disappeared because of the degeneration of
synapses, may be due to the activation of the formerly inactive
synapses, the enhancement of the effectiveness of the remaining
synapses owing to hypertrophy, the origin of the chemical hypersensitivity of the postsynaptic denervated membrane, and the appearance
of new synapses instead of the degenerated ones owing to new afferents (Stavraky, 1961; Riesen, 1975; Merrill, Wall, 1978; Zimmermann,
1979. Neurons deprived of afferent inputs can generate intrinsic
autogenic activity (see Chapter 3). The chronically deafferent
neurons can become excited when the other afferents are activated
(April, Spencer, 1969; Basbaum, Wall, 1976; Mendell et al., 1978).
Owing to their hyperactivity, these neurons generate a powerful flow
of impulses when the adjacent afferents are stimulated. All these
factors promote the formation of GPEE in the spinal dorsal horns or
in trigeminal <u>nucleus caudalis</u>. Wall (1981) believes that there are
eight factors which play a certain role in the origin of pain in
peripheral deafferentation (section of nerves below ganglia). They
include those which were considered above in the case of causalgia
and neuralgia.

Hence, the mechanism of the paradoxical phenomenon of anesthesia
dolorosa becomes clear. This phenomenon originates when the nerves
or the dorsal roots are cut in order to relieve a patient of paint.
It consists in the sustenance and even intensification of pain, while
sensitivity in the denervated region is lost. The phenomenon is
apparently based on the origin of GPEE in the spinal dorsal horns
after deafferentation.

CERTAIN PECULIARITIES OF THE PAIN SYNDROMES IN
THE LIGHT OF THE GENERATOR MECHANISM THEORY

The Origin of Pain Attacks

This is an expression of GPEE activation that can produce a
flow of impulses which hyperactivate the respective structures of the
nociceptive system. At the early stages of the formation of GPEE,
when the thresholds of the excitability of its neurons decrease
inconsiderably, while the inhibitory mechanisms are still relatively
effective, only nociceptive stimulation can (due to spatial and
temporal summations) allow the critical mass of the GPEE neurons to
originate and can thus cause a fit. In this case, nociceptive stimulation is not only an inducing, but also a forming factor. This
mechanism clearly explains the phenomenon of the intensification of
pathologic pain with the extension of the irritated area, which was
described by Noordenbos (1959). As the insufficiency of the inhibitory mechanisms develops and the thresholds of the excitability of
the neurons of GPEE decrease, rather weak and brief nociceptive
impulses can activate it and, consequently, cause pain. At this
stage, the modality of the stimuli being induced (nociceptive effects)
is still important. At the late stage of the pathologic process,
when the inhibitory mechanisms greatly weaken and the thresholds of
the excitability of the GPEE neurons decrease, GPEE can be triggered
even by slight stimulation, which may be of different modality. At
this stage, therefore, the normally indifferent stimuli (e.g. sound,
light) can be effective in provoking a pain attack. There may be
also spontaneous paroxysms due to both the activity of the trigger
neurons of GPEE (neurons of group I) and the greater reactivity of
the neurons of groups II and III (see Chapter 3). The changes in
both the organism's internal media and the external environment,
which influence the excitability of the nervous system, particularly
the nociceptive system, and the reactivity of the GPEE neurons, can
alter the readiness of GPEE to be activated, thus enhancing or reducing the probability of a seizure.

Characteristics of Central Pain

The role of consciousness, the higher levels of the central
nervous system, and the psychic and emotional spheres in the determination of the sensation of pain is a special topic (see Melzack,
1973, 1981; Sternbach, 1978) and will not be discussed here. It
should be noted that the characteristics of pathologic pain and the
peculiarities of a pain fit can be largely attributed to the activity
of GPEE and the areas of the nociceptive system that are activated by
GPEE.

The intensity and duration of pain paroxysms are determined by
the power of GPEE and the duration of its activity: the more powerful

is the flow of impulses produced by GPEE and the longer are they
produced (see Chapter 3), the more severe and more prolonged are the
paroxysms. Therefore, the intensity and duration of paroxysms grow
as the process develops. Owing to the great insufficiency of the
inhibitory process and the considerable growth of the excitability of
the GPEE neurons at the late stages, stimuli of different intensity
can produce the same effect, namely, they can trigger off the maximal
activation of GPEE. Thus, GPEE seems to act as if it is obeying the
'all-or-nothing' law (see Chapter 3). Therefore, even inconsiderable
stimuli can cause a severe fit. As we have seen (Chapter 3), when
GPEE is being formed, the latent period of its activation is reduced
and its reaction becomes more and more expolsive-like, i. e., the
power of impulse flow which is produced sharply grows. The peculiarities of the fits being induced evolve, accordingly; their latent
period diminishes and their intensity quickly (and sometimes immediately) reaches the maximum. However, slight nociceptive stimuli in
the algesic region can cause pain with a certain delay, but once it
originates, it becomes severe and lasts a long time. This phenomenon, which is known well to clinicians (Noordenbos, 1959), can also
be explained from the standpoint of the generator mechanisms: temporal summation plays a big role in the activation of GPEE, in which
the inhibitory mechanisms are still preserved and neuronal excitability is not high. After activation, GPEE can continue to be active
for a long time.

Prolonged constant pain is due to the tonic character of GPEE
activity, which s caused by the intermittent activity of the GPEE
neuronal pools (Chapter 3). Conversely, the rapid hypersynchronization of these neurons can cause acute paroxysms, instant piercing
pains, abruptly commencing and discontinuing pains, intensive prick-
ing, etc. These features of pathological pain can apparently depend
also on the apparatus of either tonic or phasic pain (Dennis,
Melzack, 1977) which is activated under the influence of GPEE.
Evidently, the peculiarities of the structural and functional organization of the formations constituting the tonic and phasic apparatus
of the nociceptive system determine to a certain extent the properties and pattern of GPEE activity that originate in these formations.
The nature of pain sensation depends on the epicritical or protopathic system (Head, 1920) which participates in the process.
Diffuse burning pain (Bishop, 1959) and hyperalgesia (Noordenbos,
1959) originate when the structures of the system of protopathic
sensitivity are disinhibited and GPEE is formed in the structures.
The diversity of the types of pathologic pain of central origin
(Melzack, Torgeson, 1971; Melzack, 1981) is due to the specifics of
the site of GPEE in the nociceptive system and the involvement of the
appropriate formations of CNS in the process.

The refractory period following neuraligc paroxysms, during
which a new fit cannot be produced and the trigger zones disappear,
can be compared with the known phenomenon of the refractoriness of

the epileptic neurons after their discharge (Chapter 2). Just as in
an epileptic seizure, it can be engendered by originating depression, the inactivation of the generator mechanism of the neurons, and
the hyperpolarization of their membrane. We have observed such
phenomena also in the above models of the central pain syndromes,
which were produced by forming GPEE in the spinal dorsal horns and in
trigeminal <u>nucleus caudalis</u>.

The <u>stability and resistance</u> of the central pain syndromes are
an important feature of pathologic pain that makes the whole problem
so important. It can be satisfactorily explained from the standpoint
of the theory of the generator mechanisms and the concepts of the
role of the pathologic system as the pathogenetic basis of those
syndromes (Chapter 4). The presence of stable GPEE, which can self-
sustain excitation and can be activated by impulses from various
sources, shows why neurotomy and rhizotomy, being produced in order
to stop the pathologically intensive noiceptive impulses going from
the periphery to the central nervous system, can be ineffective.
Attempts to get rid of the pain syndrome by eliminating the secondary
GPEE in the dorsal horns or in trigeminal <u>nucleus caudalis</u> by means
of coagulation will fail if there is still even a small part of GPEE
or there are conditions which promote the origination of GPEE
(Chapters 7, 8). It has been shown that GPEE in the dorsal horns may
not be electrophysiologically exhibited at very late stages of the
spinal pain syndrome; instead, a new GPEE originates in the thalamus.
In such cases, the coagulation of the GPEE region in the dorsal
horns, the section of the spinothalamic pathways, and even complete
cordotomy may be ineffective.

The pain syndromes are stable also because they are a clinical
expression of the pathologic systems which have been formed. These
systems originate under the influence of primary or secondary GPEE,
which act as determinant structures that sustain the activity of the
pathologic systems and determine the characteristics of this activity
(see Chapters 1, 4). The <u>pathologic system</u> of the primary pain
syndrome of peripheral origin may have a complex organization and may
include various structures, beginning with the peripheral source of
pain (neuroma, chronically traumatized degenerating and regenerating
nerves) and ending with the cerebral regions. The resistance of the
pain syndrome to pharmacological influences depends on the resistance
of the pathologic system. The longer does this system exist, the
higher is its resistance (Chapters 1, 4). A stabilized chronic
pathologic system fixed by memory imprinting is especially resistant
to therapeutic influences. In this case, there must be a set of
interventions which are intended not only to eliminate the determinant structure (this is a prerequisite), but also to destabilize and
suppress the activity of the pathologic system.

The removal of a neuroma or its inactivation by administering
either novocain or alpha-adrenoreceptor blocking agents is known to

produce a long therapeutic effect in causalgia, and also the alcohol-
ization of the trigeminal nerve branches in trigeminal neuralgia.
This means that, in such cases, secondary GPEEs and a stable pathologic system are not formed yet or the GPEE deprived of sustaining
stimulation is eliminated by the appropriate neurophysiological
mechanisms. Plasticity, which plays a big role in the development
and consolidation of pathologic changes, including the formation of
GPEE in the pathologic system (Chapter 4), is just as important in
the reorganization of the intracentral and interneuronal relations,
and it promotes the suppression and elimination of GPEE and the
pathologic system.

It follows that stimulation therapy of the pain syndromes in the
form of ES of the nerves, the dorsal columns and the cerebral antinociceptive structures has complex mechanisms which consist in not
only the modulation of sensory input (as it follows from the gate
control theory), but also the suppression of GPEE and the pathologic
system induced by GPEE as a determinant structure. These effects are
realized due to the activation of the antinociceptive system and the
reorganization of intracentral relations (see Chapter 13).

ON THE THEORY OF THE GENERATOR MECHANISMS OF CENTRAL
PAIN SYNDROMES: EVOLUTION OF CONCEPTS

According to the gate control theory in its original form
(Melzack, Wall, 1965), which is the most popular theory of pain, the
flow of nociceptive impulses is modulated in the sensory input of the
spinal cord. The inhibitory control of these impulses is realized by
gelatinous substance cells, which are activated by large fibers and
which allow the nociceptive neurons to be presynaptically inhibited
by depolarizating the endings of the small afferents. Activation of
small fibers cause the hyperpolarization of their endings, thus
facilitating the transition of excitation to nociceptive neurons.
Sensory flow can be controlled also by supraspinal influences, including influences of the cerebral cortex. The gate control theory,
which has the rational principles of the earlier theories, clearly
explained much data and many clinical observations. Its application
to the supraspinal levels (Melzack, 1971, 1973; Procacci et al.,
1974; Tsubokawa et al., 1975; Dong, Wagman, 1976) has been an important step in its development and has revealed its significance.

At the same time, the factual basis of the gate control theory
was criticized. The main arguments against it were that spontaneous
pains do not occur in many cases when the elimination of the influences of the large fibers is documentarily proven (Nathan, Rudge,
1974; Nathan, 1976), that the activation of small fibers causes the
depolarization, and not the hyperpolarization of the terminals and
can, consequently, also 'close the gates' (Frankstein et al., 1965;
Gregor, Zimmermann, 1973; Zimmermann, 1979; Smolin, 1981), and that

the control in sensory input can be realized by not only presynaptic inhibition, but also postsynaptic inhibition (Hongo et al., 1968). These and several other facts, which contradict the basic principles of the original gate control theory, have been recognized by its authors (Melzack, 1973, 1981; Wall, 1978).

However, the idea of controlling sensory input was preserved, being consolidated and developed. As noordenbos (1979) put it, the idea was "admirably expressed in a single word 'gate'." The processes involving the modulation of nociceptive flow in sensory input are now known much better. It has been established that only only presynaptic inhibition, but also postsynaptic inhibition as well as various neurons and the whole apparatus of the dorsal horns and the nuclei of the trigeminal nerve participate in the transmission and processing of nociceptive information. It has become clear that one of the functions of the C afferents is to keep the functional organization of the dorsal horns of the spinal cord on a normal level, controlling the extent of inhibition and the dimensions of the receptive fields (Wall, 1981).

The gate control theory has its shortcomings in spite of its significance: it concerns only the modulation of the sensory flow from the periphery at the input to the central nervous system. It clearly explains the phenomena of physiological pain and its regulation. As for pathologic pain, it is confined to the pain syndromes which are connected with the peripheral sources of pain or with the loss of control of nociceptive afferentation in the sensory input. This theory cannot explain the mechanism of pathologic pain of precisely central origin. The concepts of the central biasing mechanism, which is realized by the reticular formation in the form of tonic inhibitory influence (or tuning influence) on the transmission of information at all the levels of the somatic projection system (Melzack, 1971), have played a certain role in the further elaboration of the gate control theory. However, the use of these concepts was an expanding extrapolation of the theory to other areas of the nociceptive system and could not explain the mechanisms of the central pain syndromes proper and their characteristics (duration, resistance, etc.). It was too difficult for the theory to explain the phantom pain mechanisms, especially when phantom pain originated after complete cordotomy. Owing to the phantom pain syndrome, it had to be taken for granted that when the impulses no longer went to the central structures of the nociceptive system due to deafferentation, their neuronal pools can independently produce excitation, thus engendering pathologic pain. Melzack and Loeser (1978) have termed such a mechanism of pain origin 'the pattern generating mechanism'. The fact that cordotomy does not eliminate phantom limb pain or causalgia in many cases has suggested that the mechanism is realized not in the spinal cord, but in the brainstem. The mechanism is modulated by various influences (sensory inputs, descending influences, nociceptive impulses, memory, prior experience, emotions, the organism's state, etc.). It is extrapolated also to other types of pathologic pain of central origin (Melzack, 1981).

Our theory of the generator mechanisms of central pain syndromes (Kryzhanovsky et al., 1974a, 1974b; Kryzhanovsky, 1976, 1979, 1980a) was not initially connected with the gate control theory. Genetically, it is connected with the general concept of the role of the determinant structures in the pathologic state of the nervous system and the general theory of the generator mechanisms of central neuropathologic syndromes. It is based on direct experimental data on the production of GPEE in various areas of the nociceptive system. This technique has been used to create models and other neuropathologic syndromes in the central nervous system.

The theory of the generator mechanisms of central pain syndromes fundamentally differs from the gate control theory in that it deals not with the modulation of the sensory flow from the periphery and not with the loss of control of the flow at the spinal or supraspinal level, but with the origin of GPEE as a new functional formation which hyperactivates the nociceptive structures, and consequently with the origin of the pathologic system. It has been shown that GPEE can originate under various conditions due to the primary disturbance of the inhibitory mechanisms or the primary depolarization (epileptization) of the nociceptive neurons (Chapter 3). It can either be stimulated by impulses from various sources or be activated spontaneously due to the activity of its own trigger neurons. But even when GPEE is stimulated by afferentation from the periphery, its proper activity is pathogenically significant. The peculiarities of this activity depend on the properties of GPEE. In this sense, the central pain syndromes are endogenous.

It follows that the gate control theory and the theory of the generator mechanisms of central pain syndromes cannot be evaluated alternatively. Each of them covers a definite range of phenomena, and they necessarily supplement one another. Together, they form a broad conceptual basis for the elaboration of various aspects of the problem of pain.

Both of those theories have their prototypes. The summation mechanism, known in neurophysiology, was long used to explain the intensity and duration of pathologic pain. Its significance was emphasized even in the theory set forth by Goldscheider (1894), which is known as the pattern or summation theory. Livingston (1943, 1948) was the first to indicate the special role played by the abnormal summation of excitation in the centers. He assumed that this phenomenon was based on closed neuronal loops which were capable of self-excitation, like the reverberatory circuits of Lorente de Nó (1938). Another mechanism of that phenomenon has been postulated by Noordenbos (1959) in the concept of the multisynaptic system, which makes the summation of excitation possible due to the branched polysynaptic neuronal chains. Normally, the multisynaptic system in the spinal cord is controlled by special inhibitory mechanisms. When these mechanisms are upset, the system becomes hyperactivated, as a result of which long, diffuse burning pain originates. Noordenbos is also the author of the concept of sensory interaction in the input to the spinal cord. According to it, the activation of large fibers suppress input produced by small fibers. Since large fibers are less resistant and are the first to be affected, 'fiber dissociation' occurs, the influence of these fibers is eliminated, and the control of the flow of nociceptive impulses either weakens or is upset under pathologic conditions, as a result of which pain originates. Thus, Noordenbos's concepts are, in essence, a direct prototype of the theory of the generator mechanisms and the gate control theory. As Wall (1978) said, the data presented by Noordenbos had greatly influenced the formation of the gate control theory. Another mechanism of the hyperactivation of the spinal neurons has been proposed by Gerard (1951). According to it, neurons are hyperactivated due to the loss of the sensory regulation of neuronal activity when the peripheral nerve is damaged. In this case, the synchronously discharging neuronal pools can involve other neurons, their excitation can be sustained by weak stimuli, and they can generate powerful volleys of impulses with an abnormal pattern. Hebb (1949) believed that synchronized discharges in the thalamocortical pathways are most important in the pain mechanisms. These discharges can originate due to the loss of sensory regulation of the pattern of thalamocortical activity and can cause the disintegration of the processes of perception and recognition, being expressed in the form of pain. In this case, steady changes in the synapses of these pathways help to consolidate the given disturbances in the form of pathologic memory. In developing Livingston's concepts of the 'vicious circle' concerning the peripheral and central processes which sustain abnormal excitation in the spinal cord, Sunderland (1968, 1976) proposed the 'turbulent hypothesis', according to which the foci of abnormal activity that originate in the dorsal horns cause a chain reaction in the ascending pathways, even up to the cerebral cortex. As a result of this reaction, chaotic hyperactivity occurs in the whole sensory sphere. These changes, and not the peripheral mechanisms, are the main pathogenic link of causalgia. Cure and Corregal (1974) have emphasized that the uncontrolled central sensory epileptiform discharge plays the leading role in the mechanisms of chronic pain, especially paroxysmal neuralgias.

The idea of the role of liberation from inhibition in the origin of central pains is not a new one. It was proposed by Head (Head, Holmes, 1911; Head, 1920) in his teaching of both the abilities of the cerebral cortex to transform pain sensations into 'subpain' ones and the suppression of protopathic sensitivity by epicritical sensitivity. This teaching was supported and modified by Forster (1927), who attributed hyperpathy, which originates when the posterior columns of the spinal cord are cut, to the elimination of descending

influence. The concepts of the inhibiting influence of the system of
the pylogenically new spinothalamic tract (the specific system, the
lemniscus system, or the nonspecific system, the extralemniscus
system, or the polysynaptic pathway) have made it possible to explain
the origin of the central pain syndromes when the brain is damaged;
in this case, the structures of the ancient spinothalamic tract are
disinhibited because of the lower level of activity by the new spinothalamic tract (Bowsher, 1957; Bishop, 1959; Noordenbos, 1959, 1960;
Albe-Fessard, 1968; Winter, 1973; Procacci et al., 1974; Albe-
Fessard, Fessard, 1975). The clinical signs of this process are
expressed in protopathic sensitivity (Head, 1920) and the appearance
of 'slow', 'diffuse' and 'burning pain (Bishop, 1959) and hyperalgesia (Noordenbos, 1959). Similar concepts were evolved by Hassler
(1970). According to these concepts, the disturbances in the 'cortical pain pathway' system and, consequently, the disinhibition of
structures, particularly thalamic nuclei (<u>nucleus limitans</u>, <u>nucleus
intralaminaries</u>), in the system of the normally inhibited 'subcortical pain pathway' are decisive in the pathogenesis of thalamic
pains and trigeminal neuralgia of central origin. All these investigations have greatly contributed to the solution of the problems of
central pain syndromes.

SUMMARY

     The production of the generators of pathologically intensified
excitation (GPEEs) in definite areas of the nociceptive system engenders the appropriate experimental pain syndromes of central
origin: the spinal pain syndrome when GPEE is produced in the spinal
dorsal horns, trigeminal neuralgia when GPEE is produced in trigeminal <u>nucleus caudalis</u>, and the thalamic pain syndrome when GPEE is
produced in the gelatinous nucleus of the thalamus. When the lumbosacral segments are ipsilaterally deafferent (section of the posterior roots or nerves) or when a hind leg is amputated at a high level,
the production of GPEE in the dorsal horns of these segments engenders a pain syndrome with pain projection at the appropriate areas
of the hind leg which has lost its sensitivity or at the nonexisting
leg (phantom pain model). The suppression of GPEE by stereotaxically
introducing inhibitory transmitters into the GPEE site eliminates the
syndromes during the action of the inhibitors. Pain syndromes can be
produced by various factors which make it possible for GPEE to be
formed in the nociceptive system, i.e. the syndromes may be polyetiological, while the mechanisms of their development are monopatho-
genic, namely, GPEE formation in the nociceptive system. The severity, clinical characteristics and course of the syndromes are determined by the power of GPEEs, the peculiarities of their activity, and
their site in the nociceptive system. Several principal pain syndromes (causalgia, neuralgia, phantom pain, anesthesia dolorosa) and
some peculiarities of pathologic pain of central origin have been
considered from the standpoint of the theory of generator mechanisms.

10
Syndromes of caudate nuclei

The complex structural, functional and neurochemical organization of caudate nuclei (CN) is in accord with their special role in the activity of the central nervous system. It has long been known that the striatum plays the leading role in realizing the functions of the extrapyramidal system. Besides, much data indicate that CN participate in the modulation of the activity of various physiological systems, in the regulation of the reactions of reticular formation, the thalamus and the cerebral cortex to the stimuli of different modality, in the mechanisms of integrative and higher nervous activity, in the neuronal processes of the mind, and behavior (Jung, Hussler, 1960; Fox, O'Brien, 1962; Krauthammer, Albe-Fessard, 1964; Buchwald et al., 1065; Delgado, 1969, 1979; Cherkes, 1978; Karamyan, 1970; Divac, 1972; Suvorov, 1973; Arushanyan, Otellin, 1976; Cools et al., 1977; Divac, Oberg, 1979, and others). Therefore, the disturbances of the CN functions can be manifested in not only extrapyramidal disorders, but also various forms of the pathologic state of CNS, higher nervous activity, behavior, and psychic processes. hence, it seems to be rather difficult to model neuropathologic syndromes which exhibit disturbances of CN activity.

In accordance with the theory being evolved, the given problem should be solved on the basis of a single principle of producing hyperactive structures in CN; these structures act as determinants and induce the formation of pathologic systems. The peculiarities of the determinants and the respective pathologic systems depend on the specifics of the neurochemical mechanisms in CN which are involved in the pathologic process. In other words, every determinant should have its own neurochemical characteristics with which its functions are connected. Thus, opposite changes in the DA-ACh balance should result in the production of various determinants and, consequently, different pathologic systems and diverse neuropathologic syndromes. It follows that the sequence of the development of syndromes is connected with the sequence of the origin of the respective determinants. This process may depend on the distribution of the damaging factor in CN and on the different damage-resistance of the CN structures.

These methodological prerequisites also established research methods. Tetanus toxin (TT) was used as the main tool for producing the experimental models of CN syndromes. When TT is distributed in nervous tissue, it can cover diverse synapses and disturb neurotransmitter secretion (see Chapter 5). As various synapses differ in their resistance to TT, diverse neurochemical mechanisms were believed to be involved to a different extent in the pathologic process. It follows that various determinants and, consequently, the pathologic systems which they induce and the respective syndromes would probably originate at different times.

Indeed, syndromes were produced when TT was injected into the rostral part of both CN. They included stereotyped behavior, catatonia, parkinsonism, chorea-like hyperkinesis, myoclonia, and rotation. The last syndrome occurred when TT was injected unilaterally into one of the caudate nuclei. Those syndromes were formed in definite sequence and with a definite latent period, depending on the TT dose. The sequence of the origin of those syndromes can be nominally represented in the following way: after the administration of TT (most researches were carried out with doses of 1-3 DLM for rats), the first signs of stereotyped behavior originated in 18-24 hours; the first signs of catatonia originated in 36-48 hours, while parkinsonism was clearly exhibited on the fourth or the fifth day. The signs of individual syndromes were covered in this frequently observed sequence of events. This was especially true of chorea-like hyperkinesis and myoclonia, which could be occasionally observed against the background of stereotyped behavior, catatonia and parkinsonism. However, these occurrences did not greatly influence the course of other syndromes, showing that the pathologic systems which underlie various syndromes are somewhat independent.

To compare the characteristics of the models being produced, use was made of other pharmacological agents (besides TT) which have similar mechanisms of action and which produce analogous pathophysiological and clinical effects (see Chapter 5).

STEREOTYPED BEHAVIOR SYNDROME

The experimentally reproduced syndrome of stereotyped behavior (Kryzhanovsky, Aliev, 1976, 1979, 1981; Aliev, Kryzhanovsky, 1979) is exhibited in rats in regular repeated movements, including corner-to-corner or short to-and-fro runs or walks in a cage, a return to the starting point, a climb on the cage net and movement along its, slight motions of the forelegs and the head, and uniformly reproduced simple behavioral reactions expressed in intensive searching movements, the sniffing of the ground and walls of the cage, the gnawing of the cage net or constant masticatory motions, licking, overstepping with the forelegs, monotonous movements in the cage, and regular standing on the hind legs. The stereotyped behavior of some animals was exhibited in a set of movements which were regularly reproduced in a definite sequence, namely, a run to the next corner of the age, several slight search movements of the head, the digging or raking of sawdust, and then a run to the next corner, and so forth. In all the animals, the orientational search reaction was much fainter and assumed the features of stereotyped behavior.

The paroxysms of stereotyped behavior differed in their duration, some of them lasting 10-15 minutes. At the same time, stereotyped manifestations which continued without interruption for several hours were observed, and various stimuli did not affect them. Stereotypy remained in the animals for 4-7 days when small TT doses (less than 1 DLM) were administered. Large TT doses promoted intensive sterotypy with short latent periods. The main manifestations of the syndrome disappeared in 24-32 hours and catatonia developed instead of it. Individual, less pronounced stereotyped behavior (masticatory motions, slight search movements of the head) was observed against the background of catatonia.

As stereotypy developed, the animals drank less water (symptoms of adipsia or oligodipsia) and ate less food (symptoms of aphagia or oligophagia), the duration of grooming decreased, and it became irregular. When stereotypy intensified, ordinary grooming was not observed at all, or intensive sexual grooming originated in some cases.

According to modern concepts, sterotyped behavior is based on the activation of the dopaminergic nigrostriatal system (Shchelkunov, 1964; Randrup, Munkvad, 1966, 1967; Cools, van Rossum, 1970; Fog, 1972; Klawans et al., 1971). Stereotyped behavior originates when amphetamine is systematically administered as a DA agonist that enhances the synthesis and secretion of DA, reduces its uptake, and inhibits its metabolism (Fuxe, Ungerstedt, 1970; Besson et al., 1971; Teyer, Snyder, 1971; von Voigtlander, Moore, 1973; Uretsky, Shodgrass, 1977). That behavior originates also in the case of the intrastriatal injections of the stimulators of the dopamine receptors, namely, apmorphine, amantadine and bromcriptin (Ernst, 1967; Johnson et al., 1976; Clark et al., 1978; Fuxe et al., 1978), and amphetamine, dopamine and its precursor, i.e. ʟ-dopa (Ernst, Smelik, 1966; Fuxe, Ungerstedt, 1970; Fog, Pakkenberg, 1971; Costall et al., 1977).

Haloperidol Effects

The DA receptor blocking agents, e.g. haloperidol (Fog, 1967; Anden et al., 1970; Fog et al., 1971), either reduce or suppress the symptoms of pharmacologically produced stereotyped behavior, depending on their dose.

In our experiments (Aliev, Kryzhanovsky, 1979; Kryzhanovsky, 1979; Kryzhanovsky, Aliev, 1981), the intraperitoneally administered haloperidol effectively reduced the manifestation of the above syndrome of stereotyped behavior (Figure 79). The dependence of this effect on the dose was clearly revealed. When haloperidol was administered in a dose of 1 mg/kg, it greatly reduced the frequency of the recurrence of the stereotyped complexes and their intensity, lengthening the interval between them. When it was administered in large doses (2.5-5 mg/kg), it virtually suppressed the syndrome, but signs of minor catalepsy could appear in some animals.

Thus, the researches have shown that the hyperactivity of the dopamine apparatus of the neostriatum is a significant pathogenic link of stereotyped behavior in animals. Taking account of the specifics of the mechanisms of TT action, especially the fact that TT

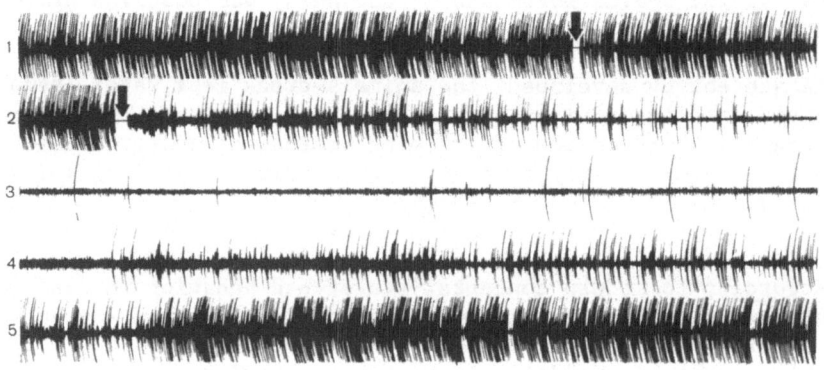

Fig. 79. Influence of haloperidol on the stereotyped behavior
         syndrome. 1: Actogram of stereotyped behavior of a rat
         before and after the solvent (control) was injected i.p.
         The moment of injection is indicated by an arrow; 2:
         gradual diminution of the frequency and intensity of
         stereotyped patterns after haloperidol was injected i.p.
         (2.5 mg/kg); 3: absence of stereotyped behavior 15-25 min
         after haloperidol was injected. Restoration of the syn-
         drome; 4: beginning of the syndrome (95-105 min after
         haloperidol was injected); 5: complete restoration of the
         syndrome (125-135 min after haloperidol was injected);
         time: 1 min.

blocks the release of the inhibitory transmitters by presynaptic terminals (see Chapter 5), it was assumed that the dopamine apparatus becomes hyperactive because its inhibition is disturbed by TT.

There is a lot of data concerning the role of inhibitory mechanisms, particularly GABA-induced inhibition, in controlling DA release in CN. Caudate nuclei contain GABAergic neurons of two types: small interneurons with branching processes inside CN and neurons whose axons form the efferent strionigral pathway (Kim et al., 1971; Otsuka, 1972). The latter represents the inhibitory neuronal feedback mechanisms, which regulate the activity of the DA nigro-striatal system (Grofova, Rinvik, 1970; Precht, Yoshida, 1971; Fonnum et al., 1973; Bunney, Aghajanian, 1976, Racagni et al., 1977). The axons of the efferent GABAergic neurons of CN form recurrent collaterals which represent proper feedback mechanisms of CN that regulate the level of neuronal activity (Szabo, 1962; Kemp, 1968; Obata, Yoshida, 1973; Richardson et al., 1977) and DA release by the caudal presynaptic terminal of the DA nigral neurons that activate GABA receptors on this terminal (Campochiaro et al., 1977). Data on the GABA effect on DA secretion in CN, being obtained in experiments with brain slices, are contradictory (Cheramy et al., 1978). But the results of the experiment carried out with an intact brain show that GABA inhibits DA release in CN (Bartholini, Stadler, 1975, 1977; Cheramy et al., 1977). DA secretion is blocked by the GABA agonist gamma-hydroxy-butyric acid (Bustos et al., 1972; Menon et al., 1974). Other inhibitory effects in CN are associated with the above-mentioned GABAergic nucleus interneurons. A part of the inhibitory CN interneurons are activated through the recurrent collaterals of efferent CN neurons (Kitai et al., 1976; Richardson et al., 1977).

Researches have confirmed the assumption that the stereotyped behavior syndrome is connected with the disturbance of GABA control of the dopamine apparatus in CN.

Effects of Penicillin and Picrotoxin

Since the stereotyped behavior syndrome is believed to be connected with the disinhibition of the dopaminergic apparatus due to the disturbance of the GABAergic mechanisms of inhibitory control under the influence of TT, experiments were carried out with other GABA antagonists, namely, penicillin and picrotoxin (Kryzhanovsky, Aliev, 1979). it was noted (Chapter 5) that penicillin not only exerts depolarizing influence, but also disturbs GABA inhibition. Picrotoxin is a specific GABA antagonist (Curtis et al., 1969; Cheromy et al., 1977b; Obata, Highstein, 1970; Precht, Yoshida, 1971). The introduction of penicillin (10-20 U) or picrotoxin ($6 \cdot 10^{-5}$M) into both caudate nuclei through microcannulae had produced a clearly expressed syndrome of stereotyped behavior with all the above-mentioned characteristic features (Figure 80). The first

symptoms originated even during the administration of the drugs and
were exhibited in brief paroxysms of intensive sniffing of the cage
and the gnawing and licking of the ground. There were jerks of the
forelegs, single myoclonia manifestations, intensive grooming of the
head, and sometimes the animals were completely motionless. Stereotyped behavior reached its peak in 10-15 minutes after the injection.
At this stage, it was expressed in licking, sniffing, search motions,
foreleg overstepping, regular passing from one corner to another,
circular motions, and backward movements. All these reactions were
very intensive. They occurred spontaneously and with definite regularity, and were observed in different combinations during diverse
periods. Such complexes originated paroxysmally. Intraperitoneally
administered haloperidol (1 mg/kg) suppressed stereotyped behavior
which was caused by either penicllin or picrotoxin.

     Thus, not only TT, but also agents which disturb GABA inhibitory
control in CN can produce a typical syndrome of stereotyped behavior.

GABA Effects

     Stereotypy was suppressed when GABA was administered simul-
taneously through implanted cannulae into both heads of CN (where TT
was administered earlier) as the syndrome was at its height (Figure

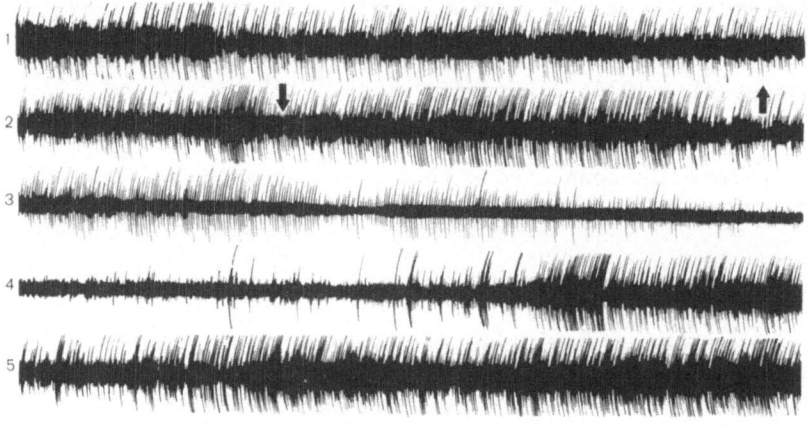

Fig. 80.   Effect of GABA injection into the rostral part of caudate
           nuclei on the stereotyped behavior syndrome. Actogram of
           the stereotyped behavior of a rat before (1) and during (2)
           the microinjection of GABA (0.2 M, 6 µl; injection·rate:
           1 µl/min). The arrows indicate the beginning and the end
           of injection; 3: first 10 min after injecting GABA; gradual
           disappearance of the syndrome; 4,5: restoration of stereo-
           typed behavior 20-30 and 30-40 min after injecting GABA;
           time: 1 min.

81) (Aliev, Kryzhanovsky, 1979; Kryzhanovsky, Aliev, 1981). In some
animals, stereotyped passages and running became less pronounced,
less intensive and less frequent, and their regularity was upset even
during the administration of GABA. At the same time, foreleg overstepping disappeared, and then the search motions of the head,
gnawing, licking, and the sniffing of the enclosure also disappeared.
Individual paroxysms of masticatory motions were observed a few
minutes after the beginning of GABA administration, but later they
were not seen.

A noteworthy fact is that grooming remained even when the symptoms of stereotyped behavior disappeared. The duration of the GABA
effect, namely, the suppression of the syndrome, varied with a change
in the GABA dose and the extent to which the syndrome developed. The
more intensive the syndrome, the more rapid was its relapse after
GABA action.

Stereotyped behavior was restored in backward order, i.e. it
began with the appearance of the paroxysms of masticatory motions,

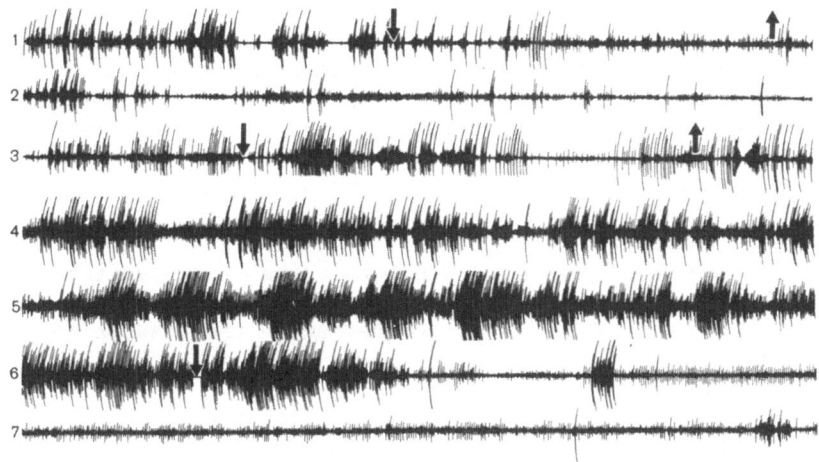

Fig. 81.   Stereotyped behavior in a rat caused by penicillin
           injection into the rostral part of caudate nuclei. Acto-
           gram of stereotyped behavior before, during (1) and after
           (2) the control microinjection of a saline (5 µl, injec-
           tion rate: 1 µl/min), and during (3) the microinjection
           of penicillin (20 IU in 6 µl of a saline, injection rate:
           1 µl/min) and 3-13 min (4) after injecting penicillin,
           when the paroxysms of stereotyped behavior appear. Pro-
           nounced stereotyped manifestations (5) 14-24 min after
           injecting penicillin (6,7). Suppression of stereotyped
           behavior by injecting i.p. haloperidol (1 mg/kg). The
           moment of injection is indicated by an arrow; time: 1 min.

which were gradually supplemented with behavioral reactions (search
motions of the head, the sniffing of the enclosure, licking, gnawing,
etc). They were followed by passages and running. The restoration
of the syndrome was, on the whole, expressed in the growing frequency
of individual steryotyped manifestations and complexes and in their
intensification.

The syndrome was suppressed whenever GABA was repeatedly administered to the same animals. The frequent administration of GABA
when the effect of its previous administration did not cease had
allowed the syndrome to be suppressed within several hours. When CN
was perfused with GABA via perfusion cannulae, the syndromes of
steryotyped behavior disappeared during the whole period of perfusion
(60 minutes) and did not originate for tens of minutes after its
completion.

The diminution of the steryotyped behavior syndrome occurred
during the intrastriatal administration of GABA whenever the syndrome
was produced by TT. The intrastriatal administration of GABA was
virtually ineffective when steryotypy was produced by penicillin or
picrotoxin.

GABA poorly passes through the hematoencephalitic barrier into
the brain. Nevertheless, it can pass to the brain in a certain
amount without losing its trophongenic and mediatory properties
(Bunyatyan, 1976; Chiflikyan, 1978). Therefore, the effect of GABA
was tested during its systemic intraperitoneal administration. In
this respect, a GABA preparation (aminalon, i.e. gammalon) was used
for systemic (intraperitoneal) administration. Investigations have
shown that GABA can suppress the steryotyped behavior syndrome produced by TT injection into both CN (Figure 82) and that the extent
and duration of the inhibiting effect depend on the dose of the
preparation.

The following peculiarity of the aminalon effect should be
noted: the animal's locomotive activity intensifies for a short time
before the symptoms of steryotypy disappear.

Diazepam Effects

Benzodiazepines are known to activate the GABAergic apparatus
and GABA control by increasing GABA synthesis (Costa, Greengard,
1975), suppressing the re-uptake of GABA by the terminal (Maisov et
al., 1976), activating the receptors and making them more affined to
GABA (Dray, Straughan, 1976; Guidotti et al., 1979), potentiating
GABA-induced inhibition (Kozhechkin, Ostrovskaya, 1977; Zakusov et
al., 1977; Gallager, 1978; Geller et al., 1978, 1980), and facilitating conduction in GABAergic synapses (Curtis et al., 1976; Polc,
Haefely, 1976). GABA increases the affinity of the benzodiazepine

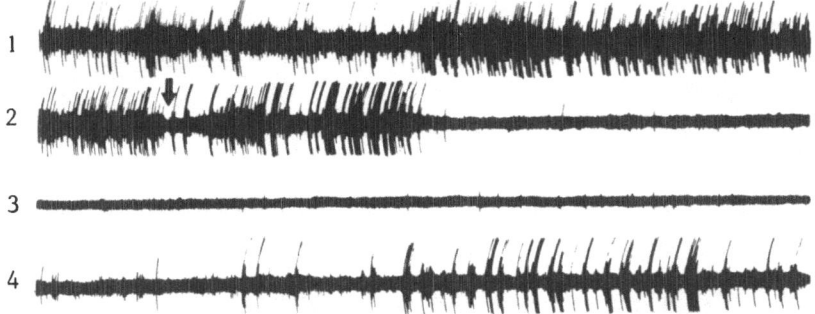

Fig. 82.   Influence of diazepam on a rat's stereotyped behavior
            syndrome. Actogram of stereotyped behavior before (1) and
            after (2) injecting i.p. diazepam (Seduxen) (2.0 mg/kg; the
            moment of injection is indicated by an arrow). Immediately
            after the injection, locomotion weakened and then briefly
            intensified, being followed by the suppression of all the
            components of stereotyped behavior. (3) 30-40 min after
            injecting diazepam, stereotyped behavior disappeared; (4)
            gradual restoration of stereotyped behavior (2.6 and 2.8 h
            after injecting diazepam); time: 1 min.

receptors (Tallman et al., 1978; Wastek et al., 1978; Gallager et
al., 1980). It was assumed that diazepam intensifies GABA inhibitory
control in the neostriatum by activating the GABAergic system.
Consequently, the hyperactivity of the dopamine apparatus becomes
less intensive, as a result of which the stereotyped behavior syndrome is eliminated. Experiments have corroborated this assumption:
diazepam (Seduxen) produced a strong inhibitory effect on that syndrome; when it was administered in relatively large doses (2m/kg), it
completely suppressed stereotyped behavior during its action (Aliev,
Kryzhanovsky, 1979; Kryzhanovsky, Aliev, 1981)
(Figure 83).

     Besides the suppression of stereotyped acts (intensive search
motions, sniffing, licking, gnawing, etc.), general locomotive
activity briefly intensified (being expressed in an increase in both
the number of passages and the walking rate) immediately after the
injection of diazepam, just as when aminalon is administered. The
intensity and duration of this effect depended on the diazepam dose.
Locomotion greatly intensified for a short time when the agent was
administered in a dose of 2 mg/kg. Later, this phenomenon disappeared. Locomotion intensified less when the diazepam doses were
smaller.

     Hence, the pathogenesis of the stereotyped behavior syndrome can
be regarded as a process which consists of the following main links:
(1) the disturbance of the GABA inhibitory control of DA secretion by

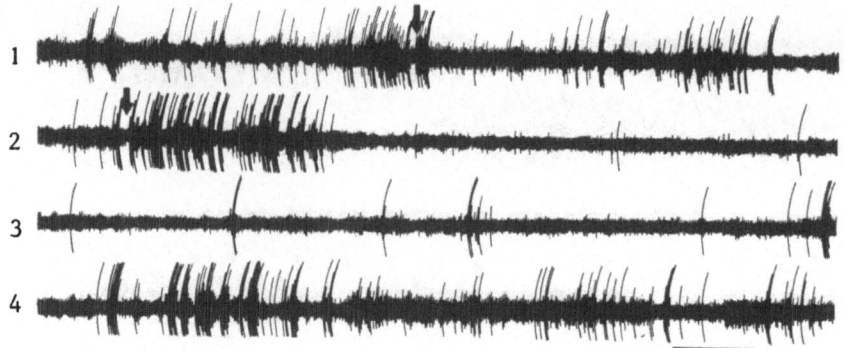

Fig. 83.   Influence of aminalone on the stereotyped behavior
           syndrome. 1: Actogram of stereotyped behavior in a rat
           before and after injecting i.p. the control solution (2%
           starch); the moment of injection is indicated by an arrow;
           2: brief intensification of locomotion directly after
           injecting aminalone (250 mg/kg) with the subsequent sup-
           pression of all the components of stereotyped behavior; 3:
           20-30 min after injecting aminalone, stereotyped behavior
           disappeared; 4: gradual restoration of stereotyped behavior
           (80-90 min after injecting aminalone); time 1 min.

the terminals of the DAergic nigral neurons in CN; (2) the disinhibition and hyperactivation of the DA apparatus of CN, pathologically enhanced secretion of DA, and (3) a change in the activity of
the postsynaptic structures, particularly the activity of the postsynaptic acetylcholine neurons, under the influence of DA whose
secretion is enlarged. The hyperactivation of the DA apparatus of
CN, i.e. the intensive, uncontrollable secretion of DA, is the
determinant structure of the pathologic system on which the syndrome
is based. The disturbance of the GABA control of the DA apparatus in
CN is a prerequisite of the origin of that determinant and the initial pathogenic link of the whole pathologic process.

     The determinant structure is suppressed or its influence atten-
uates and, consequently, the pathologic system disappears together
with the whole pathologic syndrome when that GABA control is restored
(intrastriatal administration of GABA and the systematic administration of diazepam and aminalone) or when the action of DA attenuates (systemic administration of haloperidol). This effect is
preserved until the activity of the determinant and its influence on
the postsynaptic structures are restored.

     An interesting fact is that when GABA was administered intra-
striatally, it suppressed only the form of the stereotyped behavior
syndrome which originated when TT was injected into CN. The forms of
the syndrome which were caused by the injection of penicillin or

picrotoxin into CN did not substantially change when GABA was administered intrastriatally. This fact can be attributed to the specifics of the action of the given substances: TT disturbs the release
of GABA without blocking its receptors, while the effects of penicillin and picrotoxin are associated mainly with the blockade of the
GABA receptors. Therefore, in the first case, GABA can exert its
influence, while in the second, the effectiveness of its action
substantially diminishes and depends on the amount of the free receptors which remain. Such phenomena of the different effectiveness of
the inhibitory mediators due to the neurochemical peculiarities of
the determinant are observed also as regards other syndromes (see
Chapter 13).

When diazepam and GABA were systematically administered, they
intensified the locomotive components of the stereotypy syndrome
before suppressing it. This effect was not produced when GABA was
administered intrastriatally. This phenomenon can be compared with
the data which show that small doses of amphetamine potentiate locomotive activity (Kelly, Iversen, 1975). The secretion of DA is
probably not completely suppressed at the initial stages of the
action of diazepam and GABA under the given conditions. Then the
effect of the action of its small amounts originates.

It follows from the specifics of the neurochemical organization
of CN that other mediatory mechanisms, e.g., serotoninergic and
NAergic synapses, can also be drawn into the pathologic process in
the given syndrome. This is especially evident from the data about
GABA influence on the secretion of the given mediators (Esayan et
al., 1973; Chiflikyan et al., 1978). It has been shown that the
components of stereotyped behavior originate when the serotonin
precursor, 5-hydroxytryptophan, is injected into the rostroventral
region of CN in rats (Weiner et al., 1973) and when serotonin is
injected into the rostromedial region of CN in cats (Cools, 1974).
Data show that some locomotive factors in stereotypy, like those
whose intensification was noted above when diazepam and GABA
(Aminalone) were systematically administered, are connected with the
activation of the NA mechanisms (Ungerstedt, 1970; Teyer, Snyder,
1971; Randrup, Munkvad, 1972).

CATATONIA SYNDROME

Catatonia was the next syndrome which developed after sterotypy
when TT was injected into the head of both CN (Kryzhanovsky, Aliev,
1976a). The first signs of this syndrome occurred in the form of
hypertonicity of the tail muscles, as a result of which the tail was
always either in the horizontal or vertical position. This artificial form of the tail could remain unchanged for several minutes even
when the animal moved. At the subsequent stage of the syndrome,
there were paroxysms when the animal suddenly became 'frozen' during

movement. The poses in which this occurred were separate fragments,
as it were, of the locomotive act. In the course of time, the duration of these paroxysms increased, while the intervals between them
decreased. The tests with uncomfortable postures were positive when
the animal spontaneously 'froze'. Such postures were maintained by
the animal throughout a paroxysm. Afterwards, the state of complete
catatonia developed. The animal could remain for a long time in the
position in which it was put.

The following cataleptic tests were used for estimating the
severity of the syndrome: (1) the test of standing at a horizontal
bar (the animal's forelegs were on the horizontal bar which was 12 cm
high; Figure 84 A), (2) the test of lying flat (the animal lay with
all its legs spread out; Figure 84 B), and (3) the test of an uncomfortable posture of the legs (one or two legs were on a cube which
was 3 cm high; Figure 84 C). The duration of these postures without
any change in the position characterized the severity of the syndrome. The animals could remain in these postures for a long time
(up to 10-15 minutes). The test of hanging on a horizontal bar also
characterizes the cataleptic state of animals (Figure 84 D). In
pharmacological models, an indicator of severe catalepsy is when a
rat hangs with its forelegs for more than five seconds (Temkov,
Kirov, 1971). In our model, the animal hung for more than 30 seconds. The reaction to pain stimuli (the pinching of the tail)
remained in sick animals, and they restored their normal posture when
the rolled onto their back (they preserved their vestibular reaction).

At the height of the syndrome, aphagia an adipsia developed.
The animals were given nourishment parenterally in order to prolong
their life during this period and the subsequent periods.

Different effects were produced by DA and its agonists which
were administered at the early and late stages of the syndrome. The
syndrome attenuated when ʟ-dopa (200 mg/kg) was administered intraperitoneally at the initial stage, i.e. muscular hypertonicity
diminished and the animals stood less at the horizontal bar (this
diminished from 425 ± 86.7 seconds to 239 ± 54.7 seconds; p<0.05).
At the early stages (the animal stood for 497 ± 62.1 seconds at the
bar), amphetamine (10 mg/kg) completely suppressed the syndrome in
most animals. In this case, the above-mentioned symptoms of stereotyped behavior and the phenomena of hyperkinesis in the form of
myoclonic twitches originated instead of catatonia. However, they
disappeared and catatonia was restored when the action of amphetamine
ceased. At the late stages (the animal stood for 786.8 ± 84.2 seconds at the bar), the same dose of amphetamine did not eliminate the
catatonia symptoms or engender the stereotypy symptoms, and it did
not intensify locomotive activity in any animal. However, great
myoclonic twitches occurred in all animals. At the late stages,
apomorphine (5 mg/kg) intensified the exhibition of the syndrome (the

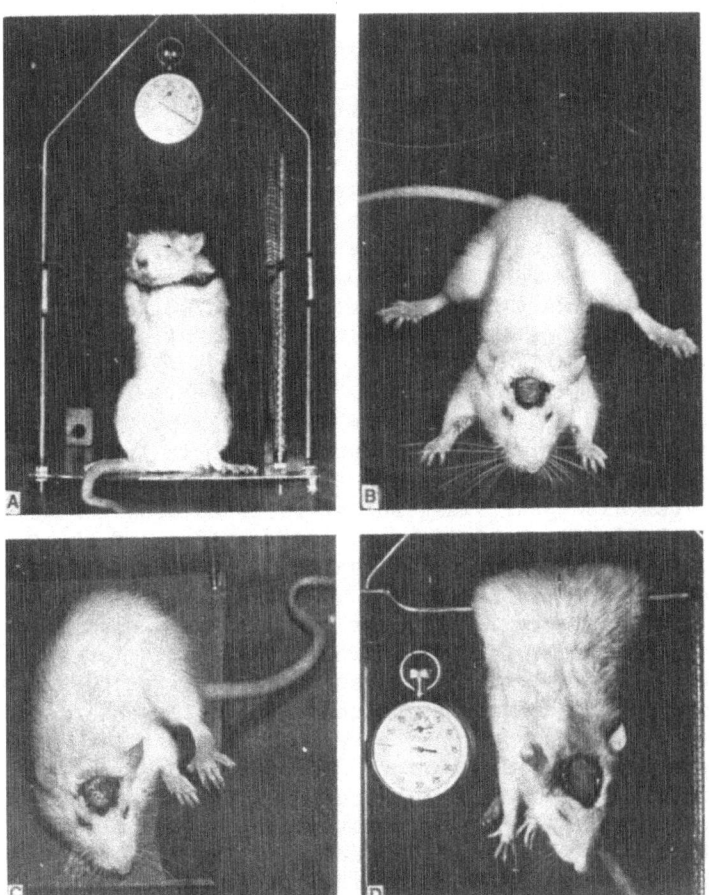

Fig. 84. Experimental catatonia in rats. A: Standing at the
          horizontal bar; B: flattening posture; C: uncomfortable
          position of the legs; D: hanging with the body on the
          horizontal bar.

animal stood at the bar 1.5 times longer and even more). The bilat-
eral intrastriatal injections of DA at the late stages (the animal
stood at the bar for 1012 ± 33.5 sec) did not greatly influence the
exhibition of the syndrome. The injection of GABA (0.2M; 6 µl)
into the rostral part of both CN and the systematic administration of
haloperidol (2.0 mg/kg) were also ineffective.

    The data above show that the given syndrome has a very com-
plicated pathogenic structure and its mechanisms apparently differ at
diverse stages of development. Catalepsy is of definite importance
in the syndrome's structure. The specifics of the clinical picture
and the possibility of using tests which are characteristic of cat-

alepsy for estimating muscular tonicity indicate that catalepsy is
present. Such tests showed that catalepsy was very severe and that
the duration of the postures in which the animal was made to stand
greatly surpassed the duration of such tests in the case of the known
pharmacological models of catalepsy. Besides the 'waxy' tonicity in
sick animals, there was rigidity which grew as the process developed.
Moreover, locomotor inhibition greatly developed and then motor
activity was suppressed and severe akinesia occurred; the animals
were indifferent to stimuli, and they refused to drink water and eat
food, etc. When a rat was taken with the hands, general muscular
tension and even rigidity could be felt; the animal could be put in
another place, and it remained 'frozen'. All these symptoms express
severe catatonia.

    As we have seen, the syndrome is suppressed by DA and its agon-
ists at the first stage of development. Hence, it is connected with
the insufficiency of the dopamine apparatus in CN. When DA is used,
the syndrome can disappear; instead, stereotyped behavior originates,
suggesting that DA may possibly be hypersecreted under the influence
of a DA injection. However, the mechanisms of the syndrome are
obviously not confined to that. The symptoms of catalepsy and its
combination with stereotyped behavior were produced when scopolamine
and DA were jointly administered intrastriatally (Fog et al., 1967)
and apomorphine was systematically administered after the animals
were treated with large doses of pilocarpine (Ahtee, Kaariainen,
1974). Thus, the DA receptors were hyperactivated together with the
postsynaptic cholinergic CN neurons. The origin of catatonic schizophrenia-like psychosis when small doses of amphetamine were regularly
administered for a long time has been described (Connel, 1958;
Warner, Pierozynsky, 1977). DA and its agonists did not reduce the
syndrome, and they could even intensify it (apomorphine) at the late
stages in our model. It has been shown that the stimulation of the
nigral DAergic neurons causes the prevalence of the stimulating
reactions in CN when CN or its stimulating inputs are activated
(Liles, 1974; Norcorss, Spehlman, 1978). Account should also be
taken of the fact that DA can exert diverse influence on the neurons
in the regions of the axonal colliculus (hyperpolarization) and the
dendrites (depolarization), that receptors which realize the inhibitory and exciting effects of DA are present in CN (Cools, van Rossum,
1976), and that different DA receptors and other specifics of DA
action exist (Iversen, 1975, 1977; Klawans, Weiner, 1976; Kebabian,
Calne, 1979; Hyttel, 1980). Hence, a set of mechanisms can participate in the pathogenesis of the syndrome at the height of its development. The hyperactivation of the dopamine apparatus is apparently
important in this respect. However, this mechanism is not the only
one and is not decisive because, as we have already seen, haloperidol
does not eliminate the syndrome. Account should be taken of the fact
that the DA effects may be distorted due to a change in the receptor
apparatus and the postsynaptic neurons. The high resistance of the
syndrome to the pharmacological drugs used coincides with the data of

the clinical experiment, which shows how difficult it is to pharmacologically treat the catatonic syndrome. An interesting fact is that the catatonic stage can be produced experimentally by injecting, into CN, substances which cause the formation of the foci of epileptic activity, i.e. the generators of pathologically enhanced excitation (GPEE). They are cobalt (Mutani et al., 1968), tungstic acid (Spiegel et al., 1965), manganese salts (Inoue et al., 1975), and cholinergic preparations in large doses. That stage disappears when the zone of chemical action is coagulated (Szekely et al., 1969). Classical bulbocapnine catatonia is also connected with the hyperactivation of the CN formations. The syndrome disappears when that hyperactivity is suppressed by intrastriatally injecting potassium chloride (Still, Sayers, 1969). These data show that the population of hyperactive neurons, which constitute GPEE, participates in the formation of the determinant in CN in the case of a catatonic syndrome.

PARKINSON'S SYNDROME

Parkinson's syndrome is manifested at the latest stages of the pathologic process which originates when TT is injected into the rostral part of both CN. The reproduction of the syndrome is influenced by the difficulties of animal maintenance that are due to aphagia, adipsia and other disturbances. Adipsia and aphagia are said to be characteristic features of experimental Parkinson's syndrome, which is caused by the destruction of DA neurons in substantia nigra (Ungerstedt, 1971b). In some animals, posture asymmetry may originate apparently due to the unsymmetrical development of the processes in both CN. Consequently, the syndrome's development is upset. Then, it becomes difficult to assess the syndrome. However, experimental parkinsonism with all the syndrome's characteristic features has been produced in most animals (Kryzhanovsky, Aliev, 1976b, 1978).

The symptoms which are characteristic of parkinsonism were at first manifested against the background of the catatonia syndrome which originated earlier on the 3–5th day, depending on the TT dose. In this case, general motor activity was substantially reduced. Movements were scarcely made during rare locomotive manifestations, being expressed in paroxysms in the form of brief walks with long pauses at every stage. Such impulsive locomotion intermingled with long pauses had either originated spontaneously or was produced by touching the animal or making a sharp sound (clicks, etc.). Most animals could not turn and bypass obstacles. When a block 1.5–2 cm high was put in front of the rats as an obstacle, many of them tried to overcome it, but once they did so, they again fell into an akinesic state. Their walking was greatly disturbed, and it was slower, the step was shorter, and movements were rigid. Later, a stage of profound akinesia developed in most rats. This stage was difficult to interrupt even by strong nociceptive stimuli (pinching the tail and applying a hot plate to the tail).

An increase in muscular tonicity during catatonia passed into rigidity which was manifested in the animals' typical posture: the back arched and gibbosity appeared (Figure 85 A). Gibbosity is regarded as a characteristic feature of experimental parkinsonism, which is produced by pharmacological agents (Jurna, 1968, 1976) (Samoilova et al., 1973). Besides gibbosity, head flexion originated in the animal. In this case, the head bent towards the body, and the legs were in an abnormal position (the forelegs were flexed and brought to the body, while the hind legs were somewhat straightened out). Rigidity was determined also by resistance when the hind legs were passively moved.

Tremor was another characteristic feature of the syndrome. At first, it involved the face and the forelegs (Figure 85 C 2). Amplitude of the tremor varied from unnoticeable tremor which was revealed when static tension was produced on an extremity to regular, sweeping visual tremor. At the early stages of the syndrome, certain sensory stimuli (touching the animal, pinching the tail, clicking which can be heard) influenced the amplitude of tremor, intensifying it. This effect was absent at the late stages of the syndrome. Tremor either diminished or even completely disappeared during sleep. Muscular rigidity, including gibbosity, also diminished.

The characteristic parkinsonian status was supplemented with heightened tonicity of the face muscles: normal movements of the vibrissae and sniffing movements were absent, the nictitating reflex was slow and, in some cases, ptosis was observed. Hence, the face expression could be equated to oligomimia or amimia in man.

Besides those symptoms, there were also severe vegetative disturbances in the form of hypersalivation, diarrhea, and dystrophic changes in the hair covering (it was yellowish and the hair intensively fell out). Vocalization was greatly disturbed, and it even completely disappeared in some animals.

The following symptom was characteristic: locomotor activity could be vehemently manifested when an attempt was made to take a rat for feeding or during other manipulations. This symptom can be regarded as an analogue of the phenomenon which is clinically known as 'paradoxical akinesia', i.e. the disappearance of rigidity and akinesia and the appearance of motor activity during strong emotions and stress influences in patients. Such a phenomenon was observed by Ungerstedt and associates (1973) in rats with experimental parkinsonism caused by 6-OHDA. They also proposed a test for that phenomenon, namely, swimming: when a rigid animal is put in water, it actively moves, and if it swims in a labyrinth, it displays sensory perception. We confirmed this phenomenon: rats with experimental park-

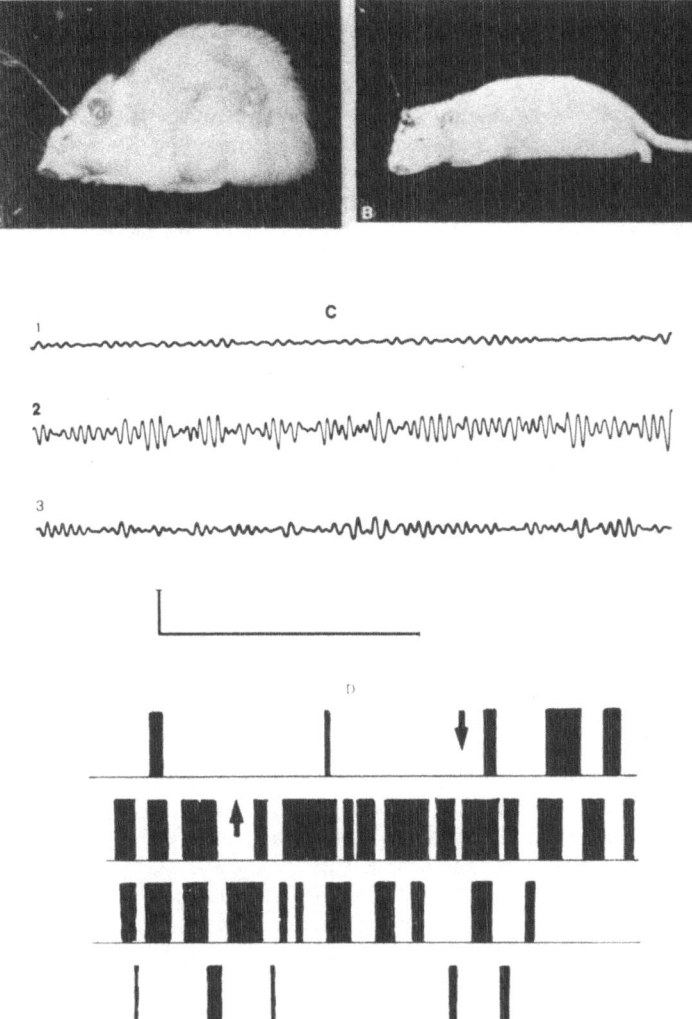

Fig. 85.   Parkinson's syndrome in rats and dopamine effects. A: Rat
           with Parkinson's syndrome: gibbosity (manifestation of
           rigidity) and akinesia; B: the same rat after injecting
           dopamine (200 μg in 5 μl; injection rate: 1 μl/min)
           into the rostral part of both caudate nuclei; disappearance
           of rigidity and akinesia; C: foreleg tremor of a control
           animal (1) and an animal with Parkinson's syndrome before
           (2) and after (3) injecting dopamine; D: motor activity in
           rats with Parkinson's syndrome before and after injecting
           dopamine. Black columns: duration of motor activity
           (movement in the cage). Arrows: beginning and end of
           dopamine injection; time: 1 min.

insonism and severe akinetorigid manifestations that were put in
water (+37°C) also began to swim, making rather coordinated and
purposeful movements.

Hence, all the above features of experimental parkinsonism
correspond to the clinical symptoms of parkinsonism observed in man.

Such features have been observed also in other experimental
models of parkinsonism, e.g. the parmacological model produced by
neuroleptics (Jurna, 1968; Steg, 1972; Buus, 1973; Samoilova et al.,
1973; Shavolina, 1978), and when the DA neurons of substantia nigra
were selectively damaged by 6-OHDA (Ungerstedt et al., 1973).

The insufficiency of the dopamine apparatus in CN is the main
pathogenetic mechanism in clinical (in man) and experimental parkinsonism, regardless of etiology. This insufficiency can be of
wither a presynaptic origin (damage done to the DA-containing nigra
neurons or their terminals and the disturbance of the synthesis and
secretion of DA) or a postsynaptic nature (DA receptor blockade).

In our parkinsonian model, the insufficiency of the DA apparatus
of CN is also an important pathogenetic factor. The introduction of
DA (150-200 µg in 4-5 µl) through implanted cannulae into the
rostral part of both CN greatly reduced the syndrome as a whole and
completely suppressed some it is symptoms (Kryzhanovsky, Aliev,
1978). A few minutes after microinjecting DA into CN, motor activity
was restored, walking was quicker, the length of the steps increased,
the length of the walk was longer (Figure 85 D), and rigidity was
greatly reduced, as a result of which gibbosity, the drawing of the
head towards the body, and the anomalous position of the legs disappeared (Figure 80 B). At the same time, tremor was reduced (Figure
85 C 3). That DA effect continued for 45-60 minutes under the given
conditions, and then the syndromes of parkinsonism gradually
reappeared: general motor activity diminished, rigidity (gibbosity)
appeared, the tremor of the forelegs and face intensified, etc.

These experiments show that the insufficiency of the DA apparatus, being the pathogenetic basis of the given model of parkinsonism,
is not connected with the blockade of the DA receptors on the postsynaptic neurons, since dopamine which was injected into CN produced
its specific effect. Taking account of the mechanism of the action
of TT, i.e. its ability to disturb the secretion of the transmitters
(see Chapter 5), it can be assumed that the insufficiency of DA in
exactly due to that mechanism, namely, to the disturbance of DA
secretion by the presynaptic terminals caused by TT.

The precursor of dopamine, ι-dopa, is known to be widely used
clinically for treating parkinsonism. Its administration (250 mg/kg
intraperitoneally) to rats with Parkinson's syndrome caused the
attenuation of akinesia and almost the complete disappearance of

muscular rigidity. The influence on tremor varied: the amplitude of tremor increased during the first minutes, but then it decreased slightly. A similar but more pronounced effect was produced by the administration of Nakom (the combined preparation of ʟ-dopa and carbidopa decarboxylase, an inhibitor of extracerebral dopa) (Cotzias et al., 1969; Papavasiliou et al., 1972). This preparation (ʟ-dopa, 100 mg/kg and carbidopa decarboxylase, 10 mg/kg) greatly attenuated akinesia and sharply reduced the tremor amplitude. The systematic administration of amphetamine (5 mg/kg intraperitoneally), which intensifies the release of DA (see above), had attenuated akinesia, rigidity and tremor, i.e. all the main components of the syndrome. At the same time, the phenomena of stereotyped behavior originated in some rats. These effects were especially pronounced at the early stages of parkinsonism. The results of these experiments show that TT caused the reversible disturbance of DA release that could be overcome by stimulating DA secretion with amphetamine. The origin of the stereotyped behavior phenomena show that either DA secretion can be substantially increased or the sensitivity of the postsynaptic DA receptors can be enhanced.

The ACh-ergic neurons of CN are controlled by DAergic neurons of substantia nigra (Butcher, Butcher, 1974; Sethy, von Woert, 1974; Agid et al., 1975; Bartholini et al., 1976). This tonic inhibitory control is manifested in a change in ACh secretion (Bartholini et al., 1976), its utilization (Stadler et al., 1973; Sethy, von Woert, 1974; Agid et al., 1975), and the turnover rate (Garenzi et al., 1975; Guyenet et al., 1975; Ladinsky et al., 1975; Trabucchi et al., 1975). DA is believed to be an inhibitory transmitter in CN, particularly as regards cholinergic neurons. The agonists of DA also inhibit these neurons, while antagonists activate them. Most of the CN neurons react by inhibition when DA is iontophoretically applied to them (Bloom et al., 1965; Hornykiewicz, 1966; McLennan, York, 1967; Felts, 1969; Connor, 1970; Gonsales-Vegas, 1974; Siggins et al., 1974; Krnjević, 1975; McCarthy et al., 1977), just as it is released by the activation of the substantia nigra neurons (von Voigtlander, Moore, 1971a, 1971b; Buchwald et al., 1973). In addition, DA seems to activate some neurons of CN (Cannor, 1970; Richardson et al., 1977). DA-activated neurons are believed to be inhibitory interneurons which limit the activity of the efferent neurons of CN (Frigyesi, Purpura, 1967; Feltz, Albe-Fessard, 1972; Richardson et al., 1977).

Taking account of these data, it could be assumed that (this was mentioned) the loss of DA influence should disinhibit the neurons and, consequently, cause the formation of GPEE from the population of these neurons. Indeed, investigations (Aliev et al., 1981) have shown that the background impulse activity of the CN postsynaptic neurons intensifies when Parkinson's syndrome develops. This effect is expressed in an increase in the number of active points (as compared with the number that usually exists) in which background ac-

tivity is recorded (Figure 86) and in the frequency of impulse activity in the active points by 4-7 times (Figure 87). The activity
of several neurons was detected and high-amplitude and high-frequency
discharges were recorded in the active points. These discharges were
not observed in the control animals (healthy rats and rats into which
inactivated TT was injected). The origin of such discharges showed
that the activity of individual neurons was synchronized and that
they formed functionally simple hyperactive micropopulations. These
phenomena were found in the zone where TT was injected.

The results of these investigations are in accord with the data
given by other authors (steg, 1969; Ohye et al., 1970; Spehlman,
1975; Ungerstedt et al., 1975; Schultz, Ungerstedt, 1978), who observed an increase in the frequency of evoked and background impulse
activity in experimental DA deficient states which are characterized
by the phenomena of parkinsonism.

The population of cholinergic neurons probably constitutes the
basis of GPEE, which originates in CN when there is an insufficiency
of DA. This explains why cholinolytics are effective, as has long
been known, in parkinsonism. Rigidity and akinesia were substantially reduced when benzohexol hydrochloride (systemic administration,
1.5 mg/kg) was used in the given model of parkinsonism. But when
cholinolytics are systematically administered, they act on different
structures of the central nervous system, and the specific value of
their influence on CN int he general therapeutic effect cannot be
established. An investigation of the effects of the intrastriatal
administration of atropine on the given model of parkinsonism has
shown that the phenomena of parkinsonism are greatly reduced when
atropine is injected into the GPEE region of both CN. The akinetorigid phenomena diminished, being expressed in an increase in motor
activity and a decrease in gibbosity, even during the injection of
atropine (200 µg in 5 µl). A little later, these symptoms of
parkinsonism completely disappeared and the animals began to walk
almost normally. The influence of atropine on tremor was not uni-
form: the amplitude of tremor increased when the drug was being
administered, and then it decreased and dropped below the initial
level, although the difference was not great. The effect of atropine
after every microinjection lasted, in general, for 1.5 - 2.5 hours.

The effects of GABA when it is administered intrastriatally have
been studied while taking account of the possibility of the inhibition of the ACh-ergic neurons of CN by GABA (Herz et al., 1970;
Curtis, Johnston, 1974; Yarborough, 1975). The microinjections of
GABA (0.2 mM in 6-7 µl) into the GPEE region of both CN produced
different results: in some animals, all the three main syndromes
(akinesia, rigidity and tremor) were expressed less, in others, only
akinesia and rigidity were reduced, and in still others, only rigidity was weakened. Tremor changed the least, and rigidity, the most.

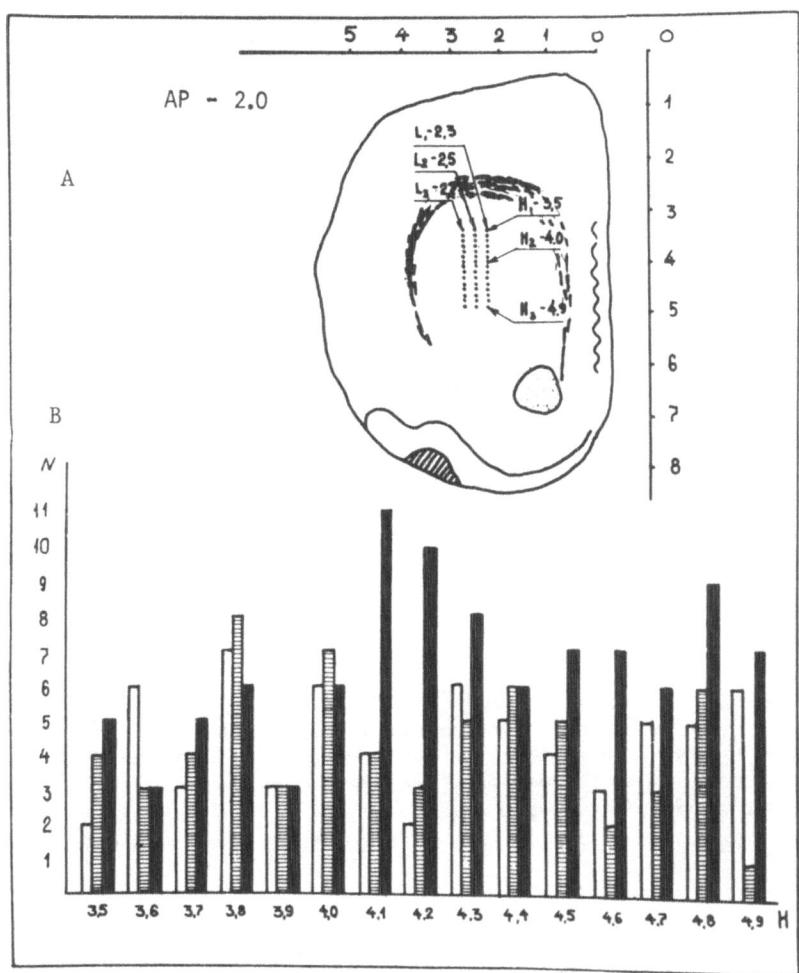

Fig. 86. Cerebral frontal cross-section passing through the head of
          the caudate nucleus. A: Electrode recording tracks; B:
          histogram of the distribution of the number of active
          points in the given region of the potential recording.
          Abscissa: track levels (mm). Ordinate: number of active
          points. Black columns: animals with Parkinson's syndrome.
          Shaded columns: control animals (injection of inactivated
          TT). Light columns: healthy animals.

     Taking account of the possible role of the disturbance of GABA
control in the pathogenesis of parkinsonism, it should be noted that
the reversible suppression of the key enzyme of GABA synthesis,
glutamate decarboxylase, in CN was detected in chronic parkinsonism
(Lloyd, Hornykiewicz, 1973). It has been shown that the disturbed
GABAergic mechanisms can be restored when ʟ-dopa is used in therapy

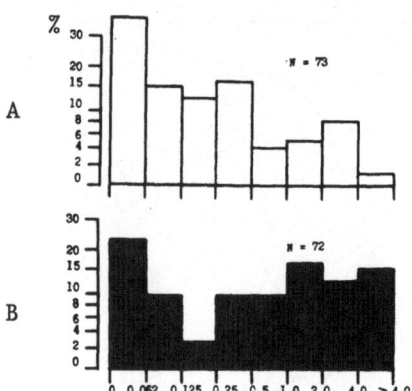

Fig. 87. Histogram of the distribution of neurons of the caudate
          nucleus according to background impulse frequency in
          healthy animals (A) and in animals with Parkinson's syn-
          drome (B). Abscissa: frequency (impulses per second).
          Ordinate: number of neurons (%); n: total number of neurons
          recorded.

(McGeer et al., 1971; Lloyd, Hornykiewicz, 1973; Achar et al., 1976).
These data suggest that the disturbance of GABA control in CN can be
one of the pathogenic links of the determinant (or determinants) in
parkinsonism. The disinhibition of the ACh-ergic neurons of CN and
the formation of GPEE from these neurons due to the deficiency of DA
may be a sign of the insufficiency of both DA control and GABA
control.

     Procaine hydrochloride was also used to suppress GPEE. Its
microinjections (20 μg/μl, 5-6 μl into the GPEE region of both CN
greatly suppressed akinesia and rigidity: gibbosity disappeared, the
number and duration of the walks increased, chewing movements appeared (they were occasionally very long and resembled stereotypy),
and the vibrissa motion was restored. However, the tremor amplitude
diminished only slightly.

     Thus, the given model of Parkinson's syndrome is pathogenetic-
ally based on (1) the insufficiency of DA and (2) the formation of
GPEE from the cholinergic CN neurons which were disinhibited. These
changes determine parkinsonism. Hence, the GPEE formed the cholinergic neurons acts as a hyperactive determinant structure.

     According to some authors (Roitrub et al., 1973; Roitrub,
Oleshko, 1977), DA activates acetylcholinesterase in CN. Therefore,
it can be assumed that the deficiency of DA in CN in parkinsonism and
particularly in this model can promote the hyperactivation of the
cholinergic neurons of CN and the formation of GPEE from the neurons.

An interesting fact in this respect is that dopa increases the activity of blood acetylcholinesterase, which is reduced in patients
with parkinsonism (Roitrub et al., 1979).

It follows from the investigations and the available experimental and clinical data that parkinsonism is not merely a complex
syndrome, but a set of syndromes, each of which has its own pathologic system. Three main features of parkinsonism, i.e. akinesia,
rigidity and tremor, are the clinical manifestations of three pathologic systems with their specifics and individual determinants,
although the above-mentioned changes, namely, DA deficiency and the
formation of GPEE from the postsynaptic CN neurons, are the primary
and general determinant of parkinsonism. Indeed, these changes
constitute the general primary determinant on which the origin of
both various secondary determinants and the pathologic systems that
are characteristic of each of them depends. This is evident from the
fact that the intrastriatal administration of DA, while specifically
suppressing the primary determinant, suppresses all the three main
features of the given model of parkinsonism at the earlier stage of
the processes. Each of these pathologic systems reacts differently
to other pharmacological influences, showing that the systems are
relatively independent. However, the system of pathologic tremor is
even somewhat antagonistic towards other pathologic systems (akinesia
and rigidity), since tremor may not only remain, but also intensity
under the action of substances which suppress the other systems
(being expressed in a diminution of both akinesia and rigidity).
Apparently, the system of pathologic tremor is also not homogeneous,
since nocturnal sleep and its influence on tremor show that there are
two pathogenetically different groups of patients with trembling
forms of parkinsonism (Wein et al., 1981).

Several clinical observations show that tremor is suppressed
under the influence of substances which block the adrenergic structures, particularly when the beta-blocking drugs are used (Strang,
1965; Abramsky et al., 1971; Dowzenko et al., 1976). Adrenaline and
noradrenaline are known to intensify tremor. According to some
concepts, parkinsonian tremor is connected with the hyperactivity of
the central adrenergic structures of the brain (Lavy et al., 1974).
An interesting fact is that our investigations have shown that lithium salt, which together with other effects, disturbs the synthesis
and secretion of catecholamine (see Chapter 13), had reduced the
amplitude of tremor in our model of parkinsonism.

The nature and specifics of GPEE activity in CN in parkinsonism
should be studied further. It is necessary to ascertain whether GPEE
is homogeneous neurochemically, i.e. whether it consists of only the
ACH-ergic neurons or constitutes a set of chemically different GPEEs
which correspond to different determinant structures.

The peculiarities of the activity of GPEE in CN and its relationship to parkinsonism at different stages of a pathologic process should be especially investigated. The number of background
active neurons in CN diminishes during the first days after the
nigral neurons are selectively damaged, but later their number grows,
reaching the maximum within a few weeks (Ungerstedt et al., 1975;
Siggins et al., 1976; Schultz, Ungerstedt, 1978), and then akinesia
and rigidity are most severe. However, heightened background activity and engendered activity of the CN neurons diminish with time
(Spehlman, 1975); such activity can reach the normal level one year
after the complete loss of the nigral neurons, although Parkinson's
syndrome is preserved (Schultz, Ungerstedt, 1978). Before drawing
any conclusions about this paradox, it is necessary to be convinced
that GPEE really disappears at the late stages of parkinsonism. Its
activity may probably assume other forms which can be revealed by
special tests. However, if GPEE disappears in CN while parkinsonism
remains, it may mean that new hyperactive structures (secondary
generators) originate in other parts of the central nervous system
under the influence of the primary GPEE, participating in the realization of akinesia, rigidity and tremor. Such a situation had
already been observed during the development of the central pain
syndrome when the primary GPEE in the spinal dorsal horn induced the
secondary GPEE in the thalamus and then disappeared (see Chapter 9).
An interesting fact is that the neuronal populations of the VL-
thalamus in parkinsonism generate discharges which are synchronized
with the frequency of tremor; these discharges disappear when the
animal falls asleep and originate when it wakes up (Albe-Fessard et
al., 1966). The coagulation of such VL zones causes the disappearance of tremor, while their electric stimulation intensifies tremor
(Cooper, 1970). Therefore, these zones are regarded as the central
link of the tremorogenous system (Raeva, 1977) or as the 'central
generator' of tremor (Calne, 1970). The change in the frequency and
amplitude of tremor in various muscles and other phenomena of the
instability of tremor cannot be regarded as a refutation of the
possible role of the main link (GPEE) in the mechanisms of tremor
since, as has been shown (Chapter 1), the functional message from the
determinant structure can be transformed in the intermediate and
terminal links of the system.

According to the general theory of the determinant structures
(see Chapters 1, 4), hyperactive structures may be formed in other
parts of the pathologic systems when the systems work for a long
time. Such structures act as new determinants which sustain and even
intensify the pathologic systems. This regularity is manifested also
in other syndromes (pathologic pain [Chapter 9] in epilipsy [Chapter
8], etc.) and is of a universal nature. It is connected with the
viability of the pathologic systems, their resistance, and the diffi-
culty of eliminating them. Therefore, just as in other syndromes,
intervention may be unsuccessful at the late stages of a process when
it is made into structures constituting the initial determinant so as

to eliminate the pathologic system. Moreover, account should be
taken (see Chapters 1, 4) of the fact that the pathologic system is
stabilized with time especially when a chronic process had been
occurring for a long time; being a very stable organization, such a
pathologic system needs far less sustaining influences of the initial
determinant.

ROTATION SYNDROME

The rotation syndrome is caused by the unilateral affection of
CN. Its different forms can be regarded as a simplified model of the
above CN syndromes of stereotyped behavior, catatonia and parkinsonism.

The models of the rotation syndrome have been produced by influences on the DA mechanisms of one of the CN (Fog et al., 1968;
Ungerstedt et al., 1969, 1978; Iversen, 1971; Heal et al., 1980;
Heikkila et al., 1981). When DA is injected into CN, rotations occur
contralaterally with respect to the injection site. Such an effect
is produced also when the neurons of substantia nigra are electrically stimulated. When these neurons are destroyed, at first contralateral rotations occur (due to the intensive release of DA at this
stage), being followed by ipsilateral rotations (caused by the cessation of DA secretion). Such ipsilateral rotations occur also when CN
is coagulated or neurologics are intrastriatally injected on one
side. The rotation syndrome can occur also when CN is directly
electrically stimulated and when substances which interfere with the
cholinergic and serotoninergic mechanisms are injected into CN
(Costall et al., 1972; Waddington, Crow, 1979).

In our investigations, the rotation syndrome originated in
animals when TT was unilaterally injected into the rostral part of
the CN head (experiments were carried out by Aliev and Guskov). The
rate at which the syndrome developed and its severity depended on the
TT doses, just as in the case of other above-mentioned syndromes.

At the initial stage of the syndrome's development, most animals
moved in the direction opposite to the TT injection site (contralateral type of rotation). These movements consisted mainly in
stereotyped walks around the pen. The direction of movements changed
in some animals: rotations were made towards the left and then towards the right. As the syndrome developed, the number of ipsilateral rotations increased, their radius diminished, and they became the
main manifestation of disturbed behavior. During this period, the
rats mainly used their forelegs for making turns, while their posterior part of the body and their hind legs could remain stationary
(Figure 88). Such rotations occurred spontaneously and were paroxysmal. They could also be provoked by different stimuli (by touching

the animal, pinching its tail, clicking, etc.). The rate of individual rotations and their total number increased as the syndrome
developed.

Asymmetry was greatly expressed during the intervals between the
animals' rotations: their head, body and tail were turned towards the
TT injection site, their fore and hind contralateral legs were drawn
apart, and the ipsilateral legs were kept flexed towards the body.
At the late stages of the syndrome's development, some animals turned
around lengthwise due to quicker rotation, rapidly returning to their
initial position. Such a phenomenon was observed when the TT doses
were relatively small. Muscular tonicity unilaterally increased, the
nictitating reflex was suppressed, and the movements of vibrissae
were either completely or partially suppressed in animals with a
strongly expressed rotation syndrome. Some vegetative disorders
which were mentioned above also occurred in this respect.

When the development of the rotation syndrome is compared with
the above syndromes caused by the injection of TT into both CN,
roughly the following relationships are observed: rotation in the
contralateral direction or alternative rotations correspond by the
time of their occurrence to the initiation of stereotypies; an increase in the number and rate of contralateral rotations corresponds
to the phase of the catatonic syndrome, and the complete disappearance of the contralateral rotations and the further increase in the

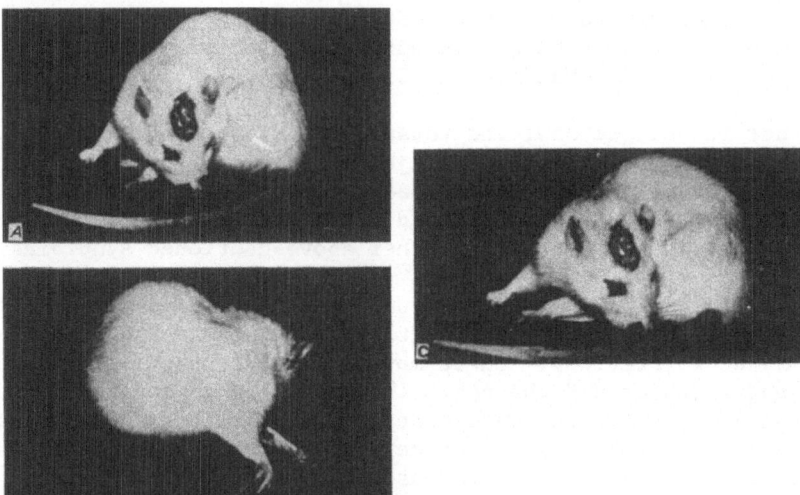

Fig. 88.   Rotation syndrome in rats. Fragments of full rotation
           performed in one place to the left; one (A), two (B) and
           three (C) days after TT was injected (1.5 DLM) into the
           left caudate nucleus.

rate of ipsilateral rotations, in the frequency of their origin and
in the duration and extent of asymmetry correspond to the initiation
of Parkinson's syndrome. Such a relative subdivision corresponds, in
general, to the pathogenic mechanisms of the above CN syndromes. It
coincides with the peculiarities of diverse forms of the rotation
syndrome reproduced by other authors by hyperactivating or hypo-
activating DA and the cholinergic mechanisms of CN.

CHOREA-LIKE HYPERKINESIS

    Spontaneous fits of peculiar myoclonia occurred in some animals
when TT was injected into the rostral part of both CN (Kryzhanovsky,
Aliev, 1976a). Three forms of hyperkinesis could be singled out with
respect to the extent and peculiarities of the manifestations. In
slight myoclonia, the animal threw its head backwards, at on its hind
legs and swung its forelegs in front of its face. When the syndrome
was of medium intensity, these phenomena were supplemented by body
movements which were like rhythmic bows, and the animal nodded as it
bowed. In strong fits, the animal fell on its back, and its hind
legs were involved in rhythmic jerks like convulsions.

    In some animals, hyperkinetic movements were manifested in
individual jerks of the head, the fore and hind limbs, the body, and
the face muscles. Tic-like jerks of the head upwards were the most
common form of hyperkinesis. These jerks could occur with intervals
of three or four seconds. The rarer throwings of the head backwards
were the most intensive movements. Tic-like movements of the head
also occasionally occurred together with the jerks of the forelegs.
In some animals, there were asynchronous jerks of the head, the legs
and the body in different combinations. This form of hyperkinesis
could originate either spontaneously or during the dynamic development of the above syndromes. At the early stages of the pathologic
process, those phenomena occurred together with the elements of
stereotyped behavior, while at the late stages, with the akinetorigid
symptoms.

    The above forms of hyperkinesis can be regarded as the model
equivalents of Huntington's chorea (the so-called rapid form), which
is expressed clinically in individual contractions of separate groups
of muscles that occur in the form of brief, rapid, uncoordinated
movements and jerks (convulsions), which involve the muscles of the
face, legs, body, etc.

    Chorea-like hyperkinesis could be reproduced by microinjecting
picrotoxin and penicillin, which disturb GABA inhibition, into both
CN. Hyperkinetic phenomena occurred rapidly in these models. Their
first features could be observed even during drug administration.
They reached the maximum a few minutes after the beginning of an
injection. The rate and intensity of the syndrome's development

depended on the doses. In these experiments, the significance of the
injection site was also clearly revealed. For instance, chorea-like
hyperkinesis occurred intensively in all animals in the form of
numerous jerks of the legs, the head, the body and the tail when
picrotoxin ($6 \cdot 10^{-5}$M; 4 µl) was injected into the medial part
(frontal plane, AP-2.0) of CN.

A pharmacological analysis of the chorea-like syndrome produced
by microinjecting TT into the rostral part of both CN has shown that
GABA (0.2 M; 5 µl) injected into both CN (in the TT injection
region) clearly suppressed hyperkinesia, i.e. the number of jerks
recorded in five minutes diminished from 234.3 ± 47.7 before the
administration of GABA to 88.7 ± 6 after its administration (in 10-15
minutes). The amplitude of jerks also decreased. Haloperidol (1.0
mg/kg intraperitoneally) likewise reduced hyperkinesis (the number of
jerks diminished from 206.6 ± 42.9 to 139.3 ± 17.7), but to a smaller
extent. This was also true of lithium chloride (50 mg/kg intraperitoneally; the number of jerks diminished from 208.0 ± 40.8 to 128.7 ±
21.7) and diazepam (0.25 mg/kg intraperitoneally; the number of jerks
diminished from 319.2 ± 41.4 to 227 ± 36.8). The combined use of
diazepam and lithium chloride in the given relatively small doses
effectively suppressed the syndrome (the number of jerks diminished
from 290.5 ± 37.1 to 25.2 ± 10.2).

These investigations clearly show that the GABAergic and DA
links participate in the pathogenesis of the given syndrome of choreiform hyperkinesis. Taking account of the mechanisms of the action
of the agents used and their effects, it can be assumed that the
disturbances of GABA control and the hyperactivation of the DA apparatus are the pathogenic links of the syndrome that constitute its
determinant structure.

In general, these mechanisms are close to those which are observed in the stereotyped behavior syndrome. The similarity of the
methods of reproducing both forms of the pathologic state indicates
this.

Morphologically, chorea is characterized by the degeneration and
loss of most of the small and part of the large CN neurons (Bruyn,
1968), and neurochemically, it is characterized by a great decrease
in both the content of GABA and acetylcholine and in the activity of
the enzymes which synthesize them (McGeer et al., 1973; Perry, 1973;
Bird, Iversen, 1974; Achar et al., 1976; McGeer, McGeer, 1976;
Iversen, 1978; Kim et al., 1980). The model of chorea-like hyperkinesis produced by microinjecting caincic acid into CN is the best
comparison of Huntington's chorea with respect to pathohistology,
neurophysiology and phenomenology (Olney et al., 1975; McGeer,
McGeer, 1976). Under these conditions, mainly small and partly large
neurons became damaged and disappeared, and the GABA content of CN
considerably decreased.

At the same time, chorea-like hyperkinesis can originate in cats when DA is microinjected into the rostral part of CN (Cools, 1972). Klawans and Weiner (1974) believe that the state produced by systematically injecting large doses of amphetamine to animals is an adequate model of choreiform hyperkinesis. The neurochemical changes detected in the brain of patients suffering from schizophrenia are similar in several respects (direction of the changes in the GABA, Ach and DA mechanisms) to the changes which were described in Huntington's chorea (Bird et al., 1977). In chorea, hyperkinesis attenuates when the DA mechanisms are suppressed and intensifies when they are activated (Divry et al., 1959; Kempinsky et al., 1960; Cohen, 1962; Gerstenbrand et al., 1963; Vaisberg, Saunders, 1963; Bruck et al., 1965; Klawans et al., 1972; Lal et al., 1973). Amphetamine may reveal latent Huntington's chorea (Klawans, Weiner, 1974, 1976). These data are in accord with the results of the employment of cholinergic and anticholinergic agents. The former weaken choreic phenomena (Klawans, 1973; Casey, Denney, 1974; Klawans, Rubovitz, 1974; Miller, 1974; Klawans, Weiner, 1976), while the latter intensify it (Crane, 1968; Klawans, McKendal, 1971).

However, when the given neurochemical mechanisms are relatively similar, stereotyped behavior and chorea as well as their experimental models are different syndromes. They differ by their clinical picture and, consequently, by their pathogenetic mechanisms. In chorea, no change has been observed in the DA content of CN (Ehringer, Hornykiewicz, 1960; Berheimer et al., 1973; Bird, Iversen, 1974; McGeer, McGeer, 1976). The DA content also does not change in the given experimental model of chorea produced by injecting caincic acid into CN. In this respect, it is believed that the effects of the relative hyperactivity of the DA apparatus can be connected with a change in the reation of the CN interneurons to DA (Klawans, 1973) or in the sensitivity of the postsynaptic DA receptors. Apparently, there are also other neurochemical and neurophysiological mechanisms in the determinant of choreiform hyperkinesis that distinguish it from the determinant of the stereotyped behavior syndrome. Small GABAergic neurons of CN are mainly affected in chorea, while the inhibitory apparatus provided by the output of large GABAergic neurons of CN is affected in the stereotyped behavior syndrome. Consequently, the differences in the functions of these neurons can be a cause of the pathogenetic distinction of the given syndromes. Apparently, the pathologic system which underlies choreiform hyperkinesis has its specifics for the syndrome. Moreover, the clinically diverse forms of chorea should differ from one another also by their pathologic systems.

SUMMARY

A complex set of syndromes is produced when hyperactive determinant structures are created by different methods in both CN. This set includes stereotyped behavior, catatonia, parkinsonism, and chorea-like hyperkinesis. When the determinant structure was created by microinjecting TT into CN, syndromes originated in definite sequence with a definite latent period, depending on the TT dose. Stereotyped behavior which originated during the earlier period is based on the hyperactivity of the dopamine apparatus of the neostriatum as a result of the disinhibition of this apparatus due to the disturbance of GABA control. The reactivation of GABA control (microinjection of GABA into both nuclei and the systemic administration of diazepam or aminalon) and the inhibition of dopamine activity (suppression of the synthesis and secretion of dopamine by lithium and the blockade of the dopamine receptors by haloperidol) eliminated the stereotyped behavior syndrome during the action of the drug. The catatonic syndrome which originated after stereotyped behavior was characterized also by cataleptic features, namely, an increase in plastic muscular tension. The syndrome could be partially reduced at the initial stage of its development by introducing dopamine into the striatum or by systemically administering amphetamine or ι-dopa. These drugs were ineffective at the height of the syndrome. Parkinson's syndrome with all its characteristic features, i.e. akinesia, rigidity and tremor as well as vegetative disorders, originated at the late stages of the process. In parkinsonism, the cholinergic neurons of the striatum are the operant part of GPEE; they are disinhibited due to DA deficiency. The restoration of the striatum (intrastriatal administration of DA and the systemic administration of ι-dopa or Nacom), the inactivation of cholinergic neurons (intrastriatal administration of atropine or the systemic administration of benzhexol hydrochloride), and the suppression of GPEE (intrastriatal administration f procaine hydrochloride) caused the disappearance or diminution of the features of parkinsonism during the action of the preparations. Myoclonia and jerks, which can be regarded as a model equivalent of chorea-like dyskinesia, originated regardless of other syndromes or were manifested against their background. The rotation syndrome was observed when CN were unilaterally affected (administration of TT only into one nucleus). The stages of its development, which are characterized by diverse clinical pictures, correspond to the appearance of the symptoms of stereotyped behavior, catatonia and parkinsonism when the nuclei are affected bilaterally.

11
Pathologically prolonged sleep

The generator of pathologically enhanced excitation (GPEE) was produced in the somnogenous regulation system in order to create a model of pathologically prolonged sleep (Kryzhanovsky et al., 1978a).

The formerly popular 'passive theories of sleep', including the theory of the deafferentation and functional deactivation of the ascending activating system (Bremer, 1935, 1954; Moruzzi, Magoun, 1949; Magoun, 1950), are now no longer convincing as new facts have been discovered and new ideas proposed. Nevertheless, the disengagement of afferentation and the blockade of the activating stimuli which go to the brain are very important in the origin and maintenance of sleep. Evidently, sleep is a physiological state which originates when a special system is activated (Hess, 1949, 1954; Moruzzi, 1962, 1972; Hernandez-Peon, 1963; Monnier, 1963, 1980; Rossi, 1963a, 1963b, Kleitman, 1964; Akert, 1965; Jouvet, 1965a, 1965b, 1967, 1978; Zancetti, 1967; Latash, 1968, 1978; Shevchenko, 1971; Wein et al., 1971; Oniani et al., 1974; Godfraind, 1976; Lemaine et al., 1977; Karmanova et al., 1978, 1981; Oniani, 1978; Webb, Cartwright, 1978; Monnier, Gaillard, 1980). This system includes formations which relate to various regions of the brain and which are connected with various structures of the central nervous system. Many experiments which were carried out under chronic conditions show that the low-frequency electric stimulation of numerous cerebral structures can produce an electrographic expression of sleep, namely, spindles and slow waves in the cortex that can be accompanied by behavioral signs of sleep. However, this does not mean that all the structures of the somnogenous system are functionally equivalent.

213

Anokhin (1958) and Akert (1965), who were among the first to conclude that a multi-unit integrative system, and not a certain center, was the anatomic substrate of sleep, admitted that specific somnogenous regions could exist. The discovery of two types of sleep, namely, slow-wave sleep and fast-wave, or paradoxical sleep (Klaus, 1937; Dement, Kleitman, 1957) and the investigation of their mechanisms have shown that those types of sleep originate when various cerebral structures are activated (Moruzzi, 1962, 1972; Jouvet, 1965, 1978). A thorough analysis has revealed that diverse components of paradoxical sleep are engendered by the activity of different structures and that various formations of the brainstem and even diverse parts of the locus coeruleus complex play a different role in the induction of sleep, in the determination of its phases and components, and in awakening. Biochemical differences correspond to these functional differences in the brainstem part of the somnogenous system: serotoninergic and catecholaminergic structures are responsible for the different phases and components of sleep (see Godfraind, 1976; Lemaine et al., 1977; Monnier, Gaillard, 1980).

The orbitofrontal cortex and particularly the cortex of the orbital gyrus play an important role in modulating the somnogenous system's activity. The excitation of this region of the cortex by acetylcholine application (Hernandez-Peon, 1963; Mazzuchelli-O'Felaherty et al., 1967) or electric stimulation (Alnaes et al., 1973) caused EEG synchronization and behavioral sleep, while its destruction reduced the duration of sleep (Villablanca et al., 1976) and caused qualitative changes in sleep (Gadea-Ciria, 1976).

The above data were the basis of the investigations of the changes in sleep when GPEE was created in the orbital cortex. GPEE was produced by microinjecting TT into those regions of the cat orbital cortex whose stimulation is believed (Alnaes et al., 1973) to reduce the arterial pressure level, the heart rate, respiration and other vegetative components which originate during sleep.

Investigations (Kryzhanovsky et al., 1978a) have shown that sleep is pathologically prolonged when GPEE is created in the orbital cortex.

Irregular peak potentials with an amplitude of 100–200 μV were recorded at the site of TT administration in the background activity of the orbital cortex by the time steady features of prolonged sleep appeared (Figure 89 A 1). Epileptiform activity periodically appeared in the orbital cortex. This activity was characterized at first by an increase in the frequency of peak potentials, the growth of their amplitude and the appearance of sharp slow waves (Figure 89 B 1). Potentials of the peak-wave type and characteristic epileptic discharges were observed at the height of such activity (Figure 89 C 1). Seizure potentials either gradually disappeared (they became rarer, their amplitude dropped and the former rhythm appeared) or

terminated abruptly. Epileptic activity was local and was recorded
only in the region of the orbital cortex where TT was administered.
Thus, a GPEE acting as a typical epileptic focus was formed in the
orbital cortex. A noteworthy fact is that the GPEE's activity was
clearly observed for the first time when stable features of prolonged
sleep appeared.

At the same time, slow waves which are characteristic of somnolence and spindle bursts (Figure 89 A–C 2) were recorded in other
cortical regions (in the visual and sensorimotor regions). The
frequency of the appearance of spindles could increase (C) at the
time when GPEE's hyperactivity originated.

The behavioral features of sleep, wakening and wakefulness were
in accord with the electrographic signs of these states. Frequent
low-voltage potentials were recorded on the EEG during wakefulness.
The electromyogram has shown that the tonicity of the antigravity
muscles, i.e. the neck muscles, was high (Figure 90 A). High-voltage
waves of sleep spindles appeared in the EEG background when animals
fell asleep. In this case, the electric activity of the neck muscles
was well defined (Figure 90 B). As the animal fell into sleep more
deeply, slow waves began to prevail in the cortical EEG, while the
tonicity of the antigravity muscles remained on a sufficiently high
level (Figure 90 C D). Paradoxical sleep (Figure 90 E) originated
after slow-wave sleep periods, which differed in their duration. Its
beginning was, as a rule, preceded by the flattening of the electromyogram against the background of the high-amplitude slow waves

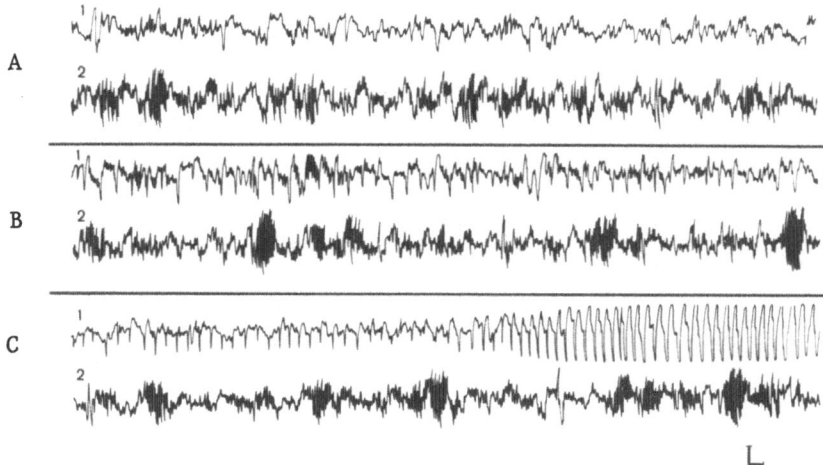

Fig. 89. Electric activity in the zone of the orbital cortex
          poisoned with TT and in the visual cortex during steady
          manifestation of prolonged sleep. A,B and C: successive
          ECoG fragments during sleep. Recordings: orbital cortex
          (1) and ipsilateral visual cortex (2). Amplitude: 100 µV;
          time: 1 s.

(Figure 90 D). Afterwards, the slow-wave rhythm was replaced by
frequent low-voltage rhythm, while the activity of the antigravity
neck muscles gradually began to disappear (Figure 90 E). Either a
gradual transition was made from paradoxical sleep to slow-wave sleep
(activity was observed in the neck muscles when EEG became desynchronized, the amplitude of the cortical potentials increased, and
slow waves appeared, Figure 90 F) or paradoxical sleep was succeeded
by brief wakefulness (10-20 sec) followed by desynchronized EEG
against the background of high electromyographic readings, after
which slow-wave sleep began.

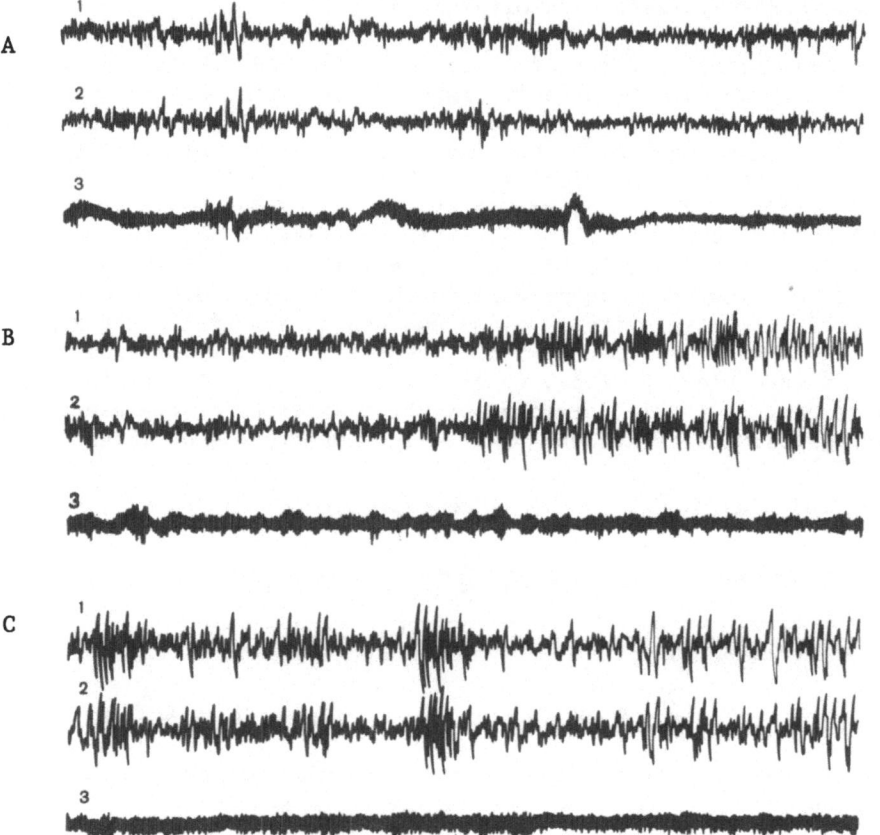

Fig. 90. Wakefulness-sleep cycle in an animal with prolonged sleep.
        Electrograms during wakefulness (A), change from wakeful-
        ness to drowsiness (B), deeper fall into slow-wave sleep
        (C,D), transition from slow-wave sleep to paradoxical sleep
        (E), and transition from paradoxical sleep to slow-wave
        sleep (F). EA in the frontal cortex (1), the visual cortex
        (2), and the neck muscles (3). Amplitude: 100 µV; time:
        1 s.

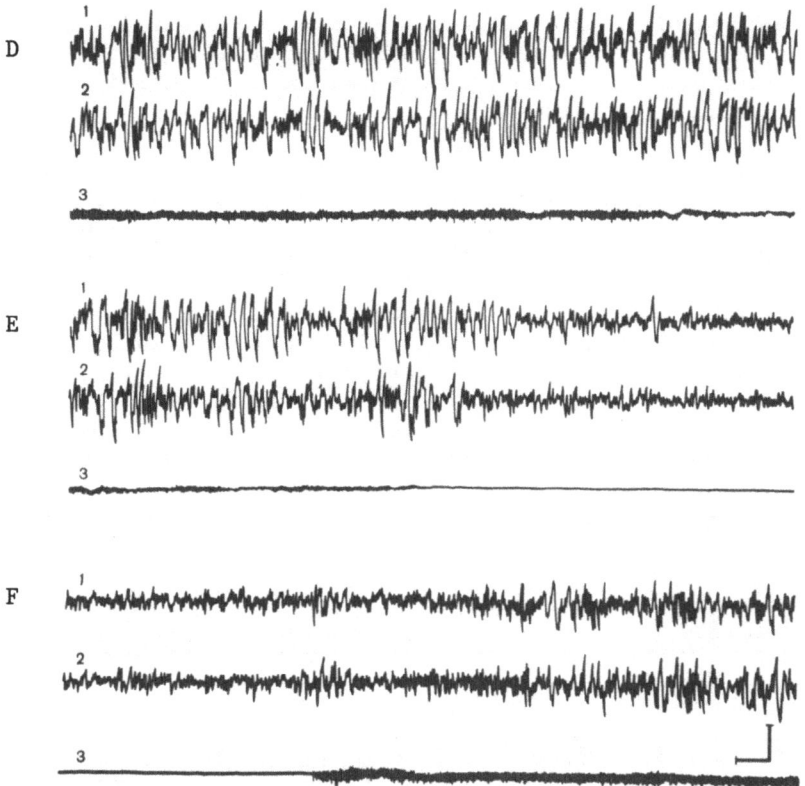

D

E

F

Fig. 90. (Continued)

Prolonged sleep was expressed in the following behavior: the
animals looked for a dark place, assumed a pose which was characteristic of sleeping cats, their respiration slowed down, and their
pupils became smaller. They did not wake up even when there was
light and noise in the laboratory. Only strong stimulation could
rouse them for a short time (1-5 min). To satisfy their physiological requirements (nutrition, defecation, etc.), the animals stayed
awake for a longer time, went over to the feed box, etc. Afterwards,
the returned to their place, sat down, lowered their head, covered
their eyes and began to drowse. Then, they rolled up and fell into
deep sleep.

This peculiarity of prolonged sleep continued for 3-4 days.
Then, the sleep-wakefulness cycle gradually became normal, i.e.
wakefulness lasted for a longer time and the animals reacted to
external stimuli more actively. Brief seizures could be observed in
some animals later (from the fourth to the ninth day).

Figure 91 shows the relationship between the periods of wakefulness, slow-wave sleep and paradoxical sleep in a healthy animal after
electrode implantation (A), in an animal during different periods
after TT administration (Bm C), and in an animal to which inactivated
toxin was administered (D).

The above data show that the duration of wakefulness in experimental animals was much shorter than that in control animals. According to Moruzzi (1962), Jouvet (1967), and Monnier and Gaillard
(1980), cats sleep for 20-70 per cent of the day. In our experiments, continuous electrographic recordings for 12 hours and observations of the cats' behavior for several days revealed that wakefulness lasted for 30-50 per cent of the day in the control groups and
88-92 per cent of the day in the test groups.

The duration of somnolence increased in absolute terms due to
the prolongation of slow-wave sleep. However, the duration of paradoxical sleep also increased. Therefore, the relationship between
slow-wave sleep and paradoxical sleep remained virtually the same as

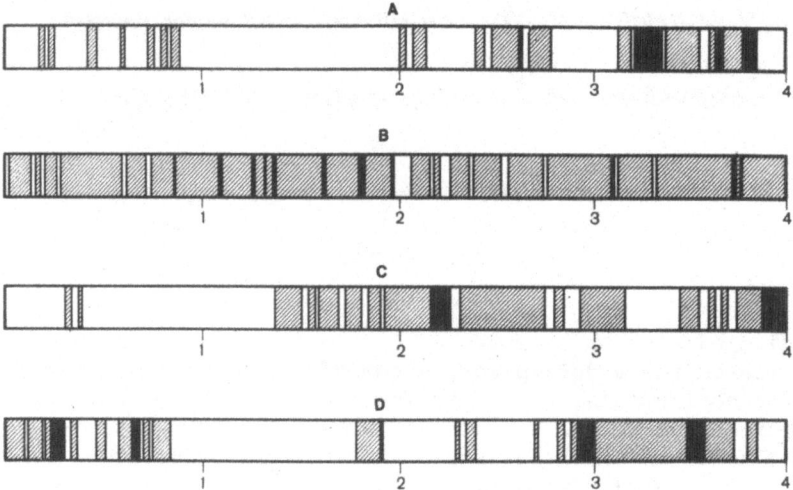

Fig. 91. Duration of wakefulness, slow-wave sleep and paradoxical
          sleep in an animal with pathologically prolonged sleep (B),
          a clinically recovered animal (C) and control animals (A,D)
          during four-hour uninterrupted recordings. Electrocortico-
          grams 24 h after the implantation of electrodes (A), 24 h
          after the injection of TT (B), 9 days after the injection
          of TT (C), and 24 h after the injection of inactivated TT
          (D). The white band fragments indicate wakefulness, the
          shaded ones, slow-wave sleep, and the black ones, paradoxi-
          cal sleep. The figures stand for the duration of continu-
          ous observation (hours).

in the control groups (the differences were not significant). The
peculiarities of continuous sleep did not substantially change. The
dynamics of the EEG changes in the case of both deeper slow-wave
sleep and the transition from slow-wave sleep to paradoxical sleep
and vice versa was the same in experimental animals as in control
animals. It was the same as the dynamics in intact animals described
by other authors (Hubel, 1960; Oniani et al., 1974). The sleep-
wakefulness cycle did not have any new components, and none of the
stages dropped out in it.

Thus, typical sleep with all the characteristic electrographic
and behavioral features was observed in experimental animals when
GPEE was produced in the orbital cortex. A pathologic feature of
this sleep was that it lasted much longer. This relates to the
determination of type III, when the hyperactive determinant merely
intensifies the system's activity, but does not change its nature
(see Chapter 1).

The results of the above investigations coincide with clinical
data (Saradzhishvili, Bibileishvili, 1975; Niedermeyer et al., 1979),
which show that sleep is generally prolonged in patients with epileptic foci in the frontal cortex. It is prolonged mainly due to the
growth of the slow-wave sleep periods.

It may seem paradoxical that GPEE, while producing epileptic
activity, increases the duration of sleep. However, this paradoxical
fact can be satisfactorily explained: the effect is due to activation
by GPEE formed in the somnogenous system, a component of which is the
orbital cortex. It should be noted that the effect is produced only
when the somnogenous system is activated, and only to a certain
extent. The creation of GPEE in the sensorimotor cortex in cats
disturbs the sleep-wakefulness cycle, increasing the duration of
wakefulness and reducing that of paradoxical sleep. These data accord
with the results of similar investigations carried out by Papuashvili
and his associates (1980). When GPEE in the orbital cortex produces
extremely enhanced excitation, such excitation becomes generalized
and the whole brain is involved in seizure activity. In this case,
the animals do not sleep, and sharp and slow waves are recorded on
their EEG. Such a state can result in a general convulsive seizure
and death.

A noteworthy fact is that the appearance of sharp waves and peak
seizure potentials in the orbital cortex was accompanied by the
origination of spindles and slow waves in other cortical zones.
Hence, the appearance of slow waves and spindles is connected with
the functional message from GPEE in the orbital cortex. The steady
spindle bursts, which occur under natural conditions when an animal
is falling asleep (Hubel, 1960; Jouvet, 1967), can be caused also by
electrically and chemically stimulating the orbital cortex, as a
result of which sleep grows (Hernandez-Peon, 1963; Alnaes et al.,

1973). The cortical regions whose stimulation reduces the frequency of respiration, lowers the blood pressure level and changes the heart rate are decisive in inducing sleep (Kaada, 1951). It has already been shown that GPEE was created precisely in these regions of the orbital cortex.

It is suggested (Hernandez-Peon, 1963; Alnaes et al., 1973) that the corticofugal impulses from the orbital cortex may regulate the subcortical and the somnogenous brainstem structures. The oligo-synaptic and monosynaptic pathways which connect the orbital cortex with the neuronal structures of the lower part of the brainstem are apparently the anatomic substrate through which that regulation is realized. Morphological (Brodal, 1971) and neurophysiological (Hun, 1976) investigations have shown that the orbital cortex is directly connected with the zones of the reticular formation of the lower part of the brainstem; these zones are believed (Moruzzi, 1962, 1972; Jouvet, 1967, 1972) to be responsible for the induction of slow-wave sleep and paradoxical sleep. The structures responsible for both types of sleep are probably activated when GPEE originates in the orbital cortex. These structures begin to operate more vigorously, but they retain their usual functional relationships (de Andres et al., 1975; Sato, Kanamori, 1975). Thus, slow-wave sleep and para-doxical sleep are prolonged.

It has been shown (Mancia et al., 1975) that the neurons of the mesencephalic reticular formation that maintain wakefulness (Moruzzi, Magoun, 1949; Magoun, 1965) are inhibited when the somnogenous struc-tures of the brainstem are stimulated. This mechanism probably participates in the above-mentioned phenomenon in which the duration of wakefulness decreases and that of sleep increases when GPEE is created in the orbital cortex.

Of course, these data do not cover the whole problem of the pathology of sleep and do not represent all the possible mechanisms of prolonging it. It can be assumed that the phenomenon of patho-logically prolonging sleep can be produced during intervention in other units of the somnogenous system apparently in the same way as the duration of sleep can be reduced when the structures responsible for wakefulness and the inhibition of the somnogenous system are activated. The above data on the fact that slow-wave sleep and paradoxical sleep are due to the activity of various structures of the somnogenous system suggest that the creation of long-term gen-erators in them will upset the pattern of sleep with the prevalence of slow-wave sleep or paradoxical sleep, accordingly. It should be taken into account that the pathologically intensified components of sleep can be reproduced when the generator is created in other cen-tral nervous system structures which are connected with the somnoge-nous system. It can be assumed from all the above data that the pathologic state of sleep is based on generator mechanisms, namely, the origination of GPEE in definite areas of the central nervous system under the influence of pathogenic conditions. Such mechanisms probably play an important role in the pathogenesis of narcolepsy.

However, physiological sleep seems also to be determined by the formation of a generator in the somnogenous system due to physiological effects which are produced in the course of wakefulness during the organism's active state. The duration of sleep, its phasic changes, the alternation of sleep and wakefulness, slow-wave sleep and paradoxical sleep, the maintenance of each type of sleep for a definite length of time, and other peculiarities of sleep can be explained by the concepts of the generator mechanisms of the origin and maintenance of sleep. These concepts do not contradict the data on the role of the humoral and other factors in the origin of sleep, since the neurophysiological mechanisms of its realization are concerned.

SUMMARY

Pathologically prolonged sleep was produced when GPEE was created in the cat's orbital cortex. The intrinsic structure of continuous sleep was not disturbed, i.e. the relationship between the overall duration of slow-wave sleep and that of paradoxical sleep was the same as in the normal state. The electrogram recorded in the orbital cortex in the GPEE zone revealed epileptic activity, while that recorded in other zones (visual and sensorimotor cortices) showed characteristic slow waves and spindle bursts. Spindles could appear more frequently in other cortical regions when epileptic activity increased in the orbital cortex. The behavioral features of sleep, awakening and wakefulness were in accord with the EEG indices of these states.

12
Models of neurotic and psychotic states

This material gives examples of some psychosis-like and neurosis-like states in animals when hyperactive structures are produced in definite areas of the central nervous system with due regard to the functional and neurochemical specifics of the systems which are drawn into the pathologic process.

SYNDROMES OF HYPERACTIVITY OF THE DOPAMINE APPARATUS

While not discussing the complicated, contradictory and largely unresolved question of the changes in the neurotransmitter systems and their metabolism and disbalance in schizophrenia, it can be concluded on good grounds that the brain's dopamine apparatus is functionally changed in schizophrenia. This mechanism is regarded as an important neurochemical link in the pathogenic structure of schizophrenia (Randrup, 1970; Klawans et al., 1972; Randrup, Munkvad, 1972; Snyder, 1972, 1976; Snyder et al., 1972, 1974a; Stevens, 1973; Costall, Naylor, 1974a, 1974b, 1978a; Matthysse, 1974; Anokhina, 1975; Arushanyan, Otellin, 1976; Meltzer, Stahl, 1976; Crow et al., 1976, 1978; Barchas et al., 1978; Berger, 1978; Carlsson, 1978; Fuxe, 1978; Stevens, Livermore, 1978; Stille, Christ, 1978; Anokhina, Gamaleya, 1979; Bird et al., 1979; Crow, Johnstone, 1978; Crow, 1980; 1982; Eggers, 1981; King et al., 1982). Amphetamine, an activator of dopamine release (Randrup, Munkvad, 1967), causes psychosis in man when it is administered in large doses. This psychosis resembles a paranoid paroxysm during schizophrenia (Connel, 1958; Bell, 1965; Ellinwood, 1967; Griffith et al., 1972; Snyder, 1972; Ellinwood et al., 1973; August et al., 1974; Woodrow et al., 1978). Amphetamine as well as other agents which potentiate dopamine activity in the brain can augment the psychotic behavior of those suffering from acute schizophrenia (Janowsky et al., 1973) when they are administered in minimum doses, which either produce an inconsiderable effect or do not produce any effect in healthy persons. The origination of schizophrenia-like psychoses in persons who took ʟ-dopa for a long time for therapeutic purposes in connection with Parkinson's disease has been described (Hartmann, 1976). The antipsychotic effect produced by reserpine, which unspecifically depletes both serotonin and catecholamine depots, is connected to the greatest extent with the diminution of the dopamine content (Carlsson, 1974). The pharmacological therapy of patients suffering from schizophrenia involves the use of neuroleptics, whose therapeutic significance is in accord with their action as the antagonists of dopamine (Fog et al., 1968; Kety, Matthysse, 1972; Snyder, 1974; Iversen, 1975; Crow, Johnstone, 1977; Crow, 1982). Dopamine hyperactivity in the animal brain was regarded as a determinant pathologic unit of the schizophrenia model (Kornetsky, Markowitz, 1978). The enhanced functional activity of the DA apparatus in schizophrenia may be connected not with the dopaminergic neurons' hyperactivity, but with the increase in the number of DA receptors on the appropriate mesolimbic neurons' postsynaptic membrane (Crow, 1982) (see Chapter 13).

It has been shown (Chapter 10) that the hyperactivity of the dopamine apparatus in the neostriatum entails the stereotyped behavior syndrome in animals. This syndrome is regarded as a model of paranoid schizophrenia (Randrup, Munkvad, 1966; Klawans et al., 1972), while some of its manifestations, as a model of chorea (Klawans, Weiner, 1976).

The clinical pattern of the above stereotyped behavior syndrome may be either very complex or relatively simple. In all the cases, however, it consists of the unmotivated, repeating and aggravated components or complexes of the species-ecologically determined behavioral acts. The exhibition of the species-specifics of animal behavior during the stereotypy syndrome is one of its characteristic features. This is indicated also by other authors who reproduced the stereotyped forms of behavior pharmacologically (Shchelkunov, 1964; Randrup, Munkvad, 1967; etc.).

The phenomena which we observed in animals displaying that syndrome include intensive genital grooming, whereby the animals ceaselessly lick their genital organs; in males, this may end in the sequestration of the penis.

As we have seen, the stereotyped behavior syndrome is caused by the hyperactivation of the dopamine apparatus of the neostriatum: the syndrome disappeared when haloperidol, which causes the blockade of the dopamine receptors, or lithium chloride, which inhibits the dopamine apparatus (see Chapter 13), was administered. The hyperactivity of the dopamine apparatus in our model of stereotyped behavior was due to the disturbance of GABA inhibitory control by GABA

antagonists (TT, penicillin, picrotoxin), which were injected into
the caudate nuclei. The stereotyped behavior syndrome was suppressed
when the insufficiency of GABA was covered by either injecting it
into both nuclei or systemically administering aminalone (gammalon)
or diazepam, which activates the GABAergic apparatus. When account
is taken of the inhibitory role of the caudate nuclei as the modulators of the cortical processes and the inhibiting influence of the
dopamine apparatus on the principal neurons of the caudate nuclei
(Chapter 10), the pathogenic chain of events, whose final clinical
expression is the stereotyped behavior syndrome, can be represented
as follows: disturbance of the GABAergic control of dopamine secretion into the caudate nuclei → increase in dopamine secretion → the
inhibition of the basic neurons of the caudate nuclei → the disinhibition of cortical motor activity in the form of species-specific
motor stereotypes.

    It is interesting to compare the given data with the results of
the investigations involving the hyperactivation of the dopamine
apparatus of the mesolimbic system which, together with a change in
the regulation of the cortical dopamine apparatus, is (as was
assumed) closely connected with the pathogenesis of certain forms of
schizophrenia (Stevens, 1973; Crow et al., 1977; Berger, 1978; Crow,
1982). This system's nuclei are the main target of amphetamine when
it induces motor hyperactivity (Iversen, 1977). Pathologic motor
activity originates in rats when dopamine is injected into these
areas (Pijnenburg et al., 1976). The affection of the limbic dopamine apparatus may be a cause of the exhaustion of motor activity and
some other symptoms in parkinsonism (Price et al., 1978a). The
hyperactivation of the mesolimbic dopamine apparatus in cats had
produced a state which was described as the experimental model of
psychosis (Stevens et al., 1974; Stevens, Livermore, 1978). The
following was observed; intense arousal, staring, fear, withdrawal,
waxy flexibility, a statuesque posture, looking, searching, sniffing,
hiding behavior, and the loss of social behavior. These phenomena
could be caused by either injecting bicuculin, a GABA antagonist,
into the ventral tegmental areas or regularly electrically stimulating those areas in accordance with the kindling type. The unilateral
administration of bicuculin caused the rotation syndrome. In the
case of regular electric stimulation, those phenomena appeared in one
combination or another with certain severity during the first week of
kindling. Then, they were augmented and could remain for several
months, even when electric stimulation was stopped within two weeks.
At the same time, slow waves, spike activity and occasionally brief
discharges were recorded in nucleus accumbens septi and sometimes in
the medial geniculate nucleus and the lateral geniculate nucleus.
Such electric activity originated also when bicuculin was administered and electric stimulation was regular; in the latter case, it
was observed for a long time even when electric stimulation was no
longer applied. This syndrome is connected with the hyperactivation
of the dopamine mesolimbic apparatus, as can be seen from the fact

that it did not originate in animals when 6-OHDA, which destroys the dopaminergic structures, was injected into the same ventral tegmental areas. The syndrome was considerably suppressed when haloperidol or closepin was systemically administered. The syndrome was greatly reduced also when <u>nucleus accumbens</u> was electrically stimulated for many days. Apomorphine intensified the syndrome and enhanced the amplitude of electric activity which was being recorded. It follows that this syndrome is due to the hyperactivation of the dopamine mesolimbic apparatus. Hyperactivation may result from either the disinhibition of this apparatus when GABA control is disturbed (the bicuculin effect) or primary activation when electric stimulation is regular. The appearance of characteristic electric activity and the long existence of the syndrome after electric stimulation is stopped shows that the syndrome has generator mechanisms.

A comparison of the above syndromes shows that both forms of the pathologic state of the central nervous system have the same pathogenic mechanism, namely, the hyperactivation of the dopamine apparatus due to the absolute or relative insufficiency of GABA inhibitory control. Some symptoms of both syndromes coincide. At the same time, the syndromes differ from one another in certain essential features. Psychopathologic phenomena are more pronounced and are more diverse when the mesolimbic dopamine apparatus is hyperactivated, while stereotyped behavior prevails when the neostriatal dopamine apparatus is activated, although some forms of stereotypy may be also due to the activation of the mesolimbic dopamine apparatus (Eichler et al., 1980). In analyzing these distinctions, account should also be taken of the species-specifics of the behavior of the test animals (rats and cats). However, these distinctions are in accord with the physiological significance of the two apparatus of the dopamine system and with their role in the pathogenesis of the disturbances of both behavior and higher nervous activity (Barches et al., 1978; Berger, 1978; Fuxe, 1978).

Thus, syndromes with common or different symptoms may originate when the same neurochemical system is drawn into a process, depending on the area of the central nervous system in which the structures of the given neurochemical system are hyperactivated. The above-mentioned syndromes clearly bear out the assumption (Chapter 4) that the specifics of the syndromes based on the hyperactivation mechanisms are determined by the physiological system in which the hyperactive determinant structure originated. Since the dopamine system includes functionally different cerebral structures (Fuxe, 1978), it can be assumed that a certain syndrome originates when the dopamine apparatus of the respective cerebral area is hyperactive. Moreover, the participation of various DA receptors may be selective, as has been shown earlier. Naturally, all that does not mean that the mechanisms of those models of the pathologic state involving the higher cerebral levels consist only in the dopamine system's hyperactivation. The given pathologic forms are an example of the inter-

connected multi-unit disbalance of the mediatory and other neurochemical systems of the brain.

The enhanced functional activity of the DA apparatus in the
corresponding forms of schizophrenia can be due not to the augmented
activity of the DAergic neurons, as was assumed earlier (Snyder et
al., 1974; Crow et al., 1976; Meltzer, Stahl, 1976; Snider, 1976;
Carlson, 1978). An increase in homovanillic acid has not been observed in the liquor of patients suffering from schizophrenia
(Bowers, 1974; Post et al., 1975), although such an increase may be
an indication of the hyperactivity of the DA neurons and the enlargement of the DA turnover; in amphetamine psychoses, this increase has
been established (Angrist, Gershon, 1970). Prolactin secretion,
which is suppressed by DA, diminishes in neither acute (Meltzer et
al., 1974) nor chronic schizophrenia (Johnston et al., 1977). A
postmortem examination of the brain of patients who suffered from
schizophrenia did not reveal any evidence of increased DA turnover
(Owen et al., 1978). However, an increase in the number of DA receptors in the brain has been detected (Owen, Crow, 1978; Owen et al.,
1978; Lee et al., 1978). This increase could not be attributed to
neuroleptics, since it was observed also when treatment did not
involve these drugs (Owen, Crow, 1978; Owen et al., 1978). Among the
two known types of receptors, only receptors labelled by the butyrophenone antagonist drugs increase numberically in schizophrenia (Owen
et al., 1980; Cross et al., 1981), while receptors labelled by both
antagonist and agonist drugs increase numerically when they are
influenced by neuroleptics (Muller, Seeman, 1977). Therefore, an
increase in the number of DA receptors in the brain can be regarded
as an essential pathologic factor of the forms of schizophrenia that
are connected with DA activity (Crow, 1982).

An interesting fact is that, in the investigations carried out
by Stevens and Livermore (1978), the electric stimulation of nucleus
accumbens suppressed the psychopathologic phenomena which originate
when the dopamine structures of the ventral tegmental areas were
hyperactivated. A favorable effect was produced also when nucleus
accumbens of a patient suffering from severe schizophrenia was bilaterally electrically stimulated. Those authors regard this effect
as an exhibition of the influence of the negative feedback from the
given nucleus to the activated structures of the tegmentum. These
data are in accord with our concept that the antisystems' activation
is one of the most important mechanisms which suppress pathologic
systems as regards various forms of the central nervous system's
pathologic state (see Chapter 13).

The enhancement of the postsynaptic neurons' sensitivity of DA
in schizophrenia can be considered in the light of the concept of the
antisystems' role in the development and suppression of the pathologic process. Such enhancement of their sensitivity may be observed
as a result of the decrease in the activity of an antisystem which is

antagonistic to the mesolimbic dopamine system (see Crow, Johnstone, 1978). An alternative hypothesis, according to which the number of postsynaptic DA receptors increases, may also be connected with the disappearance of the corresponding antisystem's inhibitory influences (see also Chapter 13).

As regards generator mechanisms, there are interesting data about recordings of pathologic electric activity in the form of paroxysmal discharges in both the mesolimbic dopaminergic structures and the regions of caudate nuclei that are adjacent to them in patients with schizophrenia (Heath, 1954). In these structures, such activity originates after injecting hallucinations (Heath, 1954; Stevens et al., 1969; Crow, 1972). It intensifies in patients during the aggravation of psychic disturbances, endogenous perception and hallucinations (Heath, 1954; Hanley et al., 1972), and is regarded as an exhibition of the group depolarization of neurons as a result of their partial denervation (Stevens, 1973), i.e. apparently the generator mechanisms are involved. Such fluctuating activity, which has also been described by Ellinwood and his associates (1974) in cats that were treated regularly with amphetamine, is thought to be connected with dopamine neural instability over a short time (King et al., 1982).

SYNDROMES INDUCED BY GPEEs CREATED IN HYPOTHALAMIC NUCLEI

Syndromes Caused by GPEE Produced by TT

Pronounced feeding behavior originated in animals when TT was injected into the central part of the lateral hypothalamus (morphologic control), whose electric stimulation caused feeding behavior (Figure 92) (Kryshanovsky et al., 1977b). It was expressed in an excessive number of 'automatic' chewing motions (the rabbits constantly 'chewed') and search reactions. the number of true alimentary instrumental reactions (pulling the ring and eating food) and the so-called secondary motivation reactions (pulling the ring without subsequently eating food) sharply increased (by 300 per cent, on the average). A well-fed rabbit could pull the ring, sometimes continuously or very frequently, for a long time, either eating or not eating food. The animals' motor orienting and search activity greatly increased.

That syndrome was absent in rabbits which received TT in the part of the lateral hypothalamus whose electric stimulation did not evoke an instrumental reaction and did not cause them to subsequently eat food. These animals' feeding behavior was greatly reduced. They were inert and inhibited, and occasionally fell asleep, staying in the corner of the experimental cage.

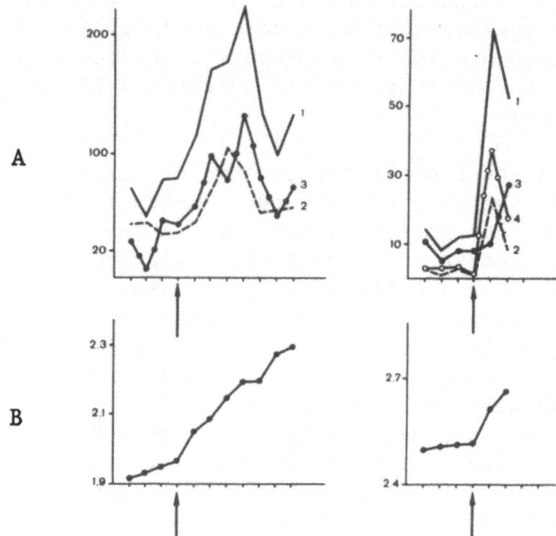

Fig. 92.   Alimentary behavior (A) and a change in the mass (B) of two
           rabbits after microinjecting TT into the lateral hypothala-
           mus.   A: 1: Total number of instrumental reactions; 2:
           instrumental reactions which were completed by taking food;
           3: instrumental reactions which were not completed by
           taking food; 4: licking, sniffing and gnawing the ring.
           Ordinate: number of instrumental reactions. Abscissa: days
           of the experiment. B: Ordinate: weight (kg). Abscissa:
           number of days of the experiment. The arrow shows the
           moment of TT injection.

     Spontaneous, occasionally continuous, chewing movements origin-
ated in rabbits when relatively large TT doses (which caused death
between the fourth and the seventh day) were administered to them.
These movements could originate even when the rabbits were in a
severe state, i.e. when they lay sidewise. The movements frequently
turned into a seizure. In many cases, epileptic fits developed in
these animals at the late stages. Even on the first day, however,
sharp permanent changes in the motivational and emotional status
occurred, and there were mixed behavioral reactions of the
'aggression-fear' type, in which fear greatly predominated. The
animals' general retardation often changed to intensive motor activity, such as 'panic running', which was easily provoked by various
stimuli (e.g. flicking, stroking). In some cases, spontaneous
aggressive reactions, expressed in the threatening tapping of the
legs, were observed. Aggressive reations prevailed and were sharply
expressed in animals to which small TT doses were administered: the
animals attacked the experimenters bit them, scratched the floor with
their forelegs, and made menacing sounds.

The exhibition of those changes in the animals' motivational and emotional behavior was paroxysmal. Fits of fear, aggression and the furious pulling of the ring could originate suddenly, and they could begin acutely. As time passed, the paroxysms became more frequent and longer; sometimes, the duration of such behavior became very long, and the fits could be induced more easily.

The recordings of electric activity in the brain showed that the epileptiform activity occurred in the hypothalamic nuclei into which TT was injected (Figure 93). This activity could spread to other cerebral structures.

Hence, the production of GPEE in the central part of the lateral hypothalamus has engendered a syndrome with characteristic signs of hyperactivity of a system which realizes alimentary, feed-behavior and 'secondary motivational' reactions. The specifics of the 'aggression-fear' syndrome, which originates when TT is administered

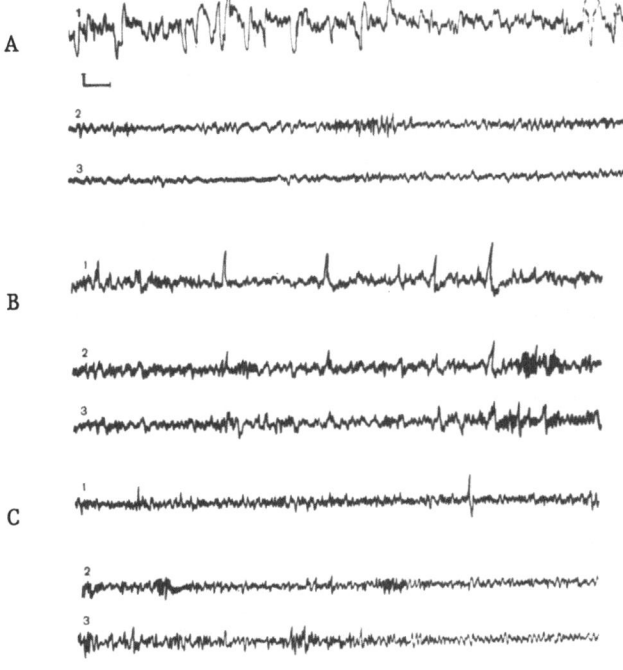

Fig. 93. Electroencephalogram of a rabbit after the formation of GPEE in the lateral hypothalamus by TT injection (A) and electric stimulation (B). Recordings: 1: lateral hypothalamus; 2: sensorimotor cortex; 3: occipital cortex. Amplitude: 100 μV; time: 1 s. B: I: during electric stimulation; II: 9 days after electric stimulation was ceased.

in relatively large doses, are apparently due to the spreading of the
toxin to the ventromedial nucleus and the formation of GPEE there.
The electric stimulation of the ventromedial nucleus is known to
cause a reaction of fear and a passive defensive reaction, and to
suppress food-obtaining behavior (Kozlovskaya, 1964; Gelgorn,
Loofbourrow, 1966; Sudakov, 1971; Waldman et al., 1976).

Syndromes Caused by GPEE Produced by Electric Stimulation

The same hypothalamic nuclei were regularly stimulated through
implanted electrodes by means of independent small electric stimulators when the animals behaved freely (Khomulo, Timofeeva, 1980).
Electric stimulation (1 V, 100 Hz, 0.5 ms, 1 s) was effected round-
the-clock with five-minute intervals for 2-4 weeks.

Alimentary instrumental activity and motor activity sharply
intensified and the orienting reaction became more rapid in the first
group of animals (in which, as histologic analysis has shown, the
electrodes were in the central part of the lateral hypothalamus) 5-6
days after the beginning of stimulation (Figure 94). That activity
was suppressed and it even stopped in the second group of animals (in
which the electrodes were between the dorsomedial hypothalamus and
the ventromedial hypothalamus). The animals of these groups rapidly
gained weight, especially during the first ten days, while the content of glucose moderately diminished in their blood. Epileptiform
activity (Figure 93) was recorded as single or group hypersynchronized discharges in the zones of hypothalamic stimulation during the
first three days. On the following days, it appeared also in the
sensorimotor cortex and then the occipital cortex. Later, it was
recorded again mainly in the hypothalamus. An interesting fact is
that epileptiform activity sharply intensified when the animals did
not eat for a day. This effect can be regarded as specific trigger
stimulation of GPEE that corresponds to the system's physiological
sign (Chapters 3, 4). All those behavioral, somatovegetative and
electroencephalographic changes remained even when electric stimulation was stopped, and they were seen throughout the whole observation period (up to four weeks) when there was no electric stimulation, while the body weight and the glucose content of blood continued to change. The thresholds of the production of the given set
of reactions, being tested by electrically stimulating the same zone
of the lateral hypothalamus, were greatly lowered in many animals.
This fact shows that GPEE neuronal excitability grows.

Comparison of the peculiarities of the above syndromes

The above data show that the syndromes caused by producing GPEE
in the lateral hypothalamus by means of TT microinjection and regular
electric stimulation are fundamentally the same: in both cases,

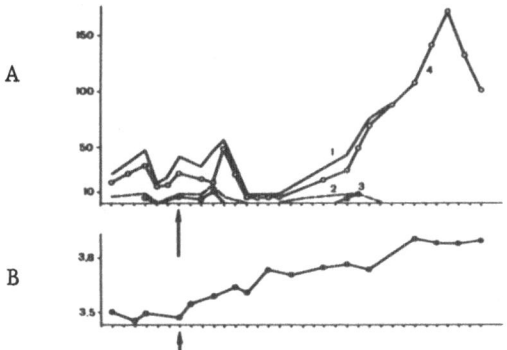

Fig. 94. Nature of alimentary behavior (A) and a change in the mass
          (B) of rabbits with chronic electric stimulation of the
          lateral hypothalamus. The notation is the same as in Fig.
          92. The arrow shows the beginning of electric stimulation.

alimentary hypermotivation occurs and, accordingly, food-obtaining
activity intensifies. However, a special analysis has shown that the
two effects greatly differ from one another (Kryzhanovsky et al.,
1980a; Kryzhanovsky, 1981b). It can be seen from Table 1 and Figure
93 that the differences are due to both the specifics of the formation of GPEE and the high level of excitation produced by it.

     The TT-induced GPEE was formed rapidly, and it could produce a
great effect, being expressed in the system's hyperactivation, already one day later. It produced the maximum effect on the second or
the third day, while pathologically enhanced excitation generated by
it was expressed in epileptiform activity. In some animals, such
activity could become generalized, being expressed in general seizures. In accordance with the rapid formation of GPEE, the latent
period of the originating syndrome as reduced and the syndrome's
signs were clearly expressed, i.e. they expressed the alimentary
motivational system's hyperactivity. An interesting fact is that
there are a large number of 'completed' instrumental reactions (which
end in the eating of food). Hyperphagia originated in rabbits with
such reactions, and they constantly ate. The whole syndrome was very
severe.

     The formation of the generator engendered by regularly electric-
ally stimulating the same area of the hypothalamus was slow, and the
generator produced less excitation in comparison with the former
GPEE. Accordingly, all the syndrome's signs originated during a long
latent period. The syndrome developed slowly, reached the maximum
after a long time, and was characterized by relative mildness. In
this case, the number of 'incomplete' instrumental reactions (pulling
the ring without eating food) increased.

Table 1. Characteristics of the Syndromes Caused in Animals as a
Result of Producing GPEE in the Hypothalamic Nuclei by
Microinjecting TT (First Group) and Regular Electric
Stimulation (Second Group).

GPEE site	Characteristics of the syndrome	
	First group	Second group
Lateral hypo-thalamus (central part)	A. 1. Sharp increase in the number of instrumental reactions, intensification of motor activity and the orienting reaction	B. 1. Sharp increase in the number of instrumental reactions, intensification of motor activity and the orienting reaction
	2. Increase in the number of instrumental reactions a day after the toxin was administered	2. Increase in the number of instrumental reactions about 5.5 days after electric stimulation was started
	3. The changes in instrumental activity are the most pronounced during the first three days after TT is administered	3. The changes in instrumental activity are the most pronounced 2-3 wks. after electric stimulation is started
	4. Great increase in the number of 'complete' instrumental reactions	4. Increase mainly in the number of 'incomplete' instrumental reactions
	5. Change in the weight a day after the toxin was administered	5. Change in the weight a day after electric stimulation was started
	6. The weight increased rapidly and intermittently	6. The weight increased uniformly and gradually with the formation of a plateau
	7. After 3-7 days, the abdomen is distended, diarrhea is observed, and some animals die*	7. The animals' state is good and is without any pathologic signs
	8. EA recording shows a long series of epileptiform discharges in the lateral hypothalamus	8. EA recording shows single epileptiform spikes in the lateral hypothalamus

Table 1. (Continued).

GPEE site	Characteristics of the syndrome	
	First group	Second group
	C. 1. Suppression of instrumental activity, motor activity and the orienting reaction	D. 1. Suppression of instrumental activity, motor activity and the orienting reaction
	2. The weight decreased in three rabbits, while it increased in one rabbit	2. The weight increased uniformly and gradually with the formation of a plateau in all the rabbits
	3. Instrumental activity is not restored even when the weight considerably diminishes	3. Instrumental activity is restored as the weight slightly diminishes
	4. Death against the background of a sharp diminution of the weight (or during seizures) in 8-10 days	4. The animals' state is good and is without any pathologic signs

* The causes of death are unclear. Besides the syndrome, tetanus in-
toxications are probably of certain significance.

An interesting fact is that such correlations were observed
during a severe syndrome also when GPEEs originated not in the lateral hypothalamus, but in the dorsomedial hypothalamus due to either
the spreading of TT or mistakes which were made when the electrodes
of electric stimulation were inserted (see Table 1, C and D).

The given data clearly show the role that the hyperactive determinant structures and generator mechanisms play in the genesis of the
neuropathologic syndromes, which are characterized by the systems'
hyperactivity. The lateral hypothalamus is regarded as the pacemaker
of food motivation (Anand, Brobeck, 1951; Anokhin, 1969; Anand, 1971;
Sudakov, 1971). By making this area of the hypothalamus hyperactive,
we produced a hyperactive determinant structure in the alimentary
motivational system, which made the system under consideration pathologic and likewise hyperactive. Such a system's activity, which is
exhibited in the above-mentioned animal behavior, is a syndrome of
alimentary hypermotivation and hyperphagia. The system's pathologic
nature is apparent: the rabbit either eats or constantly looks for

food, although its stomach is full. The result of such a system's activity is of no adaptive significance. This phenomenon is analogous to the above-mentioned constant scratching by the animal when a hyperactive determinant structure was produced in the scratch reflex system (see Chapter 4).

All the given peculiarities of the syndrome of alimentary hypermotivation and hyperphagia are observed when GPEE is the hyperactive determinant structure. However, the syndrome is not so pathologic and its exhibition may be suppressed by physiological mechanisms when the operant part of the determinant is a generator which produces relatively weak excitation and whose activity can be corrected. It has already been shown that incomplete reactions (pulling the ring without eating food) prevailed in rabbits with such a generator which was produced by electric stimulation, i.e. motivation could be suppressed at a certain stage of a reaction. This effect may be attributed to the fact that the instrumental part of the reaction, which assumed the properties of sustaining unit, had induced the feeling (Osborne, 1977). In other words, the intermediate result, being achieved at the stage of the instrumental reaction when the animals were really filled, was effective in suppressing the further course of the reaction. Hence, inhibitory control was still preserved in this case, and the reaction could be controlled yet, although, on the whole, it was already abnormal.

SYNDROMES WHICH ORIGINATE WHEN GPEE IS PRODUCED IN THE BULBAR DORSAL RAPHE NUCLEUS AND OTHER BRAIN AREAS

A complex set of pathologic reactions originates a few hours after TT is microinjected into the rats' dorsal raphe nucleus. This set of reactions begins with a paroxysm of motor hyperactivity. An animal which sat in a semidark cage suddenly jumps from its place and begins to quickly run around along the cage walls. The duration of this running may vary, and its stops just as suddenly as it begins. Then, the rat freezes, as it were, in the posture in which it stopped running. These postures may be diverse. In another characteristic phenomenon, the rat assumes a guarded posture. It tensely looks at a certain point without moving, holding its head high, and seems to watch objects which it alone appears to see. In this case, it does not react to external stimuli (knocking, light), and its orienting reaction is completely suppressed. Sometimes, the animal assumes defensive postures as if to defend itself from someone, then it screams and begins to move back. Apparently, hallucinations originate. Afterwards, the rat suddenly jumps from its place again and begins to quickly run around the cage. Then it suddenly stops, freezes in its place, etc.

The animal's behavior sharply changes if it is taken from the cage to an open illuminated place, such as a wide corridor: motor hyperactivity disappears, the rat sits in one place as if frozen and does not try to run away; external stimuli (touching) do not cause the animal to run away, and it seems to be frightened to move (unlike healthy rats, which tend to go away from an illuminated place to a dark place). However, when it is taken to a semidark cage, it begins to quickly run around the cage again, and the other above-mentioned phenomena originate.

That syndrome originates when GPEE is formed in the bulbar dorsal raphe nucleus. Recorded electric activity of PEE is exhibited in the appearance of new components in the evoked potential (EP) and in an increase in both the EP amplitude and the frequency of neuronal impulse activity.

We have detected psychosis-like disturbances when GPEE was produced also in other areas of the brain. When TT was injected into the anterior nuclei of the thalamus, the rats inadequately reacted to external stimuli, ran chaotically, jumped, and were in a state of agitated motor excitation. Besides the pain syndrome (Chapter 9), psychopathologic reactions originated in rats when TT was injected into the thalamic gelatinous nucleus: the animals rushed about the cage, jumped up high, hopped out of the cage when it was opened, leaped on the experimenter, etc. Signs which could be regarded as visual hallucinations originated in animals when GPEE was produced in their lateral geniculate nucleus: cats could continuously turn around as if to watch what was on the top of their tail. Such behavior was monotonous and could continue for a long time. Sometimes, it occurred before a seizure, being a visual aura in photogenic epilepsy, and sometimes it assumed an independent form of a pathologic state.

SUMMARY

The hyperactivity of the dopamine apparatus in both caudate nuclei, which is due to the disturbance of GABA inhibitory control, entails the development of stereotyped behavior in the form of a set of motions of different complexity that are hyperbolized constituents of the animals' species-specific behavior. According to the data found in literature, a state which can be regarded as a model of psychosis is produced in animals when the mesolimbic dopamine apparatus is disinhibited also as a result of the disturbance of GABA control. In both cases, the syndromes are eliminated during the action of drugs when the hyperactivity of the dopamine apparatus is directly suppressed or GABA control of dopamine secretion is restored. The syndrome of hyperactive alimentary motivation and hyperphagia originates when GPEE is produced in the lateral hypothalamus. In this case, the syndrome's severity, the possibility of spontaneously arresting it, and the peculiarities of animal behavior depend on both the power of GPEE and the preservation of the inhibitory mechanisms in its neuronal population. The production of GPEE in the

bulbar dorsal raphe nucleus together with analgesia causes a state
which can be regarded as a model of psychopathologic disturbances
with very complex behavioral signs. The disturbances of animal
behavior with diverse clinical pictures are due to the production of
GPEE in some other brain areas.

Part III
Basic therapeutic principles and recovery mechanisms

13
Basic therapeutic principles

The data given earlier show that the treatment of the neuropathologic syndromes should consist in the elimination of the pathologic systems on which they are based. This is evident from the fact
that the neuropathologic syndrome is a clinical expression of the
pathologic system's activity. The aim of pathogenetic therapy is to
eliminate this system.

Although our consideration of the principles of treating neuropathologic syndromes began with pathogenetic therapy, and not with
etiologic therapy, this does not mean that the latter is of no significance in the given forms of the pathologic state. On the contrary, it is necessary to eliminate the etiologic factor as a source
of the sustenance and development of the pathologic process and its
recurrence. Therefore, pathogenetic therapy should be combined with
etiologic therapy whenever the etiologic factor is established.

The therapeutic principles which will be discussed here are
those that pertain not to nosologic forms, but to neuropathologic
syndromes in various pathologic states.

The syndromes considered earlier have shown that the GPEEs which
constitute their basis can originate under different conditions and
under the influence of diverse effects. Therefore, GPEEs and the
syndromes which they induce are causally polyetiologic. However, all
the syndromes are monopathogenetic with respect to their basic mechanism, since they have the same pathogenetic basis, i.e. the formation of a hyperactive determinant structure, whose operant part
consists of GPEE or abnormal transmitter effects. Under these conditions, the center of gravity shifts to pathogenetic therapy.

There is another circumstance which accentuates the significance
of pathogenetic therapy. It has been shown that a stabilized pathologic system which acts for a long time can become relatively independent of the initial pathogenetic factor that has engendered it
(primary etiologic factor) and can exist and even develop independently under certain conditions after this factor's action ceases.
In this case, the elimination of the primary etiologic factor alone
will not produce a significant result. However, it is emphasized
that etiologic therapy is an indispensable part of a therapeutic
complex together with pathogenetic therapy. The therapeutic methods
used in modern neurology and psychiatry are, in essence, a system of
influences on syndromes and their complexes, i.e. they constitute
syndromic therapy.

The pathologic system can be eliminated mainly in three ways:
(1) by influencing the hyperactive determinant structure which forms
and sustains the pathologic system and determines the nature of its
activity; (2) by influencing the system as a whole, including the
determinant structure, the intermediate links and the efferent links,
and (3) by influencing the general state of the brain. This division
is conditional. In most cases, certain links of the pathologic
system are mainly influenced in conservative therapy. Nevertheless,
all the links are somewhat affected in any form of pathogenetic
therapy.

INFLUENCE ON THE DETERMINANT STRUCTURE

The elimination or the irreversible suppression of the hyperactive determinant structure at the stage when the pathologic system
depends on the determinant is a radical form of therapy. Such an
effect causes the pathologic system to break down and be eliminated
and, consequently, the neuropathologic syndrome to disappear.

It has been shown (see Chapter 1) on the basis of the models of
epileptic complexes in the cerebral cortex that the complex gradually
disintegrates, its foci become autonomous, their activity attenuates
and they gradually disappear when the determinant focus is surgically
removed or its activity is suppressed by locally applying nembutal.
This process has noteworthy peculiarities which reflect a general
regularity, i.e. the foci which are influenced least of all by the
determinant are the first to disappear; the complex itself remains
the longer (after the determinant structure is eliminated), the more
it has functioned under the structure's influence.

Clinical experience in treating epilepsy clearly shows that the
determinant structure must be eliminated when an epileptic system
originates; this should be done even at the late stages of the process. However, it is important for the focus being eliminated to be
the determinant one: if it is not such a focus, its elimination will

not considerably influence the epileptic system, and if it is the
dominant focus, its elimination may produce an opposite effect, i.e.
other foci which it suppresses will be disinhibited and the process
will be activated (Chapter 1).

Since the determinant structure is a key link in the pathologic
system, it should be removed also when therapeutic influences are
intended to suppress and reorganize the pathologic system (see
further). The fullest effect can be achieved at the stage when
secondary determinant structures have not been formed yet and when
connections and other parts of the system have not been stabilized.
However, even under such conditions, the determinant structure should
be eliminated. Penfield and Jasper (1954) have described a case in
the experimental epilepsy of monkeys when the elimination of the
primary focus produced a favorable effect very late.

It has been shown by taking photogenic epilepsy as an example
(Chapter 8) that the elimination (coagulation) of GPEE in LGN prevents the syndrome from developing further. Such coagulation is
effective at the early stages of the disease, when GPEE is small.
Even in this case, however, coagulation should be sufficiently
extensive. At the late stages, even greater coagulation is ineffective. We have observed such a situation also in the case of the
vestibulopathy syndrome (Chapter 7). When a small part of the population of neurons which constitute GPEE remains, a recurrence may be
observed either immediately or after some time (which is apparently
necessary for restoring GPEE). It should be taken into account that
GPEE can be formed again from intact neurons when the conditions of
its formation remain.

Interferences specific of the pathogenetic organization, i.e.
the neurochemical nature, of GPEEs are of great significance together
with the nonspecific forms of the hyperactive determinant structure's
elimination.

Since the insufficiency of the inhibitory mechanisms in the
population of the neurons which constitute GPEE is a prerequisite of
the GPEE's origin and activity (Chapter 3), this activity can evidently be suppressed by using the appropriate inhibitory transmitters. Examples were given earlier with respect to the glycine suppression of GPEE produced in various systems, namely, in the propriospinal neuronal system (Chapter 1), the giant-cell nucleus of the
bulbar reticular formation (Chapter 3), and the caudal trigeminal
nucleus of the trigeminal nerve (Chapter 9). The activity of the
neurons in the vicinity of the injection site is suppressed when
glycine is microinjected into the GPEE region, while GPEE is completely suppressed when glycine is injected into several points of
the GPEE region, as a result of which the syndrome is suppressed.
This effect is specific. It is not caused by a trauma (as is evident
from control experiments with other agents) and lasts as long as
glycine acts.

The fact that the suppression of the hyperactive determinant
structure causes the whole syndrome to disappear means that the
syndrome's basis, i.e. the pathologic system, is eliminated. Such an
effect has been observed in the case of spinal myoclonia (Figure 95),
a pain syndrome of spinal origin, the trigeminal pain syndrome
(Chapter 9), the stereotyped behavior syndrome, and parkinsonism
(Chapter 10).

Investigations have shown that the suppressive transmitter
effects are specific: they are connected with the role which a given
transmitter plays in the activity of the population of the neurons
that constitute GPEE and with the pathogenic (neurochemical) structure of the latter, i.e. with that which shows what inhibitory mechanisms are damaged or are functionally insufficient in the neuronal
populations of GPEE, what neurons are hyperactive, and what receptors
are either free or blocked on these neurons.

These conclusions are underpinned by the results of the above
investigations: glycine produces a favorable effect when GPEE is in
the medulla and the spinal cord, where it is significant as an inhibitory transmitter: the stereotyped behavior syndrome is suppressed
when GABA is injected into the rostral part of both caudate nuclei,
where the drug is significant as an inhibitory transmitter and where
its insufficiency causes dopamine hyperactivity, which is the basis
of the stereotyped behavior syndrome (Chapter 10): Parkinson's syndrome is suppressed when dopamine is injected into the affected
neostriatum, where the drug acts as an inhibitor and where its insufficiency causes cholinergic neurons to be hyperactivated, producing
parkinsonism (Chapter 10).

The given conclusions were clearly illustrated also by the
results of the experiments with the model of the pain syndrome which
was caused by GPEE in the dorsal horns of the spinal lumbosacral
segments by agents that mainly either upset the inhibitory mechanisms
(TT, strychnine, penicillin) or depolarize neurons (KCl, ouabain; see
Chapter 5). Glycine and GABA were used to suppress GPEE's activity:
they were deposited, together with the agents which engender GPEE, on
an agar plate which was applied to the dorsal surface of the spinal
cord. The idea of this experiment was to ascertain the effectiveness
of the inhibitory transmitters with respect to GPEEs of different
pathogenetic structures and neurochemical characteristics.

Investigations (Kryzhanovsky, 1976; Danilova et al., 1979) have
shown (Table 2) that glycine was effective (it prevented the pain
syndrome from developing, i.e. it suppressed GPEE's activity) when
GPEE was produced by TT, penicillin, KCl and ouabain, but it was not
effective when GPEE was produced by strychnine. GABA was effective
when GPEE was produced by TT, KCl and ouabain, but it was hardly
effective when GPEE was produced by penicillin: the syndrome's intensity did not diminish, and GABA did not prevent it from developing;

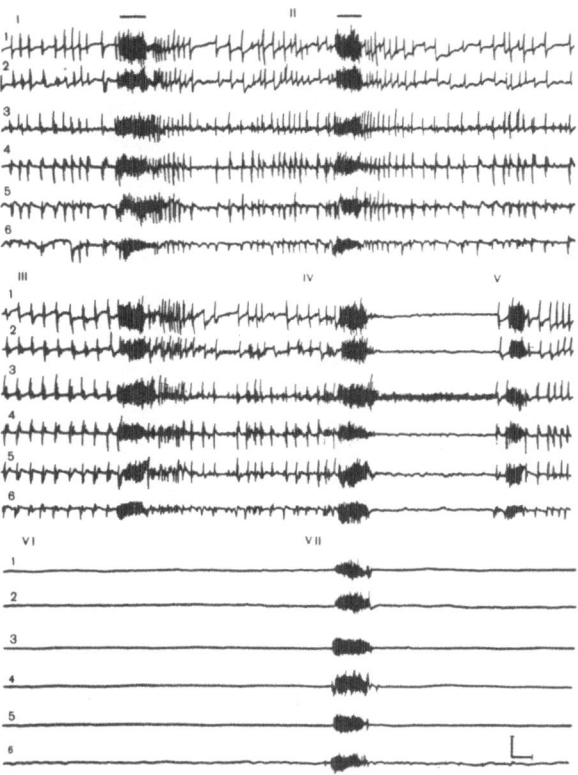

Fig. 95. Generalized spinal myoclonia ('spinal epilepsy'). EA in various muscles in the case of provoked and spontaneous activation of two different GPEEs produced on the left and the right in the lumbosacral segments; effect of glycine injected into the region of one GPEE. I, II, III: Single synchronized discharges which originated spontaneously and also after the activation of 'left-sided' GPEE; burst of tonic EA when this GPEE was activated; IV: burst of asynchronous tonic EA caused by the activation of 'right-sided' GPEE and the subsequent disappearance of spontaneous discharges; V: spontaneous burst of tonic EA and the appearance of synchronized discharges; VI: suppression of the activity of 'left-sided' GPEE after injecting glycine into the GPEE region ($1 \cdot 10^{-4}$ml of 20% solution into the central horns, L_4-S_1); disappearance of provoked and spontaneous EA; the disappearance of spontaneous discharges means that they were produced by 'left-sided' GPEE; VII: provoked burst of tonic EA when 'right-sided' GPEE was activated; apparently, this GPEE produces only tonic activity. Recordings of EA: 1,2: right and left neck mucles; 3,4: right and left back muscles; 5,6: posterior group of right and left femoral muscles. The duration of provoked stimulation (squeezing the ipsilateral foot) is indicated by a horizontal line. Amplitude: 250 µV; time: 1 s.

only its duration was somewhat reduced. When GPEE was produced by strychnine, GABA retarded the pain syndrome's exhibition but did not fully prevent it from developing.

The results obtained can be explained when account is taken of both the mechanisms of the action of the agents which engender GPEE

Table 2. Effects of Glycine and GABA in the Case of Pain Syndromes Produced by GPEEs of a Different Nature.

Agent used for producing GPEEs	Transmitters	Latent period of the origin of the pain syndrome from the moment of application of the agents, min*	Duration of the pain syndrome;h	Extent of the suppression of the pain syndrome during transmitter action
Tetanus toxin	Glycine	200±40	17±5	+++
	GABA	120±30	18±5	+++
	Control	40±10	20±5	--
Strychnine	Glycine	10±5	2±0.5	--
	GABA	40±10	1.5±0.2	(+)(--)
	Control	10±5	2±0.5	--
Penicillin	Glycine	Did not originate	--	+++
	GABA	10±5	1.5±0.5	(--)(+)
	Control	10±5	3±0.5	--
KCl	Glycine	Did not originate		+++
	GABA	Did not originate		+
	Control	40±10	1.5±0.5	--
Ouabain	Glycine	Did not originate		+++
	GABA	Did not originate		+++
	Control	40±10	1.5±0.5	--

* With regard to the duration when the agent was applied and the animal came out of anesthesia. In the case of TT application, the pain syndrome remained and became more severe until the end of observation (on the second day) due to the development of local TT intoxication. The sign (+) indicates the transmitter's favorable effect (suppression of the pain syndrome), and the sign (-), the absence of an effect; the sign (--)(+) indicates a small favorable effect after the transmitter exerted influence for a long time, and the sign (+)(--), a rapidly occurring but brief favorable effect. At least five rats were used in every series of experiments.

and the neurochemical characteristics of the latter (Figure 96).
TT acts on the presynaptic apparatus, upsetting the release of the
inhibitory transmitters, including glycine and GABA, and not blocking
the postsynaptic membrane for the inhibitory transmitters (see Chapter 5).   Penicillin blocks GABA receptors but does not block glycine
receptors on the postsynaptic membrane (Curtis, 1962; Curtis et al.,
1972; Davidoff, 1972; Roberts et al., 1976; Antonidis et al., 1980a,
1980b, 1980c; Smythies, 1980), although it can depolarize the neuronal membrane (Futamachi, Prince, 1975; Hochner et al., 1976; Heyer
et al., 1981).   Strychnine blocks postsynaptic glycine receptors
(Curtis et al., 1969, 1971; Wermann et al., 1968; Curtis, Johnston,
1974; Smythies, 1980).   Potassium and ouabain depolarize the postsynaptic membranes, but do not block the inhibitory receptors on them
(Lee, Klaus, 1971; Thomas, 1972; Khodorov, 1976).   Hence, GABA and
glycine were effective when their receptors on the postsynaptic
membrane of the GPEE neurons were free.   These transmitters were
either hardly effective or ineffective when the receptors were
blocked by the appropriate convulsants (strychnine or penicillin).

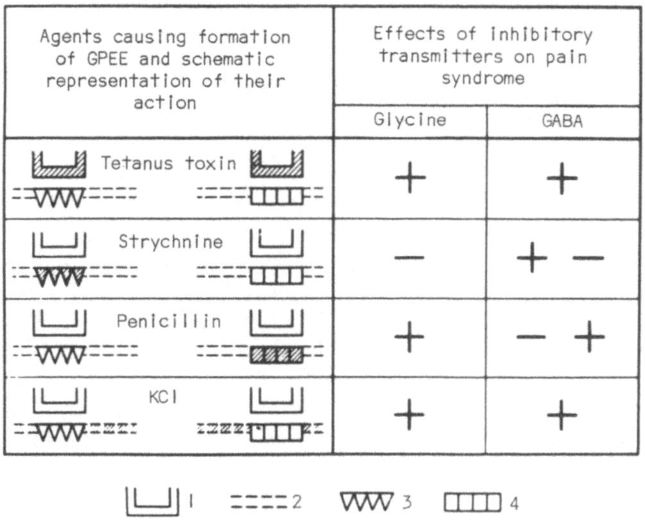

Agents causing formation of GPEE and schematic representation of their action	Effects of inhibitory transmitters on pain syndrome	
	Glycine	GABA
Tetanus toxin	+	+
Strychnine	−	+ −
Penicillin	+	− +
KCl	+	+

|⊔| 1   ====2   ⋁⋁⋁ 3   ▭▭▭ 4

Fig. 96.   Effects of glycine and GABA on GPEEs produced by various
pharmacological agents in the dorsal horns of the spinal
cord.   Left: diagram of the action of agents, which induce
the formation of GPEE, on the presynaptic and postsynaptic
membranes.   The shaded parts are the regions of the presynaptic and postsynaptic membranes which are the targets
of those agents.   1: presynaptic membrane; 2: areas of the
electrogenous membrane;   3: glycine receptors: 4: GABA
receptors.   The minus sign stands for the negative effect,
the plus sign, for the positive effect, the plus and minus
signs, for a brief positive effect, and the minus and plus
signs, for the late occurrence of the positive effect.

Those experiments clearly show the significance of the pathogenetic (in this case, the neurochemical receptive-mediatory) characteristics of GPEEs in determining the effectiveness of the agents which can be used for suppressing GPEEs in pathogenetic therapy.

The significance of the GPEE properties can be seen also when antiepileptics (e.g. diphenylhydantoin, carbamazepine and diazepam) are used. The effects of these drugs were tested by systemic (intramuscular and intravenous) administration involving a model of the spinal pain syndrome induced by GPEE in the spinal dorsal horns (Chapter 9). Investigations (Kryzhanovsky, 1976; Grafova et al., 1979) have shown that the effects of the above drugs differed in accordance with the nature of GPEE. Diphenylhydantoin slightly influenced the development of the pain syndromes engendered by GPEEs which were produced by TT, penicillin and strychnine, but it suppressed the pain syndromes engendered by GPEEs which were produced by KCl and ouabain. Carbamazepine was effective in the case of syndromes engendered by GPEEs which were produced by KCl and ouabain (Figures 97 and 98). It prolonged the latent period of the origin of the pain syndrome, partially suppressed the syndrome which already developed when GPEE was produced by TT, and reduced the duration of the pain syndrome when GPEE was produced by strychnine, but it did not influence the pain syndrome when GPEE was produced by penicillin. Diazepam suppressed the pain syndromes engendered by GPEEs which were produced by KCl and ouabain; this effect was observed when relatively small doses (2 mg/kg) of the drug were used. In the case of pain syndromes engendered by GPEEs which were produced by TT or strychnine, diazepam in small doses reduced and weakened the pain fits, while in large doses (5 mg/kg), it completely suppressed the pain syndrome during its action. Diazepam did not produce an effect even in large doses in the case of a pain syndrome engendered by GPEE which was produced by penicillin. Carbamazepine and diazepam were ineffective in the case of GPEE produced by penicillin not because this GPEE was powerful, since such a result was obtained also when GPEE was produced with small penicillin doses. Phenazepam, which is a new drug and a derivative of diazepam (Bogatsky et al., 1975; Bogatsky, 1982) and which is a more powerful tranquilizer and anticonvulsant than diazepam, produced a more powerful and an 'analgesic' effect: it suppressed the pain syndrome engendered even by GPEE which was produced by penicillin (Kryzhanovsky et al., 1982a).

The action of these drugs is complex, and they are applied to various areas of the central nervous system. Therefore, their 'analgesic' effects cannot be reduced to their influence on GPEE in the spinal cord, especially since the effects were evaluated only by animal behavior and there was no direct evidence of what was happening in GPEEs. Apparently, those agents act on various links of the whole pathologic system, and its inhibition by action on diverse links, especially at the supraspinal levels, should be of great significance in the mechanisms which suppress the pain syndrome.

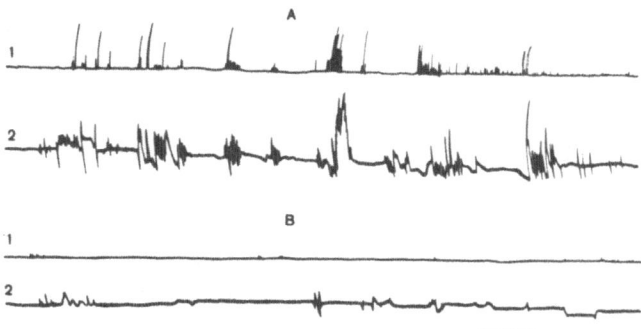

Fig. 97. Carbamazepine arrest of the pain syndrome caused by GPEE
          produced in the dorsal horns of the spinal cord by KCl. A:
          Pain syndrome in a control animal. Phonogram (1) and
          actogram (2) one hour after applying an agar plate with KCl
          (75 mM/ml of agar) on the dorsal surface of the spinal
          lumbosacral segments on the left. B: Absence of a pain
          syndrome in an animal to which carbamazepine (200 mg/kg)
          was administered <u>per os</u> one hour before KCl was applied.
          The phonogram (1) and the actogram (2) were taken one hour
          after applying KCl. The horizontal line indicates 5 min.

     At the same time, those drugs possess a common property: they
suppress seizure activity. The fact that the effects of these drugs
depend on the nature of the GPEEs suggests that they act on the
generator mechanisms of the pain syndromes and that this action is of
great importance in the general pain-suppressing effect. A noteworthy fact is that all the drugs were effective when GPEE was produced by depolarizers (KCl, ouabain), i.e. when the inhibitory mechanisms were preserved in its neuronal population. Carbamazepine and
diazepam were ineffective when GPEE was produced by penicillin.
Since the penicillin effects interfere with the GABA effects, it can
be assumed that the influence of carbamazepine and diazepam is
realized with GABA's participation. This conclusion is in accord
with the data which show that diazepam activates GABA receptors and
enhances GABA inhibitory control (see further). Thus, the specifics
of the selective effects of the drugs which possess common anticonvulsant properties are exhibited as they act when GPEEs with
different neurochemical characteristics exist.

     Anticonvulsants suppress pathologically enhanced activity in
GPEE itself and in the regions of the cental nervous system where
excitation produced by GPEE spreads and is realized. They can influence various parts of the nociceptive system and other systems which
are involved in the pathologic process. By suppressing the hyperactivity of the nociceptive system in which GPEE is formed, anticonvulsants arrest the pain syndrome, acting in this case as
analgesics. Anticonvulsants are now being used more and more fre-

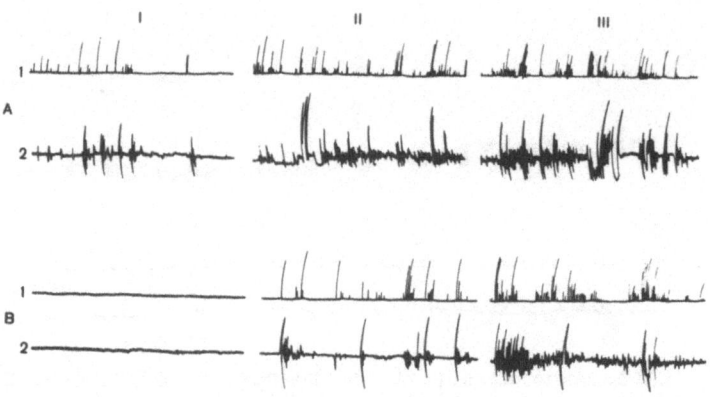

Fig. 98.   Influence of carbamazepine on the development of the pain
           syndrome caused by GPEE produced in the dorsal horns of the
           spinal cord by TT. A: Development of the pain syndrome in
           a control animal. The phonogram (1) and the actogram (2)
           were taken 40 min (I), 1.5 h (II) and 2.5 h (III) after
           applying an agar plate with TT to the dorsal surface on the
           spinal lumbosacral segments on the left. The pain syndrome
           increased with time. B: Development of the pain syndrome
           in an animal to which carbamazepine (200 mg/kg) was admin-
           istered <u>per os</u> one hour before applying TT. The develop-
           ment of the pain syndrome was delayed: its signs were
           absent during the first 40-60 min; during the second period
           (1.5 h), the syndrome was less pronounced than in the
           control animal, and during the third period (2.5 h), the
           syndrome became almost as severe as in the control animal.
           It should be taken into account that during this time, bot
           only did the syndrome intensify, but also the action of
           carbamazepine weakened. The horizontal line indicates 5 min.

quently in the case of many pain syndromes of central origin;
carbamazepine is now one of the main and most effective therapeutic
agents in trigeminal neuralgia.

    The given effects of antiepileptics, which act not only on the
pathologic system as a whole, but also on its key link, i.e. the
hyperactive determinant structure, are clearly exhibited in the
syndromes of convulsive activity, particularly in the above-mentioned
(Chapter 6) spinal myoclonia. A comparison of such effects is
interesting also because both syndromes are of spinal origin: in both
cases, GPEEs originate in the segmental apparatus. Studies (Grafova,
Danilova, 1980; Kryzhanovsky et al., 1982a) have shown that all the
antiepileptics used produce the same effect: they suppress convulsive
activity in spinal myoclonia. This effect is exhibited above all in
the suppression of clonic activity, the disappearance of spontaneous
discharges, and then in the diminution of the duration of the tonic

phase of evoked activity (Figure 99). Diazepam and especially
phenazepam were the most effective among the given drugs; in the
appropriate, relatively small doses, phenazepam suppressed even tonic
activity. An important fact is that these drugs acted also when the
spinal cord was cut, thus underpinning the conclusion that they
influence GPEE in the spinal cord. These effects of antiepileptics
were even more pronounced when the connections between the spinal
cord and the supraspinal structures remained. This fact suggests
that descending inhibitory influences participate in the realization
of the tested antiepileptics' influences. The characteristics of
electric activity change the same in all muscles, i.e. in all the
spinal motoneuronal pools, under the influence of those drugs; this
fact can also indicate that the drugs act on GPEEs in the spinal
cord, suggesting that the source of activity, i.e. GPEE, changes.

Hence, anticonvulsants, particularly diazepam and phenazepam,
directly influenced GPEE, suppressing its activity. This effect can
be understood when account is taken of even only one aspect of the

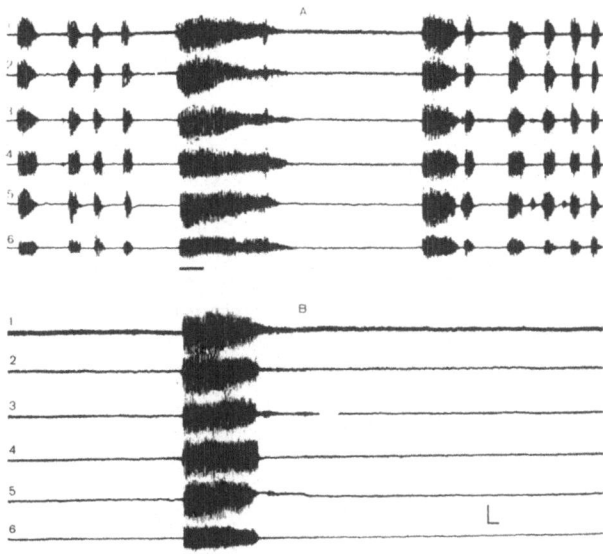

Fig. 99. Effects of diphenylhydantoin on generalized spinal
          myoclonia. Recordings of EA before (A) and 20 min after
          (B) injecting i.m. diphenylhydantoin (100 mg/kg). 1,2:
          Right and left back muscles; 3,4: right and left sacral
          muscles; 5,6: posterior group of right and left femoral
          muscles. Section of the spinal cord at the Th_1-Th_2 level
          24 h before recording EA and 96 h after injecting TT.
          Amplitude: 200 µV; time: 1 s. Stimulation (squeezing the
          toes of the left hind leg) is indicated by a horizontal
          line.

action of the given drugs on the polysynaptic connections: they can influence these connections also in the GPEE neuronal populations. An important fact is that the action of anticonvulsants on the hyper-active determinant structure is less effective than on the other parts of the pathologic system due to both the high level of exci-tation in GPEE and the peculiarities of its functional organization (Chapter 1 and 3).

Apparently, not all the GPEE neurons are inhibited to the same extent under the action of antiepileptics, since they are differen-tiated by their 'epileptic properties' and by their ability to be involved in the epileptic process and to generate pathologically enhanced excitation (Chapter 3). Apparently, the least active neurons of group III are the first to be suppressed under the action of anticonvulsants. Then, the neurons of group II are inhibited. The neurons of group I, or the so-called 'trigger' neurons, may remain unsuppressed and may continue their activity. This activity is hardly significant functionally. Therefore, it is not exhibited clinically during therapy. At the same time, such neurons constitute the basis for the restoration of GPEE and for the recurrence of the process under the appropriate conditions. In urgent cases, they can apparently be suppressed under the action of drugs in very large doses, when toxic effects may originate.

Several mechanisms of the resistance of neuropathologic syn-dromes, which are connected with a pathologic system's biological peculiarities, have already been discussed. Account should also be taken of some other factors with which not only the resistance of syndromes, but also the ineffectiveness of their therapy may be connected. One such factor, which is very important, is the nature of the pathologic determinant and the neurochemical organization of GPEE. It has already been shown that the determinant structure may be based on either presynaptic or postsynaptic mechanisms of func-tional hyperactivity. The neurochemical organization of GPEE differs when it is produced by agents with different mechanisms. However, the pathologic systems induced in such cases and the respective neuropathologic syndromes may be clinically similar. It has been shown by taking the example of the pain syndrome of spinal origin that the therapeutic effectiveness of drugs depends on the disturb-ance of the type of inhibitory control with which the formation of GPEE is connected. Drugs are ineffective when they act via the activation of the inhibitory mechanisms which were damaged.

In comparing the results of the action of drugs of two different classes, i.e. specific inhibitory neurotransmitters and nonspecific antiepileptics, it should be taken into account that these different substances played the same role, although the role was unusual for them, under the given conditions: they suppressed the pain syndromes in spite of the fact that they were not analgesics in the strict sense of the word. This effect is connected with their suppressing

influence on the nociceptive system when it is hyperactivated and
with the suppression of the generator mechanisms of the pain syndromes. An important fact is that the effects of these substances
depended on the neurochemical organization of GPEEs. This may
explain the well-known clinical observations when the same drug can
be effective or ineffective as regards neuropathologic syndromes with
seemingly the same clinical manifestation. In such cases, the term
'individual sensitivity' is used, but it does not have a definite
meaning. It may be believed that the differences in the pathogenetic
(pathochemical) organization of GPEEs are among the causes of the
given phenomenon, since such an organization establishes the different relation of the pharmacologic drugs to GPEEs and, consequently,
the different effectiveness of the drugs.

In this connection, the following question inevitably arises: to
what group do the substances which suppress convulsive activity in
some cases and pathologic pain in others belong? It should be noted
in this respect that the given agents are also used in treating the
so-called internal diseases, e.g. cardiovascular diseases, paroxysmal
arrhythmias, and various dyskinesias of the cavitary organs. These
and other similar agents are used whenever the functional structures
are hyperactive, when GPEEs originate.

INFLUENCE ON THE PATHOLOGIC SYSTEM

It has already been shown that all the anticonvulsants suppress
the pathologic system's activity, acting on not only GPEE, but also
other parts of the pathologic system. The action of drugs on other
links of the pathologic system is a question which should be
specially discussed.

A series of experiments were carried out to study the influence
of general anesthesia on the complex of epileptic foci which was
produced in the cerebral cortex and which is a distinct model of the
pathologic system (Chapter 1). The anesthetics used were ether and
halothane, which act mainly on the cerebral cortex. Studies
(Kryzhanovsky et al., 1977c, 1978c) have shown that ether (Figure
100) and halothane (Figure 101) suppress the epileptic complex in the
cerebral cortex and cause it to break down; in both cases, the complex was eliminated as a system in accordance with the same regularity: the process began with the suppression of the dependent
foci's activity; the activity of the focus which was influenced least
of all by the determinant focus was suppressed at first. Such a
sequence was especially clearly exhibited in experiments with
ethereal anesthesia, which develops more slowly than halothane
anesthesia. This made it possible to trace the process as it successively developed. The determinant focus is the last to remain.
To suppress it, anesthesia should be relatively deep.

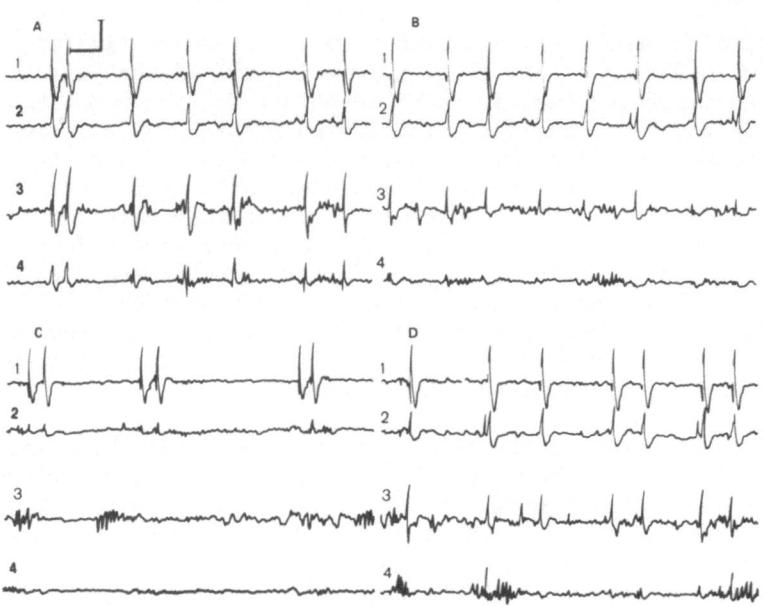

Fig. 100.   Influence of ether infalation on the complex of epileptic
            foci in the cerebral cortex.   State of the complex before
            (A), and 4 min (B) and 30 min (C) after ether inhalation;
            D: 30 min after inhalation was ceased.   The foci were
            produced by applying strychnine.   The determinant focus is
            in zone 1.   The details of producing such complexes were
            given in Chapter 1.   1: Orbital contex; 2: coronary gyrus;
            3: anterior sigmoid gyrus; 4: posterior sigmoid gyrus.
            Amplitude: 500 μV; time: 1 s.

        That distinguishing feature should be mentioned also in another
respect: electrographically, the determinant and dependent foci could
often be characterized by seemingly the same activity.   This fact is
in accord with the clinical data which show that the main epileptic
focus cannot always be ascertained by electrographic features alone
(Ugryumov, Zotov, 1971; Rasmussen, 1975; Rossi, 1975; Sarajishvili,
Geladze, 1977; Ojemann, 1980).

        The determinant focus is 'exhibited' in anesthesia, when the
dependent foci of the complex are the first to be suppressed.   Such
differential diagnosis based on the functional heterogeneity of the
foci can be significant with respect to not only multifocal epilepsy,
but also other forms of the pathologic state of the central nervous
system, where several hyperactive structures, one of which is the
determinant, may originate.

        The epileptic complex was restored when anesthetics were no
longer given.   In both cases, it was restored by the same regu-

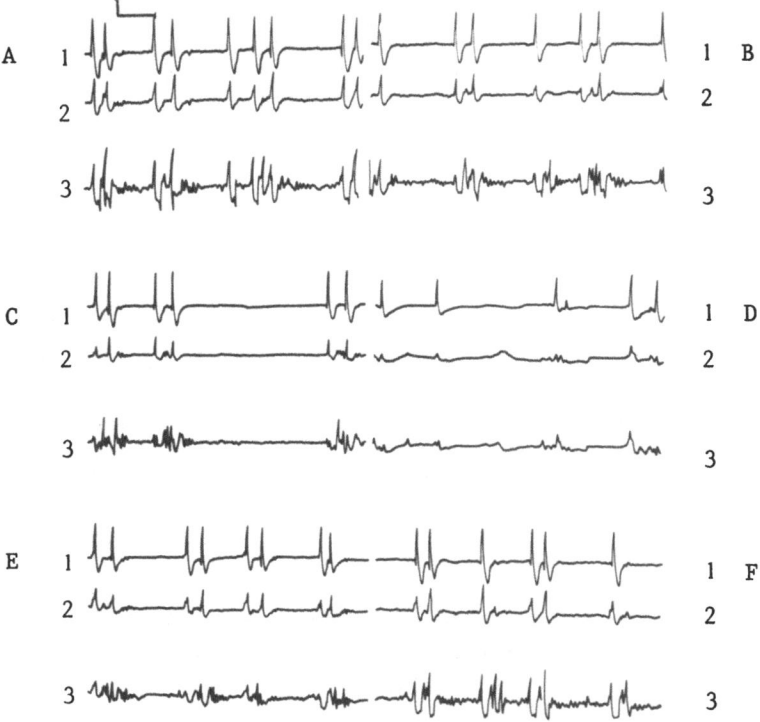

Fig. 101. Influence of halothane inhalation on the epiletic activity
           complex in the cerebral cortex. State of the complex
           before (A), and a few seconds (B), 1 min (C) and 2 min
           (D) after the beginning of halothane inhalation; E: 3 min
           after halothane inhalation was stopped. The determinant
           focus is in zone 1. The indications are the same as in
           Figure 100.

larities as if in the reverse sequence of focal suppression: slight
activity in the foci which were influenced most of all by the determinant focus appeared at first, and then it intensified and was
simultaneously exhibited in other foci. The activity being restored
was synchronized from the very beginning with the activity of the
determinant focus. Hence, the determinant focus which is deprived of
its 'stations of designation' (dependent foci) in anesthesia no
longer acts as a determinant structure, but it can act as such a
structure again as soon as the conditions under which the foci are
suppressed are removed.

A similar effect was produced when diazepam was being tested
with respect to its influence on the complex of epileptic activity in
the cerebral cortex. Studies (Kryzhanovsky et al., 1980e) have shown
that the epileptic complex is suppressed under the influence of

diazepam; when small doses are used, the process usually occurs in
the same above-mentioned sequence: activity is at first suppressed in
the dependent foci, and it disappears last in the determinant focus
(Figure 102). Such an effect was observed when phenazepam was also
being tested, the only difference being that phenazepam is more
active and acts in smaller doses than diazepam (Shandra et al.,
1982).

Experiments with diazepam have another noteworthy feature. As
it can be seen from Figure 102 F, the repeated application of strychnine to the determinant focus region with a view to activating the
complex had activated the determinant focus, but the complex was not
restored; among all the foci, only the focus which depended most of
all on the determinant one was slightly activated. Hence, diazepam
not only suppressed the activity of the foci of the complex, but also
made it difficult for the influences to spread from the determinant
focus. The determinant focus may remain when epileptic activity is
on a sufficiently high level. Diazepam 'cuts off', as it were, the
dependent foci from the determinant focus, suppressing them and
isolating the determinant focus.

The sequence in which the structures of the pathologic system
were functionally eliminated was studied on the basis of two more
syndromes, which are also characterized by epileptic activity but
which pertain to other areas of the central nervous system, namely,
spinal myoclonia ('spinal epilepsy') and photogenic epilepsy. It has
already been shown that GPEE formed in the propriospinal neuron
system of the ventral horn of the lumbosacral segments causes spinal
myoclonia (Chapter 6), and GPEE formed in LGN induces photogenic
epilepsy (Chapter 8).

In the case of spinal myoclonia, the activity of the formations
which realize the functional message from the hyperactive determinant
structure were suppressed in definite sequence under the influence of
sodium amobarbital. Electric activity in various muscles, which
reflects the activity of the corresponding segmental motoneurons and
the interneurons, shows that the motoneurons of the brainstem and
those of the proximal and then the distal segments of the spinal cord
were successively eliminated from the reaction under the influence of
sodium amobarbital; neuronal activity remained at a definite stage in
the segments where GPEE was produced, but later it also disappeared
(Figure 103). Hence, the pathologic system is suppressed in the same
sequence in this form of the pathologic state, too: the activity of
the parts which are the furthest from the hyperactive determinant
structure, i.e. those which are the least influenced by the determinant, is the first to be suppressed; then, the activity of the parts
which are closest to it is suppressed, while GPEE is the last to be
suppressed. The sequence of the elimination of the pathologic system
as its parts were 'cut off' from the hyperactive determinant was
clearly exhibited on the basis of that model due to the peculiarities

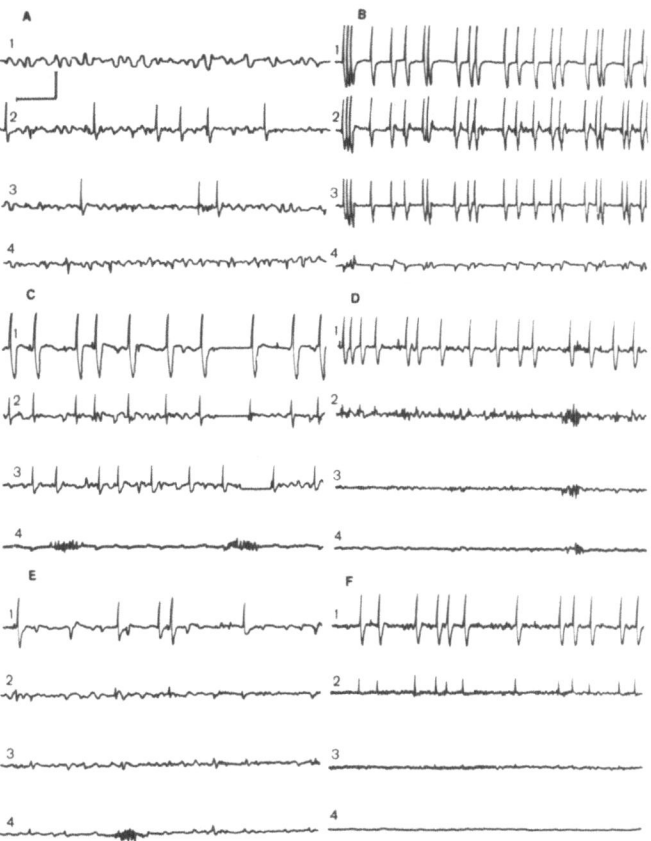

Fig. 102.   Influence of diazepam on the epileptic activity complex in
the cerebral cortex. A: Formation of foci in zones 2 and
3 after applying 0.1% strychnine; strychnine was removed
when epileptic activity appeared; B: synchronization of
epileptic activity in all the foci and the formation of an
epileptic complex under the influence of the determinant
focus (zone 1), which was produced by applying 3% strychnine. The recording was made 14 min after the determinant
focus was formed. C: 2 min, D: 12 min, and E: 24 min
after diazepam (3 mg/kg) was injected. F: 8 min after
strychnine was applied to zone 1 again in order to restore
the complex. 1: Orbital cortex; 2: coronary cortex; 3:
posterior sigmoid gyrus; 4: anterior sigmoid gyrus.
Amplitude: 500 μV; time: 1 s.

of the conduction pathways (short-axonal propriospinal connections)
and the development of the process along the spinal cord.

Studies of the influence of diazepam on the syndrome of photogenic epilepsy (Kryzhanovsky et al., 1979a; Rekhtman et al., 1979)

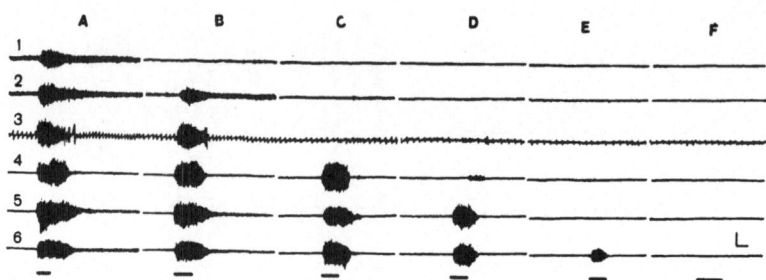

Fig. 103.   Effect of sodium amobarbital on generalized spinal
            myoclonia. Electric activity in the muscles before (A),
            and 1 min (B), 3 min (C), 5 min (D), 10 min (E), and 15
            min (F) after injecting i.v. sodium amobarbital (3 mg/kg).
            Recordings of EA: 1: right mastication muscles; 2: right
            neck muscles; 3: right intercostal muscles; 4: right
            dorsal muscles; 5,6: posterior groups of right and left
            femoral muscles. Seventy-two hours after TT was injected
            into the mucles of the left hind lef of a rat (4 DLM for
            the rat, and 0.025 IU of antitoxin i.v.). The horizontal
            line indicates stimulation. Amplitude: 20 µV; time: 1 s.

showed that diazepam in selected doses completely suppresses hypersynchronized activity in all the cerebral structures which were
examined, but it suppresses this activity far less in LGN, where GPEE
is formed (Figure 104). Under the action of diazepam, compulsive
photogenic reactions characteristic of the given syndrome disappeared
and the evoked responses in the visual cortex sharply diminished, but
they were more pronounced in LGN on the side of GPEE than on the
opposite side. Noteworthy phenomena were observed when the action of
diazepam weakened at the initial stage of withdrawal from its influence: characteristic pathologic focal autorhythmicity appeared during
this period in LGN on the side of GPEE. Spikes which sporadically
originated in it could be evaluated as interictal spikes. They
terminated in spontaneous generalized discharges, and then originated
again with greater amplitude (Figure 104 C). Later, the origin of
the initial epileptic discharges was connected with the greatest
epileptic activity in LGN, where GPEE was sited (Figure 104 D).
Hence, GPEE was suppressed less than other cerebral structures when
the epileptic system produced under the influence of GPEE in LGN was
being suppressed. It was the first to restore its activity when the
action of diazepam weakened. The origin of generalized epileptic
activity was connected with that restoration.

A study of the peculiarities of the restoration of the pathologic system in spinal myoclonia as the action of diazepam weakened
has shown that the process occurs according to a definite regularity:
at first, evoked activity originated in the segments which were close
to GPEE; afterwards, it successively originated in segments which

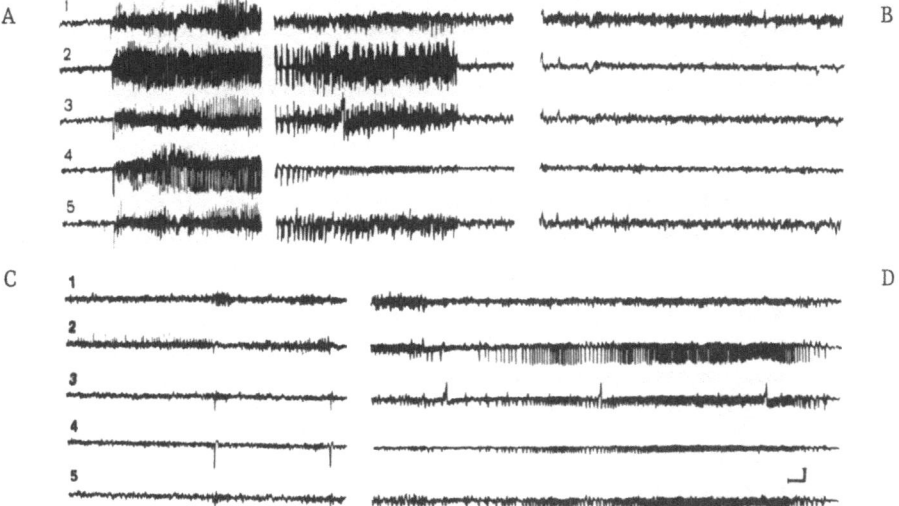

Fig. 104.   Change in EA in various structures of the brain in an
             animal with photogenic epilepsy under the action of
             diazepam. A: The beginning and the end of generalized
             seizure activity before the injection of diazepam
             (4 mg/kg). B: EA 5 min after injecting diazepam
             (4 mg/kg). C: EA 12 min later; interictal discharges in
             the right LGN. D: The epileptic after-discharge is pro-
             nounced in the right LGN 70 min after injecting diazepam.
             Recordings of EA: 1: right sensorimotor cortex; 2: right
             LGN (poisoned with TT); 3: right VC; 4: left LGN; 5: left
             VC. Amplitude: 200 μV; time: 3 s.

were further away. At the same time, the amplitude and duration of
the discharges in other segments grew, indicating that GPEE became
more and more powerful.

     Thus, the same regularity was observed in diverse models of
epileptic activity, which was induced by GPEE in various systems,
under the action of different anticonvulsants: the pathologic system
was suppressed nonuniformly under the influence of these agents.
When all other conditions were equal, the activity of the parts of
the system which were influenced least of all by the determinant
structure was the first to disappear, while GPEE which makes the
determinant structure hyperactive was the last to be suppressed.
The determinant disappears last. These data are explained well by
the concepts of the role of the hyperactive determinant structure in
the formation and maintenance of the pathologic system (Chapters 1
and 4).

The process of the pathologic system's general suppression when the anticonvulsant's action is generalized can be represented in the following sequence: the suppression of the activity of the parts of the system that are less dependent on the determinant → the suppression of the activity of the parts that are more dependent on the determinant → the disappearance of the determinant as a controlling structure, but the sustenance of its neurophysiological basis, namely, local GPEE → the suppression of the activity of the GPEE neurons, at first those of group III → then those of group II → finally those of group I. In urgent cases, GPEE can be suppressed in the event of the general suppression of the pathologic system by using large doses of drugs, when (depending on their properties) general relaxing, anesthetic or toxic effects originate. It may be useful to employ drugs in doses which are relatively small but which are enough to generally inhibit the pathologic system due to the disturbance of the facilitating connections between the parts of the pathologic system that sustain and enhance the system. If the physiological defense mechanisms and the systems which sustain the functional homeostasis of the central nervous system and which counteract the pathologic system (antisystems, see further) are not suppressed in this respect, such therapy can ultimately produce a favorable result when it is used for a long time due to the activity of the given natural sanogenic systems, the employment of the reserves of the central nervous system, and its plasticity reorganizations. The prolonged treatment of epilepsy by sustaining doses of barbiturates and other anticonvulsants is an example of the favorable effect of such therapy.

Various parts of the pathologic system can be influenced by appropriate therapeutic action (Figure 105). The determinant structure's effects are prevented from being finally realized when the activity of the effector structures is suppressed (Figure 105 B). Consequently, the syndrome clinically disappears. However, its pathogenic basis, i.e. the pathologic system with the determinant and intermediate links, remains. Many drugs of such pathogenetic therapy, which is in essence symptomatic therapy, produces an effect that can be called the 'effect of the effector link'.

Of course, the use of such drugs is far from being the best therapeutic method. They can be used from the standpoint of the concepts of the pathologic process as a process of only the breakdown of the physiological systems and the disorganization of the regulatory mechanisms. From that standpoint, this therapy is considered to be the most effective and even the only possible therapy. Its disadvantage was revealed by the concepts that neuorpathologic syndromes are based on pathologic systems. The tendency to normalize the activity of the system's effector link, i.e. the target organ, by separately influencing it without influencing the pathologic system's other links, particularly its determinant, is like treating epilepsy with curare-like agents. A physician who acts in this way is like a

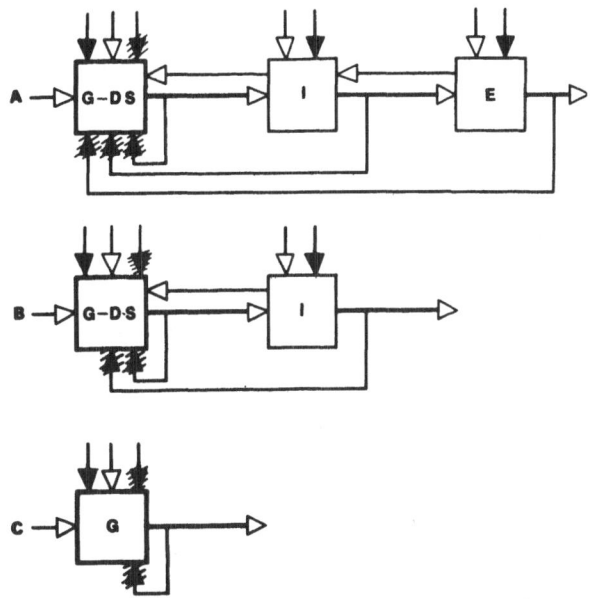

Fig. 105. Successive stages of the suppression and breakdown of the
           pathologic system under various therapeutic influences.
           G-DS: generator - determinant structure; I: intermediate
           links of the system; E: effector links of the system;
           G: generator. White triangles indicate exciting connec-
           tions, dark ones, inhibitory connections, and shaded ones,
           disturbed inhibitory connections.

person who tries to protect the floor in his house or to make it more
resistant to rain by improving its quality instead of repairing the
roof. Therapy that normalizes only the target organ's activity,
which was disturbed under pathologic central influences, can produce
a clinical effect, i.e. the syndrome's manifestations may disappear,
but it does not eliminate the pathologic system; consequently, there
may be a recurrence. Under these conditions, infinitely long sustaining therapy is required, and the doses should, in some cases,
gradually increase due to the formation of new receptors (on either
the effector link or the target organ's structures), the mechanisms
of pathologic plasticity, metabolic changes, etc.

     However, such therapy can sometimes produce a favorable result.
This result can be produced in the event of the visceral organs'
so-called functional disturbances, which are included in the wide
range of pathologic processes that we termed 'regulation diseases'
(Kryzhanovsky, 1978a, 1978b, 1981c, 1981d). Since the positive
feedbacks of various parts are important in enhancing the pathologic
system's activity and resistance, the pathologic system weakens and

becomes destabilized, and its resistance diminishes when, as we will
see later, it is reduced at the expense of its various parts, including the effector links. It has been experimentally shown on the
basis of pathologic states, such as conditioned-reflex hyperkineses
and hyperkineses in the event of neuroses, that feedback afferentation from the central effector links and from the target organs,
which are the 'outlets' of the pathologic states, can not only act
inhibitively (Khananashvili, 1978), but also sustain the pathologic
system and promote its stabilization (Danilov, 1968). The more
important the pathologic system's effector link in sustaining the
system, the more significant is its elimination from the system.
Then, it is easier for the plasticity and homeostasis mechanisms of
the nervous system to suppress or get rid of the pathologic system.
A definite length of time is required for realizing these processes
and consolidating the new state. Therefore, such sustaining therapy
should be sufficiently long.

     Pathogenetic therapy produces a greater effect when the system's
intermediate links are influenced. In this case, the intermediate
parts of the pathologic system as well as the effector link are
switched off. Then, the pathologic system becomes greatly reduced.
Being deprived of the dependent structures, the determinant loses its
own governing role. Only its neuropathophysiologic basis, i.e. a
generator (which is weakened and functionally limited), remains
(Figure 105). Such a local formation is less resistant to the natural mechanisms of suppression and to therapeutic influences (see
further). Under these conditions, the pathologic system as a whole
can be eliminated by natural homeostatic mechanisms more rapidly and
more effectively than in the former case, although a definite length
of time is required in this case, too. However, there may be a
recurrence while the generator remains (even if it is not manifested
clinically). This recurrence may take place under conditions which
are favorable to the activation of the generator and the restoration
of the determinant and other parts of the system (see Chapter 14).

     An analysis has shown that the specificity of the effect of
drugs increases as regards their action from the efferent links to
the determinant link of a pathologic system.

     The peculiarities of some therapeutic methods used in resistant
cases can be explained when account is taken of the data on the
pathologic system's properties. It has already been noted that the
pathologic system can be stabilized and can become more stable as
time passes. Like any biological system, it can become adapted to
the influences directed at it, as a result of which its resistance to
them grow. This is promoted especially when therapeutic methods are
inadequate, the doses of medicine are too small, therapy by the same
drugs is prolonged and hardly effective, etc. In some cases, the
pathologic system's resistance may be very great. Apart from the
cases of resistance that are connected with the drugs' enhanced

metabolism, their accelerated excretion and insufficient entry into
the brain, it should be noted that favorable results produced by
special therapeutic methods are due to the overcoming of the pathologic system's resistance. These methods include, for instance, the
treatment of the resistant forms of schizophrenia with very large
doses of neuroleptics, 'zig-zag' therapy, and the alternation of
large and usual doses (Donlon, 1976; Aubree, Lader, 1980; Belyakov,
1981; Dencer et al., 1981; Avrutsky et al., 1983). Many such doses
are in essence aimed at the pathologic system. They cause its destabilization, reduction or breakdown. Therapy by various types of
shock (insulin, electric, hyperthermic and other shocks) as a definite procedure is an extreme expression of such a method.

INFLUENCING THE PATHOLOGIC SYSTEMS BY ACTIVATING THE ANTISYSTEMS

It has been shown (Chapter 1) that two independent pathologic
systems, each of which could function when the other was destroyed,
had originated when two GPEEs were produced on both sides of the
spinal lumbosacral segments. In this respect, the following question
arises: what result will be produced when both GPEEs are activated?

Studies (Kryzhanovsky et al., 1977a) have shown that the GPEE's
effect will be suppressed when it is activated after the activation
of another GPEE (Figure 101 IV). This is evident from the fact that
the activity of the first (in this case, 'left') GPEE, which is
recorded in the form of hypersynchronous discharges in all the
muscles, is not observed when the second (right) GPEE is at first
activated. The result cannot be attributed to the depression of the
motoneurons due to the small interval between the stimulations of
both GPEEs, since the given suppressing effect was not produced when
each GPEE was doubly activated: the same burst of activity originated
in the same muscles when the 'left' generator was activated again
(Figure 101 II). Thus, the second GPEE's effect was actively suppressed when both GPEEs were successively activated. In short, the
other system's activity is suppressed when one system is activated.
This is an example of reciprocal or dominant relations (Chapter 1).
A noteworthy fact is that although both activated systems are pathologic, they can produce an inhibiting effect and can be inhibited.

The above data are noteworthy in many respects. Firstly, they
show that the pathologic system can produce a powerful inhibitory
effect. Such an effect should be really powerful in order to suppress the activity of the other system which is just as hyperactive.
It follows from this fact that the pathologic system strongly influences also other cerebral structures with which its disorganizing
influences and aggressiveness are connected (Chapter 4). Secondly,
the given active suppression of one pathologic system by another
shows that the inhibitory influences, which can suppress the activity
of the whole system and that of GPEE when activation is sufficiently
great, are somewhat preserved in the pathologic system and in GPEE.

Taking that example, every given system can be regarded on the model plane as an antagonistic system or an antisystem with respect to the other corresponding system. This fact suggested that antisystems, which can suppress the pathologic systems' activity on the basis of physiological mechanisms, should exist or originate in the central nervous system when pathologic processes develop. Such a conclusion was based on the physiological principle of regulating the activity of various structures of the central nervous system by activating the functional antagonists. The conclusion was verified on the basis of two different forms of a pathologic state that originated when GPEEs were produced in different systems.

There is now much data which show that analgesia is produced when some cerebral structures are electrically or pharmacologically stimulated. The analgesic effect was produced by stimulating the hypothalamus (Balagura, Ralph, 1973), the ventral posterior lateral nucleus of the thalamus (Hosobuchi et al., 1973; Gyvels et al., 1976; Mazars et al., 1976), the caudate nucleus (Lineberry, Vierck, 1975), catecholaminergic nuclei (Segal, Saudberg, 1977; Saudberg, Segal, 1978), the septa (Gol, 1967), and the reticular formation (Akaike et al., 1978; Zorman et al., 1981, 1982). The analgesic effect is especially pronounced when the periaqueductal gray substance and the dorsal and major raphe nuclei are stimulated (Reynolds, 1969; Mayer et al., 1971; Oliveras et al., 1974; Oliveras et al., 1975; Fields et al., 1976, 1977; Morrow, Casey, 1976; Olesson, Liebeskind, 1976; Hosobuchi et al., 1977; Fields, Anderson, 1978; Igonkina, Kryzhanovsky, 1977, 1979; Olesson et al., 1978; Behbehani, Fields, 1979; Bennett, Mayer, 1979; Carstens et al., 1979; Jurna, 1980; Mohrland et al., 1982). These structures constitute the antinociceptive system (Mayer, Liebeskind, 1974; Melzack, Melinkoff, 1974; Akil, Liebeskind, 1975; Liebeskind, 1976; Mayer, Price, 1976; Waldman, Ignatov, 1976; Basbaum, Fields, 1978; Fields, Anderson, 1978; Kryzhanovsky, 1979, 1980a; Mayer, 1979).

We carried out similar investigations as regards a pain syndrome of spinal origin that was produced by GPEE in the dorsal horns of the spinal lumbosacral segments created by TT. As it has been shown (Chapter 9), this syndrome is very severe, and its severity grows as GPEE's power increases. Studies (Igonkina, Kryzhanovsky, 1977; Kryzhanovsky, 1979) have shown that analgesia with respect to not only physiological pain, but also pathologic pain originates when one of the antinociceptive structures (the dorsal raphe nucleus) is electrically stimulated. The sufficiently intensive electric stimulation of the given zone stops the spontaneous or induced pain fits even when the pain syndrome has developed considerably (Figure 106).

Studies (Kryzhanovsky, Igonkina, 1978; Igonkina, Kryzhanovsky, 1979; Kryzhanovsky, 1979) have shown that the inhibiting analgesic effect can be produced by injecting TT into the dorsal raphe nucleus (Figure 107). To maximalize the exhibition of this effect, TT was

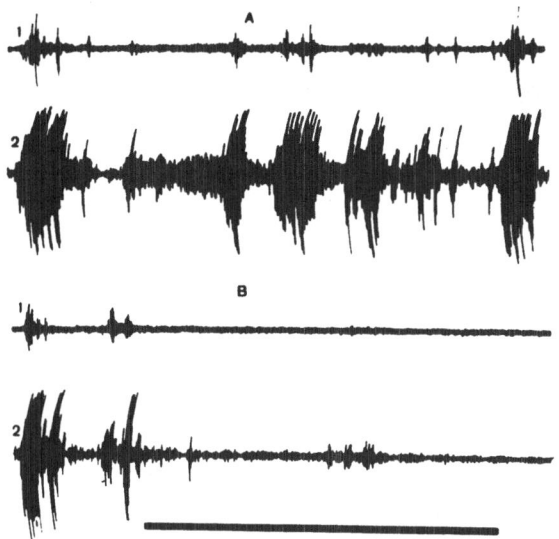

Fig. 106.   Influence of ES of the dorsal raphe nucleus on the pain
            syndrome of spinal origin. A: Exhibition of the pain
            syndrome (1: phonogram, 2: actogram) before ES. B: Sup-
            pression of the pain syndrome during ES of the dorsal
            raphe nucleus. The duration of ES is indicated by a
            horizontal line. Time: 10 s.

injected into the raphe nucleus beforehand, i.e. a few hours before
TT was applied to the spinal cord's dorsal surface. This was done
because a sufficiently powerful generator must be formed in the
dorsal raphe nucleus in advance so as to suppress the effects of a
powerful GPEE in the spinal cord.

     A paradoxical situation originates: the effect of TT in the
spinal cord is suppressed by the effect of the same agent in the
dorsal raphe nucleus. However, such a situation can be easily explained: the cause lies not in TT, but in GPEEs, which originate
under its influence. TT can be replaced by another agent which
engenders GPEE, and the same effect will be produced. Such an effect
was produced in the experiment, in which the dorsal raphe nucleus was
electrically stimulated in order to produce an artificial generator
in the nucleus.

     Experiments in which the electric activity of the structures of
the dorsal raphe nucleus was studied have shown that GPEE really
originates in this nucleus under the influence of TT (Igonkina,
Kryzhanovsky, 1979; Kryzhanovsky, 1979). It was established in these
experiments that the evoked potentials greatly changed in the dorsal
raphe nucleus after TT was injected into it: the first negative wave
considerably increased, new components originated, the first positive

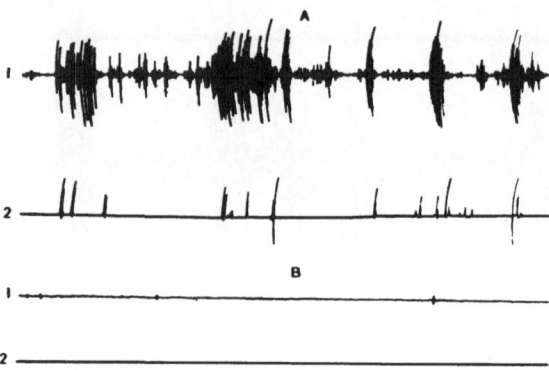

Fig. 107.   Preventive suppression of the pain syndrome of spinal
            origin with GPEE in the dorsal raphe nucleus. A: Exhi-
            bition of the pain syndrome (1: actogram; 2: phonogram)
            produced by applying TT to the dorsal surface of a rat's
            spinal lumbosacral segments. The recording was made 4 h
            after TT was applied. B: No exhibition of the pain syn-
            drome when GPEE was produced in the dorsal raphe nucleus
            by microinjecting TT 8 h before producing GPEE in the
            dorsal horns of the lumbosacral segments (by applying TT
            to the spinal dorsal surface). The actogram and the
            phonogram were taken 4 h after applying TT to the dorsal
            surface of the spinal cord. Time: 5 s.

fluctuation diminished, and spontaneous potentials of a high amplitude originated more frequently; all these changes increased with
time (Figure 108). Background impulse activity of the neurons of the
nucleus greatly increased. As it has already been shown (Chapters 2
and 3), these features are characteristic of the formation of GPEEs.
An important fact is that the origin of analgesia after the injection
of TT into the dorsal raphe nucleus coincided as regards time with
the origin of the above changes in the evoked potentials in the
nucleus.

    Hence, the production of GPEE in the dorsal raphe nucleus, being
part of the 'antinociceptive' system, causes profound and virtually
permanent analgesia, stopping the development of the pain syndrome
engendered by GPEE in the dorsal horn.

    The role which the antisystems play in suppressing pathologic
hyperactivity can be shown by taking the example of epilepsy. Epileptic activity can be suppressed by electrically stimulating some
cerebral structures: the cerebellum (Dow, 1961, 1965; Dow et al.,
1962; Kreindler, 1962; Mutani et al., 1969; Hutton et al., 1972; Babb
et al., 1974, 1977; Dolina, Arshavsky, 1975; Maiti, Snider, 1975;

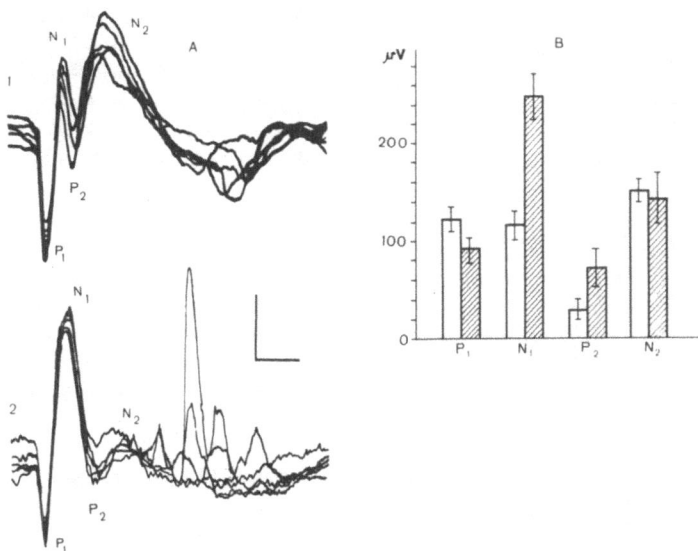

Fig. 108.   Change in EP in the dorsal raphe nucleus after TT
            injection. A: EP before (1) and 8 h after (2) TT was
            injected into the nucleus. The potentials were evoked by
            nociceptively (electrically) stimulating a region of the
            skin on the tail. Amplitude: 100 μV; time 50 ms. B:
            Amplitude of the compounds of EP in the dorsal raphe
            nucleus before (white columns) and 8 h after (shaded
            columns) TT injection. The ordinate indicates the ampli-
            tude. P_1, N_1, P_2 and N_2 are the negative and positive
            components of EP.

Hablitz, 1976; Hablitz, Rea, 1976; Majkowski, 1980; Kryzhanovsky et
al., 1981c, 1983a), the caudate nucleus (Mutani, 1969; Mutani,
Fariello, 1969; La Grutta et al., 1971; Arushanyan, Otellin, 1976;
Oakley, Ojemann, 1980), locus ceruleus (Khanababyan, 1980), certain
thalamic nuclei (Smith, Purpura, 1960; Wagner et al., 1975; Oakley,
Ojemann, 1980), and the caudal reticular pontis nucleus (Okujava,
1969; Okujava et al., 1979; Kryzhanovsky et al., 1980c). There are
data on the clinical effectiveness of the electric stimulation of the
cerebellum (Cooper et al., 1973, 1980; Cooper, 1978; Bidzinsky et
al., 1980; Ryabokon et al., 1980; Davis et al., 1982) and the caudate
nucleus (Sramks et al., 1976; Ojemann, 1980).

    It was shown in our researches (Kryzhanovsky, Russev, 1976) that
the strychnine foci in the cerebral cortex are suppressed when the
zone of the forebrain medial band and the preoptic region are damaged. Further researches have shown that this effect is connected
with the activation of the caudate nucleus.

The peculiarities of the suppression of the epileptic system, its parts and the determinant when the antisystem is activated are especially noteworthy. Investigations in this respect were carried out on the basis of an epileptic complex in the cerebral cortex. It was shown that definite electric stimulation of the posterior hypothalamus causes the epileptic complex to break down and be suppressed. This process occurs in the above sequence in which the dependent foci are 'cut off' as the determinant focus remains (Figure 109). Electric stimulation had to be performed several times in order to suppress the epileptic complex. The inhibiting antiepileptic effects of electric stimulation can greatly vary, and they depend on many conditions; a change in the conditions of electric stimulation can considerably alter the effect. This distinguishing feature has been indicated by other authors, too (Gellhorn et al., 1960; Gellhorn, Loofbourrow, 1963).

The effects varied as regards epileptic complexes in the cerebral cortex when the cerebellar structures, i.e. the vermis (Shandra et al., 1982), the dentate nucleus (Godlevsky, Shandra, 1983) and the fastigial nucleus (Shandra, Godlevsky, 1983), were also electrically stimulated. Such specifics were observed by other researchers, too (Babb et al., 1974). Researches (Kryzhanovsky et al., 1983a) which were specially carried out in this respect have shed light on some conditions and mechanisms of variability of the given effects. It has been established that when the activity of the primary, determinant epileptic focus in the cerebral cortex is not relatively great, the electric stimulation of the contralateral dentate nucleus of the cerebellum suppresses it. At the same time, epileptiform activity induced in other areas of the cortex by the determinant focus is also suppressed. When epileptic activity in the determinant focus is originally very high (e.g. when the focus is produced by a concentrated penicillin solution), electrical stimulation with the same parameters of the contralateral dentate nucleus may enhance it, i.e. the frequency of the potentials increases (Figure 110 B, zone 1). Mixed inhibiting and stimulating effects are simultaneously observed in the focus of induced epileptic activity (Figure 110 B, zone 2): the frequency of the discharges increases (due to the influence of the determinant focus), while their amplitude decreases. When epileptic activity in the determinant focus attenuates (in the course of time or under the influence of the continuing electric stimulations of the contralateral dentate nucelus), the electric stimulation of the nucleus no longer enhances, but inhibits activity in the determinant focus (Figure 110 C, zone 1); in the focus of induced epileptiform activity, the latter disappears during electric stimulation (Figure 110 C, zone 2). A noteworthy fact is that activity in zone 3 remains, being due to the greater and direct influence of the determinant focus on the 'mirror' focus. Later, activity in all the foci, including the determinant focus, is suppressed and disappears when the nucleus is stimulated (Figure 110 D, zones 1,2,3). This effect may remain even when the nucleus is not stimulated (Figure 110 E).

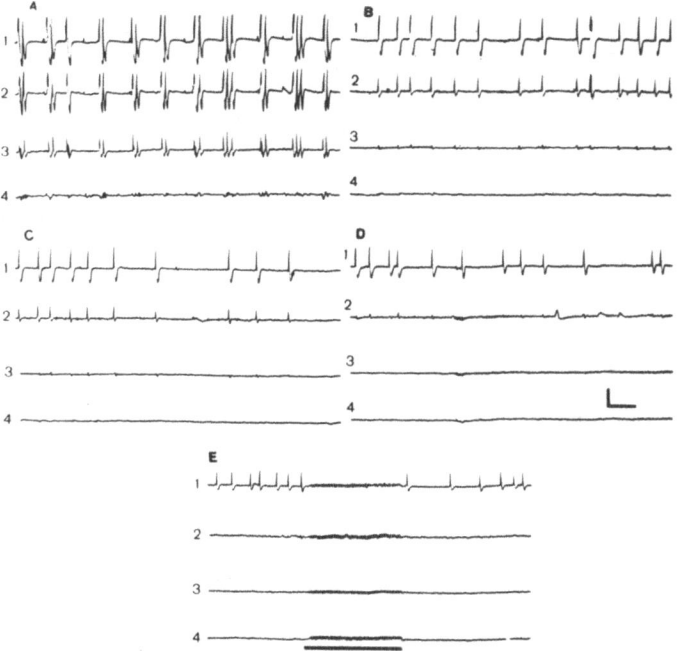

Fig. 109.   Effect of ES on the posterior hypothalamus on the
            epileptic complex in the cerebral cortex. Activity of the
            complex before ES (A), after three ES seances (B); six ES
            seances (C), seven ES seances (D), and during ES
            immediately after D (E) (25 min after the beginning of
            stimulation treatments). Parameters of every ES seance:
            duration: 10-20 s; frequency: 100 impulses/s; duration of
            the impulse: 0.5 ms; amplitude: 4 V. The determinant
            focus was produced by applying crystal strychnine to zone
            1 (orbital cortex); other foci were produced by applying
            0.1% strychnine; zone 2: coronary gyrus; zone 3:
            ectosylvian gyrus; zone 4: sigmoid gyrus. When the
            complex was produced, strychnine was removed from zone 1
            (determinant focus). E: the duration of ES (with the same
            parameters as those given above) is indicated by a
            horizontal line. Amplitude: 500 μV; time: 1 s.

Under these conditions, the coagulation of this nucleus causes a
recurrence of epileptic activity in the previously inhibited foci
(Figure 110 F, zones 1,2,3).

     The dual effect of the electric stimulation of the dentate
nucleus, i.e. the intensification of epileptic activity in the powerful focus and the suppression of this activity in a weaker focus, can
be explained by the concepts of the dual nature of functional transmission produced by the structure being stimulated in the central

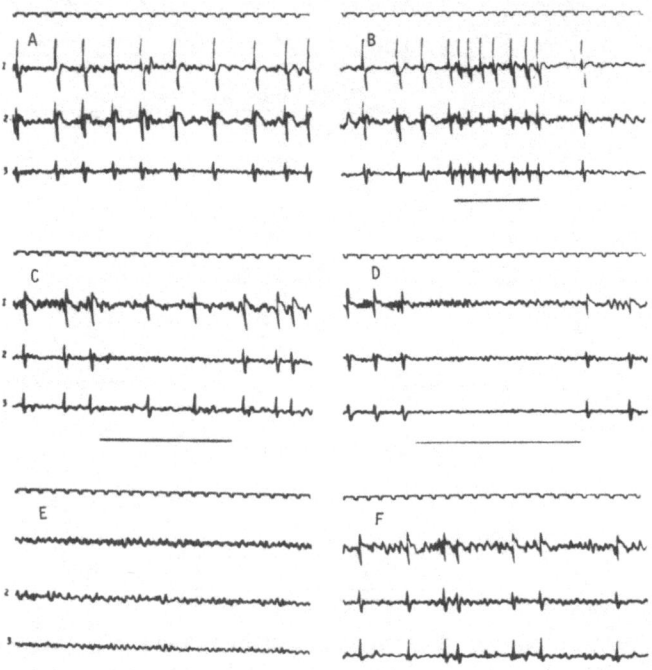

Fig. 110.   Influence of ES of the dentate nucleus on a powerful
            epileptic focus in the anterior sigmoid hyrus, a secondary
            induced focus in the contralateral hemisphere, and the
            'mirror' epileptic focus. A: 9 min after applying a
            penicillin solution (40,000 U/ml) to zone 1; penicillin
            was removed when discharges appeared; B: increase in the
            frequency of discharges in zones 1 and 3 and a decrease in
            the amplitude of discharges in zone 2 during 5 min of ES
            of the dentate nucleus after the cessation of penicillin
            application; C: diminution of the amplitude and frequency
            of discharges in zones 1 and 3 and the suppression of
            discharges in zone 2 in repeated ESs of the nucleus 12 min
            after the beginning of ES; fifth ES seance; D: suppression
            of epileptic discharge in zones 1, 2 and 3 during ES of
            the nucleus after seventh ES seance (17 min after the
            beginning of ES); E: complete suppression of the dis-
            charges in all the zones following ninth ES seance (20 min
            after the beginning of ES); F: 3 min after the electric
            coagulation of the dentate nucleus. 1: Anterior sigmoid
            gyrus of the contralateral hemisphere; 2: posterior
            sigmoid gyrus of the contralateral hemisphere; 3: 'mirror'
            focus. ES parameters: 300 Hz, 0.25 ms, 3.5 V. Amplitude:
            500 μV; time: 1 s.

nervous system (see Chapter 3). Exciting and inhibiting functional
messages originate when the dentate nucleus is stimulated. The
preservation of the inhibitory mechanisms in the structures which
receive the message that induces an inhibitory effect is a prerequisite for the realization of such an effect. When the inhibitory
mechanisms are more or less preserved (e.g. in the determinant focus
with low epileptic activity, as has been shown in the researches by
Hutton and others [1972], and in the induced foci which are free of
epileptogen action), the electric stimulation of the dentate nucleus
suppresses activity in the given foci. When the inhibitory mechanisms are upset (e.g. in the powerful focus engendered by a large dose
of penicillin), the inhibiting message is not realized, while the
exciting message, which is not exhibited under normal conditions, is
realized. This fact is expressed in the intensification of the
epileptic activity of the focus which was influenced by the dentate
nucelus being stimulated. Such relations were observed also in other
forms of the pathologic state of the central nervous system
(Chapter 3).

Investigations involving the use of benzodiazepine preparations
when the cerebellum is electrically stimulated have illustrated the
significance of the level of both epileptic activity and the preservation of inhibitory mechanisms in the cortical epileptic foci as
regards the effects produced by stimulating the cerebellar structures. These investigations showed that the additional administration of benzodiazepines, which enhance GABAergic inhibition,
reduces the exciting effect and enhances the inhibitory effect when
the vermis is electrically stimulated (see further).

Hence, the variability of the inhibiting effects concerning
epileptic activity and even their replacement by exciting effects,
which have been observed under definite conditions when some specialized cerebral structures (posterior hypothalamus, dentate and fastigial nuclei, the vermis, the reticular caudal pontine nucleus
[RCPN], and caudate nuclei) were stimulated, do not discredit the
concepts of the role of these cerebral structures and others as parts
of the antiepileptic system which maintain functional homeostasis in
the brain under the action of epileptogenic factors. When these
structures of the antiepileptic system are functionally destroyed, it
is easier for epileptic activity either to originate and be generalized or to reappear after it is suppressed.

Our researches and the data provided by other authors (Okujava,
1969) have shown that epileptic activity in the cerebral cortex is
suppressed when RCPN is electrically stimulated as the central gray
substance is being destroyed. This effect is clearly expressed when
there is one epileptic focus (Kryzhanovsky et al., 1980c) (Figures
111 and 112). The suppressing effect can be so great that single
electric stimulation causes the focus to disappear, and it will not
be exhibited for a long time (Figure 111 C). Its suppression is

maintained by the activity of the stimulated nucleus: the coagulation
of this nucleus causes epileptic activity to be restored (Figure
111 D). Hence, a state of prolonged excitation was produced in the
nucleus when it was electrically stimulated. This state was significant functionally, since it suppressed the epileptic focus. In other
words, electric stimulation engendered a temporary generator of
sufficiently powerful excitation in the nucleus being stimulated.
This generator could then operate for some time without sustaining
stimuli.

Experiments were also carried out under specially selected
conditions, when electric stimulation of RCPN was stopped after a
single focus was suppressed (see Figure 112 A), as a result of which
focal activity was restored. Against this background, additional
foci were produced by means of relatively weak epileptic influence
(by locally applying strychnine in low concentrations). All the foci
combined into an epileptic complex, which operated in a unique mode;
the first, most powerful focus in zone 1 was the determinant one. In

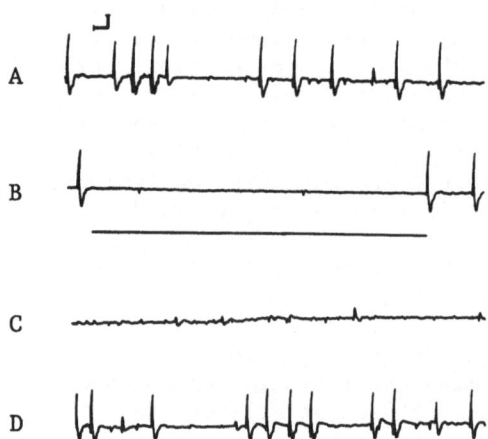

Fig. 111.   Influence of ES and coagulation of the reticular caudal
             pontine nucleus (RCPN) on the epileptic focus in the
             cerebral cortex. A: Production of the focus of epileptic
             activity in the cerebral cortex (middle sigmoid gyrus) by
             applying 1% strychnine. B: Suppression of the focus of
             epileptic activity during ES of RCPN. Parameters of ES:
             4.5 V; 220 impulses/s; 0.5 ms; the duration of stimulation
             is indicated by a horizontal line. C: Disappearance of
             epileptic activity under the influence of prolonged ES of
             the nucleus; 28 min after ES. D: 3 min after the coagu-
             lation of the nucleus: restoration of epileptic activity.
             Parameters of electric coagulation: current: 5 mA;
             duration: 2 min. Amplitude: 500 μV; time 1 s.

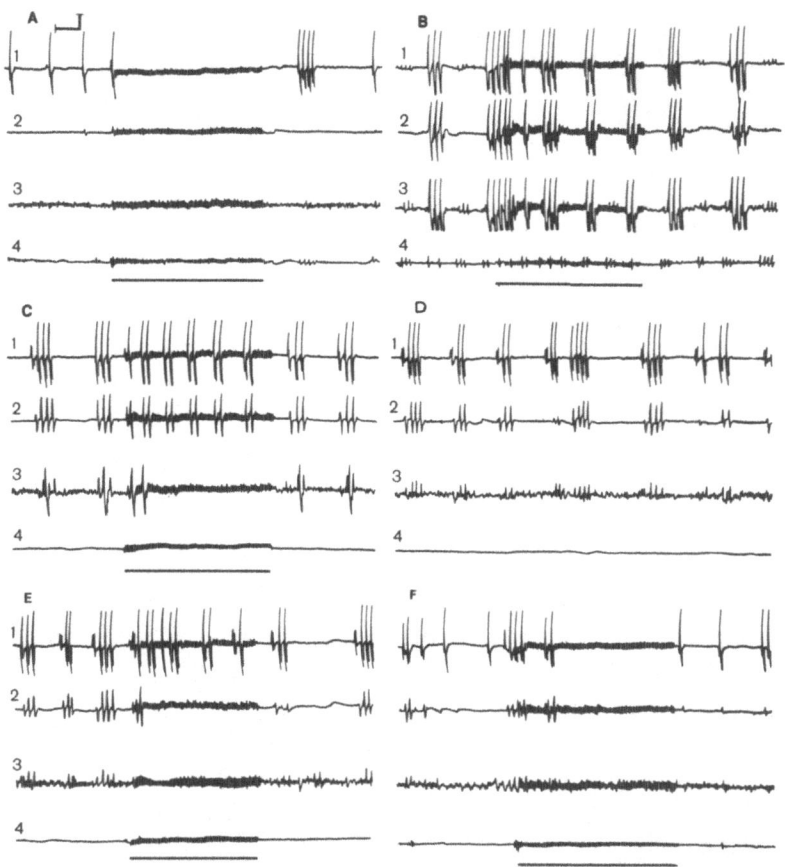

Fig. 112.   Influence of ES and coagulation of RCPN on a separate
            epileptic focus and a complex of epileptic foci in the
            cat's cerebral cortex. A: Suppression of activity in the
            epileptic focus produced in zone 1 (middle sigmoid gyrus)
            by applying 1% strychnine (strychnine was removed after
            the appearance of epileptic activity) when RCPN was elec-
            trically stimulated. The duration of ES in this and other
            fragments is indicated by a horizontal line below. Para-
            meters of ES: 4.5 B; 220 impulses/s; 0.5 ms. B: Formation
            of a complex of epileptic foci with a unique pattern of
            activity which is determined by the focus in zone 1. The
            foci in zone 2 (anterior sigmoid gyrus) and zone 3 (pos-
            terior sigmoid gyrus) were produced by applying 0.1%
            strychnine. Strychnine was removed after epileptic ac-
            tivity appeared in them. Zone 4 (coronary gyrus) is
            intact. The determinant focus in zone 1, the complex as a
            whole and dependent foci were not suppressed when RCPN was
            electrically stimulated with the same intensity. C:
            Diminution of epileptic activity in the dependent foci
                                                   (Continued overleaf)

this case, the electric stimulation of RCPN by the same parameters (or even stronger stimulation) suppressed neither the first focus in zone 1 nor weaker foci in other zones (Figure 112 B). Continuing electric stimulation suppressed the activity of the complex and caused it to gradually disintegrate. This process occurred more rapidly than under ordinary conditions (without electrically stimulating the nucleus). The complex broke down according to the already known regularities: the foci which were less dependent on the determinant were the first to attenuate (Figure 112 C and D). At this stage, the suppressing influence from the stimulated nucleus was realized on the fading focus, while activity remained in the other foci. Such an inhibiting effect was produced by stimulating the nucleus on the next focus as its activity attenuated, while the determinant focus remained (Figure 112 E). At the stage when the complex almost disappeared, the electric stimulation of RCPN suppressed epileptic activity of also the primary (the former determinant) focus in zone 1 (Figure 112 F).

Several very important conclusions can be drawn from that experiment. Just as the previous experiment, it shows that the hyperactive pathologic system should be destabilized, or 'loosened up', in order to suppress it. It can be destabilized when the antisystem continues to be active.

The phenomenon of the pathologic system's destabilization is apparently a universal regularity. It is widely known in clinical practice involving therapeutic electric stimulations. Various types of therapy aimed in essence at destabilizing the pathologic system (therapy involving different shocks, 'zig-zag' therapy, 'shock' therapy involving large doses, etc.) were discussed earlier.

(zones 2 and 3), especially in zone 3, due to the continued ES of RCPN (the duration of every ES seance was 10-15 s, and the interval between them was 2 min). The fragment was recorded 6 min after ES began; suppression of epileptic activity in zone 3. D: Further diminution of epileptic activity in foci 2 and 3; discharges in zone 3 almost disappeared; during this period, ES suppressed epileptic activity in zone 2 (12 min after ES seance began) (E). F: Substantial reduction of epileptic activity in zone 2. During this period, ES suppressed epileptic activity not only in this zone, but also in the determinant structure (zone 1). The determinant focus completely disappeared after another three ES seance of RCPN. Before the experiment, the central gray substance was coagulated. The coagulating current was 4.5 mA, and it was applied for 30 s. The animal was not anesthetized. Amplitude: 500 μV; time 1 s.

The destabilization of the intracentral relations which were formed is a prerequisite for overcoming the stable pathologic state and producing new, biologically favorable intercentral relations (Bekhtereva, 1974, 1980; Bekhtereva et al., 1972, 1978). Experiments involving the suppression of epileptic foci when RCPN is electrically stimulated have also shown that the resistance of the epileptic focus differs, depending on whether it exists alone or is in a system. The focus in zone 1 could relatively easily be suppressed as RCPN was electrically stimulated when it existed as a separate GPEE. In some cases, it could not have been restored further when electric stimulation was performed for a long time. However, such a focus became especially resistant when, being even somewhat weak as time passed after strychnine was washed away, it had turned into a determinant structure in the system. Moreover, other foci produced by relatively weak strychnine solutions were resistant to inhibiting influences as regards the electric stimulation of RCPN when they became part of the epileptic system. But when this system began to disintegrate under destabilizing influence, its foci lost their resistance and could be suppressed again when the antiepileptic structure was activated. In particular, the determinant focus lost its resistance when it no longer played a determinant role as the system disintegrated and when it became a separate generator of excitation again. This example clearly shows once more the difference between the hyperactive determinant structure as a governing structure (systemic category) and the generator of excitation as a local functional structure (neurophysiologic category). The pathologic system's resistance to the anti-system's influences depends on not only the determinant structure's power, but also the interaction between all the system's links (in this case, by all the epileptic foci), i.e. the system as a whole. Studies have shown that the more powerful the pathologic system, the more links it has and the more resistant it is.

This conclusion can be illustrated by the results of one of the experiments which were carried out in order to study the epileptic complex's resistance (Kryzhanovsky et al., 1980c) (Figure 113). A powerful focus which acts as the determinant structure was produced in zone 1, while a relatively weak focus, whose power increased under the influence of the determinant focus and together with which it constituted the epileptic complex, was produced in zone 2. Such a bifocal complex is suppressed when RCPN is electrically stimulated (Figure 113 A). As one more focus (in zone 3) was produced, the complex consisting of three foci (in zones 1, 2 and 3) was not suppressed when the nucleus was electrically stimulated, but the focus in zone 4 (not in the complex) was suppressed in this case (Figure 113 B). When the electric stimulation of the nucleus was ceased and the activity in the foci was restored, the elimination of one of the foci which constitute the complex (zone 3) by locally applying nembutal resulted in the fact that the complex was made up of only two foci, while the focus in zone 4 became autonomous (Figure 113 C). Under these conditions, all the foci of the reduced complex and the

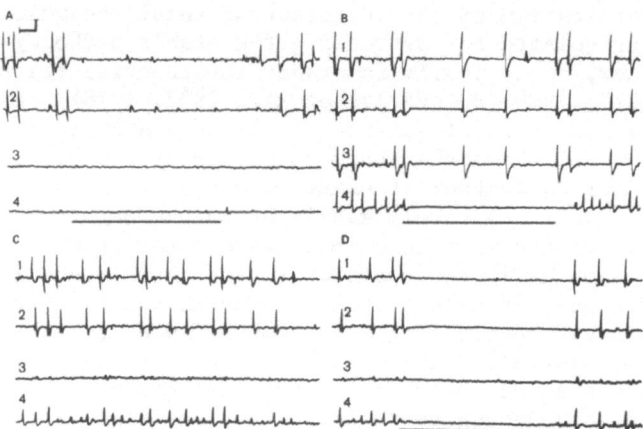

Fig. 113. Resistance of epileptic complexes consisting of a
           different number of foci to the inhibiting influences of
           ES of RCPN in a cat. A: The focus in zone 1 was produced
           by applying 1% strychnine, and in zone 2, by applying 0.1%
           strychnine. The activity of both foci which constitute
           the complex was suppressed when RCPN was electrically
           stimulated. B: Two additional foci in zones 3 and 4 were
           produced by applying 0.1% strychnine; a complex was formed
           out of three foci (zones 1, 2 and 3); the focus in zone 4
           was not included in the complex. ES of RCPN suppressed
           the activity of the focus in zone 4, but did not suppress
           the activity of the foci of the complex. C: The exclusion
           of the focus in zone 3 from the complex by locally apply-
           ing 6% nembutal (2 min after its application). D: Sup-
           pression of epileptic activity in all the foci. The
           indications are the same as in Figure 112.

autonomous focus were completely suppressed when RCPN was electrically stimulated (Figure 113 D).

     The significance of the pathologic system's power in establish-
ing its resistance to the antisystem's influences is also illustrated
in Figure 114. A complex consisting of relatively weak foci is
suppressed when RCPN is electrically stimulated (Figure 114 A). When
only the determinant focus in zone 1 is strengthened, epileptic
activity grows not only in it, but also in other foci of the complex.
Then, none of the foci of the epileptic complex is suppressed when
RCPN is electrically stimulated with the same force (Figure 114 B).
A noteworthy fact is that, under these conditions, a focus of even
low activity (zone 4) in the complex is not suppressed. This experiment shows that the whole pathologic system's resistance grows as
its power increases; that effect can be produced by enhancing the
determinant structure's power.

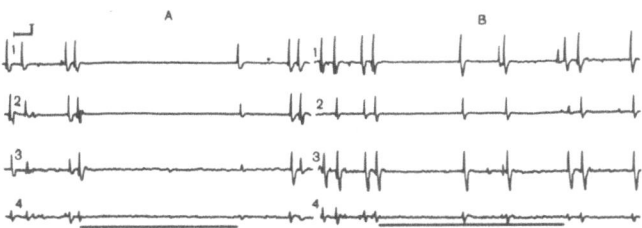

Fig. 114.   Increase in the resistance of the epileptic complex to the
             inhibiting influences caused by ES of RCPN with the growth
             of the power of the determinant focus. A: A complex of
             foci of relatively poor activity which were produced by
             applying both 1% strychnine to zone 1 and 0.1% strychnine
             to zones 2-4; suppression of activity in all the foci
             during ES of RCPN.   B: Additional application of 3%
             strychnine to zone 1; increase in the activity of the foci
             in zones 1, 2 and 3; ES of RCPN did not suppress the
             activity of the foci of the complex. Indications are the
             same as in Figure 112.

Hence, the pathologic system's formation is an essentially new
stage of the pathologic process. The system which originated, unlike
each of its parts, acquires new properties; as it is formed and
stabilized, it becomes very resistant to the influence of the physiologic mechanisms of inhibition. The nature of this intrasystemic
interaction, which enhances the resistance of both the whole system
and its parts, is a new phenomenon and should be specially analyzed.
As regards application, it is expedient to draw conclusions which can
be used: the pathologic system and its hyperactive determinant structure are suppressed more easily when the system is destabilized and
disintegrates, when it is spatially and functionally reduced, and
when its parts are 'cut off' and its power diminishes. It follows
that the influences which cannot eliminate the determinant structure
but which overwhelmingly affect the pathologic system, destabilizing
it in one way or another and causing it to disintegrate, can be
useful in the syndrome's general therapy. It also follows that if
symptomatic therapy weakens the syndrome by suppressing the pathologic system's individual links, this therapy can be considered to be
also partially pathogenetic therapy, which helps eliminate the system
in the case of specific pathogenetic interventions.

Hence, by activating the antisystem, the respective neuropathologic syndrome can be arrested, because the pathologic system which
constitutes the syndrome's basis is suppressed. These effects are
apparently realized by physiological mechanisms. For instance, the
suppression of epileptic activity in the cerebral cortex when RCPN
has been electrically stimulated is connected in one way or another
with the activation of the desynchronizing reticular formation. An

interesting fact is that additional effects in the form of switching off definite structures can greatly influence the effect of the antisystems' activation. For instance, the effect of the electric stimulation of RCPN was enhanced and became more stable when the central gray substance was preliminarily coagulated. The above-mentioned effects of the electric stimulation of the lateral hypo-thalamus or the GPEEs produced in it were enhanced when the medial hypothalamus was coagulated (Chapter 12). Thus, the therapeutic effect of the pathologic system's suppression can be enhanced by suppressing or eliminating the structures which may be physiological antagonists with respect to the structures that are electrically stimulated in order to activate the antisystem. Conversely, it has been observed in many researches that the pathologic process can sharply intensify and even rapidly develop when the antisystems' structures are damaged. For instance, the complex of epileptic activity is formed more rapidly in the cerebral cortex, and convul-sive activity becomes greater and longer, when a part of the vermis is removed (the experiments were carried out by Makulkin, Shandra and Lobasyuk). Such an effect was observed also in an experiment with cerveau isolé when the brainstem structures which control cerebral excitability were cut off (Kryzhanovsky et al., 1980b).

The antisystem can be activated not only by electrically stimu-lating its structures. Such an effect can be produced also pharma-cologically. What is borne in mind in this respect is not the drugs' generalized influence, but their effect on definite regions of the central nervous system and even separate structures in the system. The widely known action of chlorpromazine can be given as an example. This agent effectively suppresses hyperactivity in the spinal cord, particularly hyperactivity in the spinal motor nuclei. When GPEE is being produced in the spinal cord in the efferent output system (Chapter 2), different stimuli either on the segmental level (elec-tric stimulation of the cutaneous nerve) or by suprasegmental struc-tures (electric stimulation of the bulbar structures) cause enhanced electric activity in the appropriate muscles of the hind leg on the side of GPEE. Such activation of GPEE is suppressed by chlorproma-zine. However, this inhibiting effect is produced only when the connections between the spinal cord and the suprasegmental levels are preserved. It is not produced when the spinal cord is cut (Figure 115). In this case, an inhibiting effect cannot be produced even when large doses of chlorpromazine are used, although the drug is effective in relatively small doses when the spinal cord is intact. It is well-known that chlorpromazine acts indirectly on the spinal structures through the supraspinal apparatus. However, this effect cannot be attributed only to the inhibition of the descending facili-tating influences, as was the case with some data obtained from the experiments which were carried out under ordinary conditions. As we have already seen (Chapter 2), GPEEs produced in the spinal cord can operate without any supraspinal stimulation when the spinal cord is cut. The effect of chlorpromazine on GPEE in the efferent output system in the spinal cord is the result of active descending inhibition which originates when the supraspinal structures are activated by chlorpromazine.

There are more and more neuropharmacological and neurophysiological data which show that the effects which appear to be universal are due not to the generalized, but to the selective or main action of the agents on the structures that activate the appropriate systems.

An example in this respect is the effect of morphine, which causes general analgesia when it is injected in minimum doses into the antinociceptive regions, particularly the dorsal raphe nucleus or the periaqueductal gray substance. As it comes into contact with the opiate receptors, it activates their cells (Urca et al., 1977;

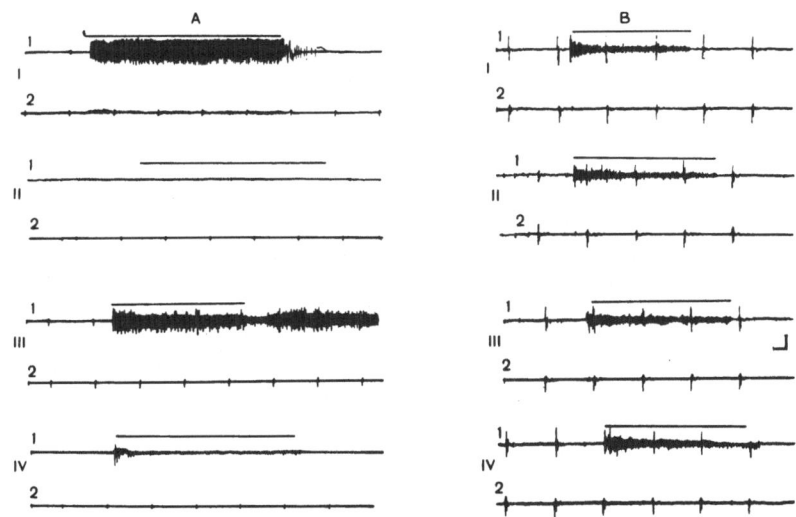

Fig. 115. Effects of chlorpromazine on EA of the muscles in cats with an intact spinal cord (A) and a cut spinal cord (B) in the case of GPEE in the ventral horns of the spinal lumbosacral segments. A: I,II: ES in the left (a) and right (2) gastrocnemial muscles during ES of the raphe nuclei before (I) and 15 min after (II) chlorpromazine was administered (0.5 mg/kg); III,IV: EA in those muscles when the sural nerve was stimulated before (III) and 15 min after (IV) the administration of chlorpromazine. B: EA in the left (1) and right (2) gastrocnemial muscles before (I) and 5 min (II), 10 min (III) and 20 min (IV) after chlorpromazine (7 mg/kg) was administered when the spinal cord was cut. The horizontal line indicates stimulation. Amplitude: 200 µV; time: 0.2 s.

Fields, Anderson, 1978; Oleson et al., 1978; Urca, Nahiu, 1978; Urca, Liebeskind, 1979). Our studies (Igonkina, Kryzhanovsky, 1979) have shown that the injection of morphine into the dorsal raphe nucleus in doses which are far less than those used systematically causes deep and prolonged analgesia as regards not only physiologic, but also pathologic pain (the pain syndrome of spinal origin). The effects of endogenous analgesics like opioid peptides are very complex. They inhibit the neurons of the nociceptive system at diverse levels. Encephalins also inhibit neurons in the supraspinal regions of the antinociceptive system. The inhibition of neurons in the formations of the antinociceptive system under the influence of encephalins, being noted by many authors (see Gebhart, 1982), may activate the antinociceptive system apparently by releasing the appropriate neurons from inhibition (Gebhart, 1982). The analgesic effects of acupuncture are also connected with the activation of various forma- tions of the antinociceptive system (Kaada, 1976; Mayer et al., 1976; Mayer et al., 1977; Melzack et al., 1977; Takeshige et al., 1979; Bragin et al., 1982). Such mechanisms of analgesia constitute the basis of the electric stimulation of the nerves and the posterior columns.

The effects of other drugs which are widely used clinically can also be analyzed in this way. Besides its general ability to enhance GABAergic control, diazepam activates the cerebellar structures which suppress epileptic activity; the activation of these structures correlated with the antiepileptic effect (Julien, 1972). A similar effect was observed when diphenylhydantoin was used (Julien, Halpenn, 1972). It has been shown that diazepam enhances the activity of the basket cells of the hippocampus, being followed by the suppression of the activity of the pyramidal cells (Lee et al., 1979). Barbiturates and diazepines enhance the recurrent inhibition of the pyramidal cells of the hippocampus (Wolf, Haas, 1977).

Of course, all this does not mean that the given drugs do not act on other areas of the central nervous system. It has been shown in many works that morphine and diazepam influence various structures of the central nervous system. However, these and similar drugs can produce a selective activating effect on the formations which play an important role in the functioning of systems that act as antisystems in a given pathologic process. These formations become determinant structures which activate the antisystems and determine their behavior.

Thus, just as the pathologic system is formed under the determi- nant's influence, the sanogenic antisystem is either formed or acti- vated under its influence. Therapeutic measures, whether they be electric stimulations of the appropriate cerebral formations or specific pharmacological influences, consist in the creation of sufficiently powerful and long-lasting determinants which can produce a state of the antisystem's tonic activity. Effects must be pro-

longed in order to produce such determinants. Therefore, the necessary effect is as a rule produced by permanent electric stimulations
of the nervous structures and long-lasting drug treatment.

The prolonged influences of the drugs can be regarded as some-
thing of pharmacological kindling. Models of pharmacological kindling have already been produced. Our researches (Shandra et al.,
1983) have shown that the brain's epileptic readiness and its sensitivity to an epileptogen increase when pentatetrazole is administered
daily in subconvulsive doses for a long time. An important fact is
that such a state lasts a long time after the drug is no longer
administered. In this respect, it would be very interesting to
produce a state of the sanogenic antisystem's long-lasting tonic
activity similar to pharmacological kindling.

The presence of receptors of certain drugs in the central nervous system shows that the organism either should have or can produce
similar substances which, whenever necessary, stimulate an effect
resembling the one induced by the given drugs. This was how endogenous neuropeptides were discovered; they can bind the opiate receptors, with which morphine is also bound, and, like morphine, can
produce the same or even greater analgesia. The discovery of both
benzodiazepine receptors in the brain tissue and their functional
connections with the GABA receptors and the GABAergic processes
(Möhler, Okada, 1977; Squires, Braestrup, 1977; Gallager, 1978;
Iversen, 1978; Martin, Candy, 1978; Tallman et al., 1978; Tallman,
Gallager, 1979; Gallager et al., 1980; Geller et al., 1980; Speth et
al., 1980) not only shows that diazepines can produce an effect, but
also suggests that agents of this class, which are capable of producing diazepine effects, may exist in the nervous system. According
to Möhler and others (Möhler, Okada, 1977; Squires, Braestrup, 1977;
Möhler et al., 1979), nicotinamide can act as an endogenous ligand
which can bind diazepine receptors and produce a diazepine-like
effect. Indeed, our researches have shown that nicotinamide can
suppress some types of epileptic activity (see further).

Hence, the organism has another very powerful means of activating the antisystems by such endogenous agents. Probably this
mechanism is specially activated under pathologic conditions, and the
'antisubstances' are rapidly produced when antisystems with a
sanogenic effect are activated. Moreover, such agents may probably
produce a direct suppressing effect on certain links of the pathologic system.

All this has given rise to a fundamentally new problem. The
matter in question is whether specific agents, which are humoral
equivalents of the specific state of both the pathologic process and
recovery, including the sanogenic systems' activation, can either
have preformed existence or be produced. A well-known example in
this respect is endogenous analgesics, i.e. neuropeptides which

activate the antinociceptive system. 'Antisubstances' which are
specific of every syndrome should either exist or be produced at the
required moment in the organism. The pathologic system's formation
should be the trigger mechanism of their metabolism (the intensification of their synthesis, release or de novo formation). The activation of the antinociceptive system by nociceptive stimuli is an
analogy of that phenomenon. It has been shown (Igonkina,
Kryzhanovsky, 1979) that nociceptive stimulation acts as trigger
stimuli for the generator of excitation which is produced in the
antinociceptive system and which activates the latter.

The origin of enhanced electric activity in the structures of
the antisystems which directly pertain to a given form of a pathologic state shows that the antisystem is activated when pathologic
processes develop. It has been established in our researches that
high-amplitude potentials appear in the cerebellar cortex at a definite stage of the development of the epileptic focus in the cat's
sensorimotor cortex produced by applying penicillin. Their origin in
the cerebellar cortex is accompanied by the development of the depression of electric activity in the neocortex; cerebellar aspiration
restores the ictal discharges in the sensorimotor cortex. It has
been observed (Fernandez-Guardiola et al., 1962) that an increase in
the impulse activity of both Purkinje's cells and the neurons of the
cerebellar nuclei coincides with the termination of epileptic activity in the cortex. An increase in the activity of Purkinje's
cells has been observed during both the depression in the cortex
(Julien, Laxer, 1974) and paradoxical sleep (Hobson, McCarley, 1972),
when epileptic activity in the cortex is known to be suppressed. The
diminution of the convulsive electric-shock after-discharge in the
cat's cerebral cortex under the influence of the cerebellar surface's
local hyperthermia was accompanied by an increase in the impulse
activity of both Purkinje's cells and the neurons of the cerebellar
nuclei (Snider et al., 1978; Snider, 1979). According to Okujava
(1969), enhanced activity in the form of frequent changes in the
potential or high-amplitude regular activity of a frequency of 3-5
per second is recorded in the caudal reticular pontis nucleus when
depression develops in the forebrain structures as the after-
discharge, being produced by electrically stimulating the cortex,
ceases. Epileptiform activity in nucleus accumbens septi has been
recorded during the dopamine psychosis-like syndrome in cats caused
by the hyperactivation of the tegmental ventral nucleus. (Chapter 12).
The syndrome is suppressed when the nucleus is electrically stimulated (Stevens, Livermore, 1978). The paroxysmal spike discharges in
various subcortical structures, including nucleus accumbens septi,
which have long been observed in schizophrenia (Heath, 1954), may be
an expression of not only the pathologic process, but also the activation of the structures which are in the antisystems.

Findings which indicate that the antisystem is activated (e.g.
electrograms) are wanting especially because purposeful research

based on the idea of the role of the antisystems as a sanogenic
factor was not carried out in this respect. A special approach
should be taken in such research, because it is necessary to prove
that the structures in the sanogenic antisystem are activated and
that the suppression of the pathologic system's activity is connected
with this activation. Besides, the defensive physiological mechanisms (antisystems) can be suppressed when the pathologic system is
hyperactive.

Apart from activating the natural physiological systems which
act as antisystems, there is another way of destabilizing the pathologic systems and suppressing their activity. This involves the
artificial production of new antisystems or new inputs into the
existing antisystems. In essence, the above methods of physiologically electrically stimulating the cerebral structures in various
ways (Bekhtereva et al., 1972; Bekhtereva, 1974, 1980; Smirnov, 1976;
Smirnov, Borodkin, 1975, 1979) pertain to that type of therapy.

The production of artifical stable functional connections (ASFC)
for therapeutic purposes (Smirnov, 1976; Smirnov, Borodkin, 1975,
1979) is interesting as regards the formation of systemic relations.
Stable functional connections are established between the cerebral
structures which are stimulated with the same frequency. When one
cerebral structure produces an inhibiting effect on the pathologic
state, the other structure whose stimulation is added becomes capable
of realizing similar suppression. In this case, the first structure
acts as a rigid unit, or a 'training' (matrix) point, while the
second structure, as the (daughter) point 'being taught'. The connection formed is strictly selective; it is specifically activated
only by the impulse frequency which was used to form it. ASFC can be
enlarged and made more complex by adding to it new structures being
synchronously stimulated, thus producing a multi-unit system. The
therapeutic effect of such ASFCs has been observed in many forms of
the brain's pathologic state (syndromes of extrapyramidal disturbances, pain syndromes, etc.).

The mechanisms of ASFC formation and the specifics of their
activity are very similar to those which were considered when other
systems were formed: the 'training' (matrix) structure acts as the
determinant in ASFC, while the system's units 'being taught' are
'tuned' to the matrix structure's activity pattern. An interesting
fact is that any link of the ASFC chain can be eliminated by electrically stimulating the caudate nucleus synchronously with the link's
stimulation. In this case, other links of the given ASFC channel
remain, and ASFCs programmed for other frequencies are also not
affected. Hence, ASFC is selectively suppressed. These specifics
of ASFC suppression somewhat simulate the specific antisystems'
activity.

The methods of special adaptive bioregulation training (Miller, 1969, 1971; Chernigovskaya, 1978; Chernigovskaya et al., 1982), the biotechnical regulation system (Khananashvili, 1978), and the new functional integration of cerebral structures when they are stimulated together (the method of systemic activation according to Khananashvili, 1978) are apparently among the methods used for producing artificial antisystems in order to destabilize and eliminate pathologic states. The last method can be regarded as the potentiation of the antisystems' activity when they are influenced in a complex manner. The therapeutic effect's potentiation concerning parkinsonism was shown when <u>centrum medianum</u> and the ventrolateral nucleus of the thalamus were synchronously electrically stimulated (Smirnov, Borodkin, 1979). The pathologic systems can be suppressed by producing a special type of ASFC by means of light and acoustic stimuli (Smirnov, Borodkin, 1979) and by creating a temporary dominating physiologic system, thus imposing enhanced locomotion on an animal (Delgado, 1969; Khananashvili, 1978). It is known that epileptic patients can prevent the seizures themselves by means of strong exteroceptive and nociceptive stimuli.

The antisystems' activation as a therapeutic principle which is intended to suppress the pathologic systems should be especially significant in the pathologic state's polydeterminant forms, when either it is practically impossible to surgically eliminate the hyperactive determinant structures or such an operation is ineffective. The clinical and neurophysiological experience gained in stimulative neurosurgery concerning multifocal epilepsy, central pain syndromes, parkinsonism and other forms of the nervous system's pathologic state has shown that the given approach is very promising.

Some specifics of the effects produced by suppressing pathologic syndromes are noteworthy. Clinical practice and experiments have revealed that a syndrome may be sometimes arrested amazingly fast when the cerebral structures are electrically stimulated; it terminates soon or immediately after electric stimulation begins. Such an effect indicates that a specific antisystem is rapidly and vigorously activated. It originates when the structures which are the most significant to a given antisystem are electrically stimulated. In this case, the determinant produced by such stimulation immediately becomes decisive, while the antisystem activated by it becomes the dominant one. The termination of the pain fit and the arrest of the pain syndrome when the dorsal raphe nucleus is electrically stimulated vividly exemplify such an effect (Figures 112 and 113). This effect is observed also during epileptic activity when some structures of the antiepileptic system are electrically stimulated with specially selected parameters (this was mentioned earlier). There are many examples of the dramatic changes in the behavior of man and animals when the respective cerebral structures are stimulated.

Cases of the virtually instantaneous disappearance of the pathologic syndrome under certain extreme conditions are very noteworthy
in this respect. Classical examples of the cases of the disappearance of catatonia which lasted for many years and which could not be
cured medicinally are connected with a situation which threatened a
patient's life (a fire in the hospital, etc.), with very disturbing
news, and so forth. These facts show that it is fundamentally possible to suppress the pathologic system by certain cerebral mechanisms
as regards a given form of the pathologic state, i.e. by an antisystem's influences when it is activated in extraordinary circumstances. They also indicate that a given antisystem was suppressed
during the whole period of a disease, as a result of which the pathologic system and syndrome lasted.

Very frequently, however, stimulation performed during every
treatment (as we have seen when the complexes of epileptic activity
in the cerebral cortex were suppressed) and even during a series of
treatments, which are conducted in clinics as chronic neuropathologic
syndromes are suppressed, should be prolonged in order to suppress a
syndrome. It is necessary to produce a stable determinant and
stabilize the antisystem which must be in a state of tonic activity.
In this case, the phenomenon involved is apparently kindling which
engenders a stable generator of excitation in the structure being
stimulated.

An amazing fact is that complex, polysystemic syndromes, which
involve many cerebral structures, can be suppressed by precisely
influencing separate structures. For instance, the whole syndrome
is suppressed when <u>nucleus accumbens septi</u> is electrically stimulated
in experimental psychosis caused by the mesolimbic dopamine system's
hyperactivity due to either the disinhibition of the tegmental ventral nucleus or its direct excitation (Chapter 12; Stevens,
Livermore, 1978). Stevens and Livermore have also observed the
favorable effect produced by electrically stimulating the given
nucleus in the case of human schizophrenia. According to the
authors, this inhibiting effect is believed to be caused by the
antagonistic relations between the given nucleus and the tegmental
ventral nucleus: the electric stimulation of <u>nucleus accumbens</u> inhibits the tegmental ventral nucleus (i.e. the determinant structure
in that form of psychosis), causing the dopamine mesolimbic system to
be inhibited and the syndrome to be suppressed. An interesting fact
is that the development of experimental information neurosis either
stops developing or attenuates when the septal nuclei are electrically stimulated (Khananashvili, 1983).

Complex multisymptomatic neuropathologic syndromes of a poly-
determinantal nature may be based on not one, but several pathologic
systems. These systems are interconnected, constituting an intricate
pathologic systemic complex. One of these systems can act as a
determinant, i.e. it can be a determinant structure of this complex,

while other pathologic systems are dependent ones. It has been shown (Chapter 1) that simple systems can produce a determinant or a dominant effect with respect to other systems. Therapeutic influences due to which the determinant pathologic system is suppressed cause the whole set of pathologic systems to disintegrate. Since the leading pathologic system has its determinant, the elimination of the latter causes a chain reaction in which the leading system and the whole set of pathologic systems disappear. This explains why a complex polysystemic neuropathologic syndrome can be suppressed when a cerebral structure having antagonistic relations with the leading pathologic system's determinant is locally stimulated.

The following fundamental question arises in this respect: if there are antisystems which produce sanogenic effects, maintaining functional homeostasis and suppressing pathologic systems, why do pathologic systems originate and why is it necessary to specially stimulate the antisystems in order to produce a therapeutic effect?

It should be noted first of all that pathologic systems by no means always originate. Most frequently, they do not originate in spite of the pathogenic agent's action. This fact is connected with the antisystems' activation. If there were no such activation, pathologic systems would always originate and the process would always be fatal.

However, when conditions are unfavorable and pathogenic agents continue to act, the pathologic system can develop extensively and intensively, acquire aggressive properties and suppress the activity of the physiological systems, including the mechanisms of homeostasis and compensation. Then, the physiological antisystems, which promote the processes of recovery and the pathologic systems' suppression, are suppressed. In such cases, it is necessary to additionally stimulate the antisystems. This may be done by electrically stimulating the appropriate cerebral structures or using special pharmacological agents. When there is no such stimulation, the antisystems cannot overcome the pathologic systems' inhibiting influences and become dominant.

Hence, the antisystems' continuous, tonic activity is both a prerequisite and a mechanism of keeping healthy. The task in pathogenic prophylaxis is to maintain the antisystems' tonic activity by the appropriate methods. It follows that a disease may originate due to the antisystems' initial damage either under the pathogenic factors' influence or as a result of the antisystems' genetically determined defect. The antisystems' relative insufficiency, being compensated under normal conditions, is expressed in the predisposition to a disease, while insufficiency during a pathologic process which has already developed causes it to become chronic. The above-mentioned formation of GPEE, when inhibitory control is upset as a result of pathogenic influences, is a universal example of the origin

of pathologic systems and the syndromes connected with them due to
the initial damage of the antisystems' local mechanisms. An analysis
has shown that many hereditary forms of the pathologic state of the
nervous system in man and animals (special strains) can be regarded
as the appropriate systems' genetic defects. These defects can
affect various molecular, membranous, synaptic and metabolic mechanisms of the formation of GPEE as well as the general mechanisms of
cerebral homeostasis. For instance, a diminution of the GABA content
and an excess of acetylcholine have been observed in EL mice, in
which convulsions are inherited according to the dominant type
(Averill, 1979).

When the disturbances of the antisystems' structures are deep-
going and irreversible, they may not be activated by either drugs or
electric stimulation. Hence, it is impossible to therapeutically
treat the syndromes whose pathologic systems have originated and are
preserved as a result of either the antisystems' great damage or
their genetic defect. It has been observed in some experiments that
the electric stimulation of the cerebellum is ineffective in some
forms of cortical seizures (Hablitz et al., 1975; Strain et al.,
1978; Lockard et al., 1979; Lockard, 1980). In this case, the number
of inhibitory cortical interneurons seems to diminish (Harris, 1980).

Substitution therapy should be used to treat such forms of the
central nervous system's pathologic state that are engendered by
deep-going and irreversible changes in the cerebral tissue. The
reconstruction of the cerebral structures by planting the appropriate
embryonic tissue may be especially significant in this respect. The
nigrostriatal dopaminergic pathway was reconstructed by intracer-
ebrally transplanting the embryonic tissue of substantia nigra to an
animal with parkinsonism caused by the destruction of substantia
nigra (Björklund, Stenevi, 1979). This direction of research can be
regarded as organic engineering.

COMPLEX PATHOGENETIC THERAPY

It has been shown that pathogenetic therapy of the neuropathologic syndromes, which are characterized by the systems' hyperactivity, is aimed at suppressing and eliminating the pathologic
systems on which the syndromes are based. The whole pathologic
system, the determinant, and the system's intermediate and effector
links may be its target. The peculiarities, effectiveness and disadvantages of such therapy have been considered earlier. Owing to
the pathologic system's properties, relatively large doses of drugs
should be used in such therapy. In this case, however, the drugs may
produce side effects and the activity of the physiological systems
and antisystems which provide sanogenic effects may be suppressed.
Such therapy is, as a rule, nonspecific, since diverse anticonvulsants are used to combat hyperactivity.

However, increasing doses of a certain drug are often required when its influence is aimed at the determinant structure or at the links which are directly connected with it. This fact is vividly illustrated by the use of neuroleptics when the above-mentioned experimental dopamine psychosis-like states are considered (Chapter 12) or when some forms of schizophrenia are clinically treated. To either effectively block DA receptors or disturb their ability to bind DA, large doses of neuroleptics should be used when dopamine secretion is pathologically enhanced due to the disinhibition of the release mechanism (Chapters 10 and 12) or when the number of post-synaptic DA receptors is enlarged (Chapter 12). However, the post-synaptic structures in essence become pharmacologically denervated as such a blockade is effected. Consequently, sensitivity grows and new DA receptors originate according to the law of denervation (Cannon, Rosenbluth, 1949). According to some data (Martres et al., 1977), hypersensitivity may originate even a few days after a single admin-istration of haloperidol. To prevent the pathologic state from being aggravated, the neuroleptic dose is enlarged with a view to more fully blocking DA receptors. Consequently, the denervation effect becomes more profound, etc. In this case, a vicious circle orig-inates and, in addition, the pathologic state becomes aggravated with respect to its basic mechanisms. Paradoxically, this aggravation is caused by therapeutic pharmacological influence. Such therapy pro-duces side effects and complications: it has been observed clincially that extrapyramidal disorders occur and even severe parkinsonism develops when neuroleptics are used regularly in large doses. That conclusion is not refuted by the present use of new neuroleptics which do not cause extrapyramidal disturbances (e.g. klozapin and thioridazine), since these drugs produce a dual effect on dopamine and cholinergic structures, precluding extrapyramidal disturbances (see further).

Taking account of the specifics of the pathologic systems' functional organization and activity, it is obviously necessary to use complex pathogenetic therapy which covers various pathogen-etically connected links of the determinant structure and a given pathologic system. Such therapy makes it possible to potentiate the action of every drug used and to produce a more significant thera-peutic effect. Complex therapy must be used also because recovery, i.e. the elimination of the pathologic system and the restoration of the disturbed functions, is a complex process with many links (see Chapter 11), and many sanogenic mechanisms should be activated in order to completely realize it. In this respect, it is likewise necessary to combine various drugs. This conclusion, just as other conclusions concerning complex pathogenetic therapy, is valid also with respect to neurosurgical interventions. The experience gained in neurosurgically treating epilepsy has shown that a greater effect is produced when the ablation of the determinant epileptic focus is combined with pharmacological influences and other methods.

Complex therapy is not a new method. Several drugs have long been used simultaneously or successively for therapeutic purposes. However, the method was often empirical, and it could not be strictly substantiated as a special method due to the prevalence of the notion about the nature of the pathologic process as a process involving only the disorganization and disintegration of the central nervous system and the disintegration of the physiological systems. The new concepts of the role of the pathologic systems as the pathogenic basis of the neuropathologic syndromes in essence represent this method in a new form, i.e. as a therapeutic principle. They also define the concrete ways of working out this principle.

Various combinations can be used for diverse methods, which help eliminate the pathologic systems and restore the disturbed functions by engaging the main mechanisms of pathogenetic therapy, i.e. the inhibiting influences on the pathologic systems, the activating influences on the sanogenic antisystems, the influences on the modulating homeostatic systems of the brain, and the influences on the trophic and plastic neural processes. Each of these forms of intervention, which are included in general complex pathogenetic therapy, may be complex itself.

Complex specific pathogenetic therapy (CSPT) is a special form of complex pathogenetic therapy. This term stands for therapy which consists in a combined specific effect on not only the pathologic system, but also especially its most important part, namely, the determinant structure. Therefore, CSPT is not nosologic, but syndromic therapy. CSPT is based on the following assumption: since the drugs used act selectively on the determinant's respective links which are pathogenetically interconnected, CSPT should produce a greater effect of the pathologic syndrome's suppression than each of the drugs used, and the doses of each compound can be reduced (Kryzhanovsky, 1980b, 1981d, 1981e). This assumption has been confirmed experimentally when our model of the stereotyped behavior syndrome was used (Kryzhanovsky, Aliev, 1979, 1981).

The determinant structure on which that stereotyped behavior syndrome (Chapters 10 and 12) is based includes the disinhibited hyperactive dopamine apparatus (enhanced dopamine release) in the caudate nuclei due to the insufficiency of inhibitory GABAergic control. The syndrome could be suppressed by influencing each link of the determinant structure. GABA was used to restore GABA control, either injecting it directly into the caudate nucleus through implanted cannulae or administering it systemically. In the latter case, the GABA compound (Aminalon, an analogue of the Japanese Gammalon) was used. For that purpose, Seduxen (a Hungarian analogue of diazepam; systematic administration) was used to enhance GABAergic inhibition.

Benzodiazepines are known to facilitate GABAergic transmission, activating GABA receptors and promoting the binding of GABA with the receptors; consequently, they enhance Cl^- conductance as the Cl^- ionic channel opens (Costa et al., 1975; Haefely et al., 1975; Zakusov et al., 1975, 1977; Dray, Straughan, 1976; Briley, Langer, 1978; Costa et al., 1978; Karobath, Sperk, 1978; Straughan, 1978; Costa, Guidotti, 1979; Polc et al., 1979; Cananzi et al., 1980; Costa, 1980; Gallager et al., 1980; Guidotti et al., 1980; Geller et al., 1980; Mezzari et al., 1981; Olsen, 1981; Olsen, Leeb-Lundberg, 1981; Study, Barker, 1981, 1982).

Lithium and haloperidol were used to suppress dopamine hyperactivity. Lithium reduces the catecholaminergic systems' activity due to the activation of monoamine oxidase and the intensification of the intraneuronal metabolism of catecholamines (Schildkraut et al., 1969), the greater uptake of catecholamines (Colburn et al., 1967; Pomeroy, Rand, 1971; Ahlawalia, Singhal, 1981, the diminuation of both their release (Blinder et al., 1971; Sheard, 1980; Beaty et al., 1981) and their content in the brain (Eroglu, Atamer-Simsek, 1980), the inhibition of adenylate cyclase activity (Dousa et al., 1970; Schildkraut, 1973; Belmaker, 1981), and the decrease in the sensitivity of adrenergic and DA receptors (Flemenbaum, 1977; Pert et al., 1978; Treiser, Kellar, 1979; Susan, Kenneth, 1980). When lithium is systematically administered, the content of GABA increases in the brain (Gottesfeld, 1975), while the content of noradrenaline and dopamine decreases in it (Eroglu, Atamer-Simsek, 1980). The ability to prevent DA receptors' supersensitivity during treatment with neuroleptics, particularly haloperidol (Burt et al., 1977; Klawans et al., 1977; Müller, Seeman, 1977; Pert et al., 1978, 1979; Allikmets et al., 1979; Verimer et al., 1980; Meller, Friedman, 1981) is an important property of lithium; Lit exerts its influence on the excited neuronal membranes through the Na^+ channels and inhibits Ca^{2+} input (Richelson, 1977).

Haloperidol suppresses stereotyped behavior very effectively (Fog, 1967, 1972). It is a very powerful pharmacological antagonist of dopamine, and its great effectiveness as a neuroleptic and an antipsychotic is connected with that property. It blocks DA receptors (van Rossum, 1967; Anden et al., 1970) and changes the activity of DA-sensitive adenylate cyclase (Kebabian et al., 1972; Clement-Cormier et al., 1974; Miller et al., 1974). Since haloperidol slightly competes with DA for the binding site but has great affinity for a certain region of the neuronal membrane (Burt et al., 1975; Creese et al., 1975, 1976; Carlsson et al., 1977; Koneiskey et al., 1978; Riddal et al., 1978; Titeler et al., 1978), it can be assumed that DA receptors are blocked by a certain indirect mechanism, probably by the membrane's conformation changes. The blockade of DA receptors in the mesolimibic system (nucleus accumbens and related structures) correlates with the therapeutic effectiveness of haloperidol (Crow et al., 1977). The fact that the therapeutic effect of

neuroleptics is connected with the blockade of DA receptors can be
seen from the specificity of that process: only the alpha-isomer
theiaxanthene flupenthixol, which causes the DA receptor to be
locked, is therapeutically active; the beta-isomer does not possess
this effectiveness (Johnstone et al., 1978). To understand the
significance of complex therapy, it is very important to take account
of the fact that the action of haloperidol is connected also with the
effects of GABA and is partially realized through them (Ostrovskaya,
Molodavkin, 1980); the effects of both drugs are synergic (Maruyama,
Kawasaki, 1975), and haloperidol suppresses GABA uptake (Iversen et
al., 1971; Maisov et al., 1975).

A set of the given drugs constituted the basis of CSPT. The
dose of every drug was reduced, while the separate use of each of
them was comparatively ineffective: certain constituents of the
stereotyped behavior syndrome were only partially suppressed and the
effect was brief (Figure 116). The effectiveness index (EI), which
is the product of the extent (in points) of the symdrome's suppression (diminution or disappearance of certain features) multiplied
by the duration (in minutes) of this suppression, was maximum when a
set of drugs was used; this is in conformity with the syndrome's
suppression for a long time (Figure 117). However, it was inconsiderable when drugs were used separately (Figures 118 I II III).
It greatly increased when two drugs were used (Figure 118 IV V).
The maximum effect was produced when all the three drugs of the set
were used (Figure 118 VI). This effect was not simply an additive
one, since then the summary EI would have been 59.1. However, it
reached 575.0, i.e. it was almost ten times higher than the summary
EI. Hence, the effects of all the drugs were really potentiated when
lithium, Seduxen and haloperidol were used together. Moreover, the
general clinical effect had a qualitatively new expression, i.e. all
the syndrome's constituents were suppressed.

The stereotyped behavior syndrome (Chapter 10) is regarded as an
expression of pathologic psychomotor excitation and as an experimental model of the respective forms of this complex pathologic
state of the brain. Pathogenetically, this syndrome (hyperactivity
of the DA caudate nucleus apparatus) is like experimental psychosis
which is based on the hyperactivity of the mesolimbic DA apparatus
(Chapter 12). These data were a prerequisite for examining the
effectiveness of the given type of complex pathogenetic therapy in
cases involving schizophrenia (Mosketi et al., 1982).

The following composition was used in clinical research: phenazepam (daily dose, 1-2 mg) was used instead of diazepam, as it is
more effective than the latter as regards several parameters (see
previous pages); haloperidol (daily dose, 3 mg), and lithium carbonate (daily dose, 0.9 g). All previous therapy was cancelled. At
first, patients were given phenazepam and lithium for 7-14 days.
Haloperidol was added when such a combination did not produce an
effect. Clinical and catamnestic examinations of 120 patients with

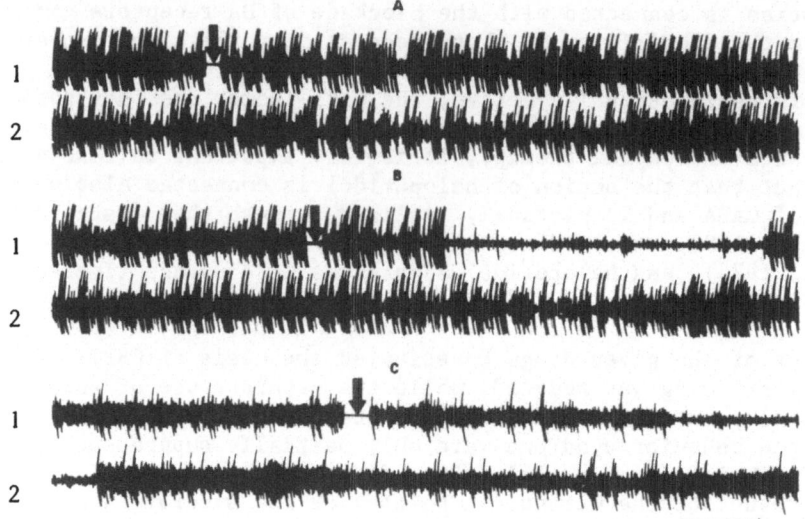

Fig. 116. Separate effects of lithium chloride, diazepam and
           haloperidol on the stereotyped behavior of a rat. A:
           Actogram of stereotyped behavior before (1) and immedi-
           ately after (1,2) i.p. injecting lithium chloride
           (50 mg/kg). Brief diminution of the intensity of stereo-
           typed behavior from the first to the fourth minute and
           from the eleventh to the fourteenth minute after the drug
           was injected. B: Actogram of stereotyped behavior before
           (1) and after (1,2) i.p. injecting diazepam (0.1 mg/kg).
           Diminution of the intensity of stereotyped behavior, the
           intensification of the locomotive components of behavior
           immediately after the drug was administered, and the
           subsequent brief suppression of stereotyped behavior.
           Restoration of stereotyped behavior (2) with an incon-
           siderable diminution of the intensity of the syndrome from
           the eleventh to the sixteenth minute after the drug was
           injected. C: Actogram of stereotyped behavior before (1)
           and after (1,2) i.p. injecting haloperidol (0.2 mg/kg).
           Diminution of the intensity of stereotyped behavior
           immediately after the drug was injected, the subsequent
           deep suppression of the syndrome, and the restoration (2)
           of stereotyped behavior. Brief diminution of the inten-
           sity of the syndrome from the tenth to the fourteenth
           minute after the drug was administered. The arrows indi-
           cate injections. Time: 1 min.

different forms of schizophrenia, most of whom were resistant to the
previous methods, have shown that this therapy was more effective
than the diverse variants of active antipsychotic therapy used
earlier. During treatment with this composition, affective, neur-

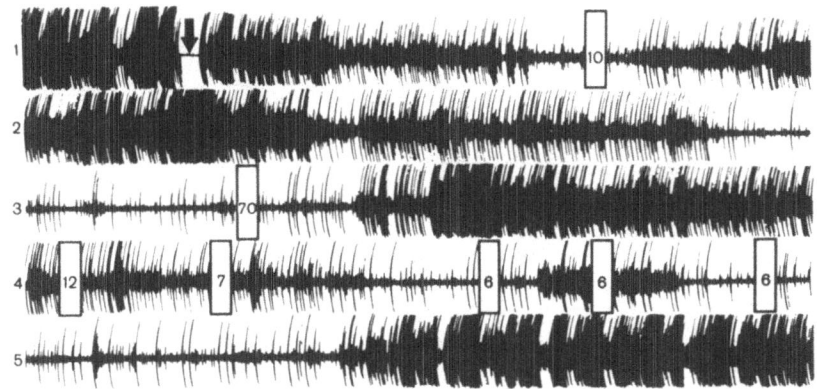

Fig. 117.   Effects of the combined administration of lithium
            chloride, diazepam and haloperidol on the stereotyped
            behavior of a rat. Actogram of stereotyped behavior
            before (1) and after (1-5) the complex (lithium chloride,
            50 mg/kg; diazepam, 0.1 mg/kg, and haloperidol, 0.2 mg/kg)
            was i.p. injected. Immediately after the complex was
            injected, the intensity of stereotyped behavior gradually
            diminished, and the syndrome was then completely sup-
            pressed within 12 minutes (the figures in the columns are
            minutes), during which the respective effects were ob-
            served. Gradual restoration of stereotyped behavior (1-2)
            to the initial level and the subsequent diminution of its
            intensity (2), and its complete absence for more than
            70 min (3). Subsequently (3-5), the periods of the com-
            plete absence of stereotyped behavior paroxysmally alter-
            nated with the periods of different severity of the syn-
            drome. Restoration of stereotyped behavior 175 min after
            the complex was injected (5). The arrow indicates the
            injection. Time: 1 min.

otic-like and psychopathic-like, paranoid and catatonci disturbances
(especially in recurrent schizophrenia) either disappeared or considerably diminished, and apathic and abulic symptoms were corrected.
Neuroleptic side effects (rigidity, tremor, dyskinesia, etc.) either
considerably diminished or were absent.

    Hence, favorable results were produced with respect to several
syndromes when benzodiazepines (diazepam, phenazepam), lithium salts
(chloride, carbonate) and haloperidol were used together in the case
of the experimental stereotyped behavior syndrome and some forms of
schizophrenia. The doses of the given drugs were rather small and,
as a rule, side extrapyramidal disturbances were absent. Moreover,
when such disturbances were considerably great prior to complex

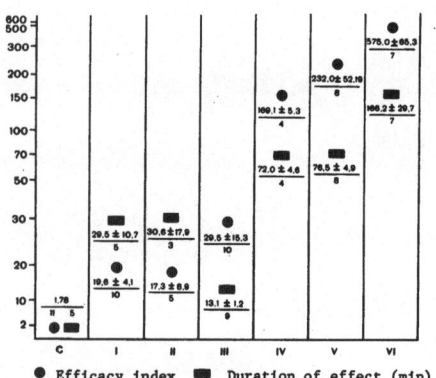

● Efficacy index   ■ Duration of effect (min)

Fig. 118.   Effects of various combinations of lithium chloride,
idazepam and haloperidol on the stereotyped behavior
syndrome. C: Control injections of a saline; I: lithium
chloride (50 mg/kg); II: haloperidol (0.2 mg/kg); III:
diazepam (0.1 mg/kg); IV: haloperidol (0.2 mg/kg) + diazepam (0.1 mg/kg); V: diazepam (0.1 mg/kg) + lithium chloride (50 mg/kg); VI: lithium chloride (50 mg/kg) + diazepam
(0.1 mg/kg) + haloperidol (0.2 mg/kg). The figures in the
denominator indicate the number of animals. The black
circle stands for the effectiveness index, and the black
rectangle, for the duration of the effect, min.

therapy, they either were reduced or disappeared when the patients
underwent that therapy.

It seems expedient to add GABAergic compounds to the composition
of phenazepam, lithium and haloperidol. The data on the interaction
between benzodiazepine receptors and GABA receptors show that the
former become more active and benzodiazepines are bound far more
effectively when GABA receptors are activated by GABA or GABA-like
drugs (Martin, Candy, 1978; Wastek et al., 1978; Tallman et al.,
1978; Karobath, Sperk, 1979; Supavilai, Karobath, 1979, 1980a;
Tallman, Gallager, 1979; Gallager et al., 1980; Geller et al., 1980;
Speth et al., 1980; Gavish, Snyder, 1981; Olsen, 1981).

An important fact is that the binding of benzodiazepines and
GABA is mutually facilitated, since benzodiazepines activate GABA
receptors. Practically, this means that it is expedient to either
use benzodiazepines together with exogenous GABAergic drugs or combine their use with the stimulation of the synthesis of endogenous
GABA. This conclusion is underpinned also by data on the insufficiency of GABA in the brain during schizophrenia (Perry et al.,
1979). In addition, haloperidol effects are connected with GABA
effects (see the previous pages). Clinical data show that therapy
becomes more effective when the derivative of GABA, i.e. sodium

hydroxybutyrate, is added to that form of CSPT. In other clinical investigations (Shaposhnikov et al., 1982), good results were produced when glutamic acid, pyridoxine and pyracetam were added to the main form of CSPT (diazepam, lithium, haloperidol). Besides producing other effects, these compounds make it possible to count on the intensification of GABA synthesis.

That form of CSPT can be improved by using specific drugs which are more active and by increasing their potentiation. For instance, the discovery of the physicochemical and functional heterogeneity of the benzodiazepine receptors (Klepner et al., 1979; Squire et al., 1979; Siegart, Karobath, 1980; Supavilai, Karobath, 1980b; Yokoi et al., 1981; Lo Mathew et al., 1982) can be conducive to the production of effective compounds which are more specific and whose action is precise.

Of course, complex pathogenetic therapy in the case of different pathologic states of the nervous system can be improved when the mechanisms of the development of these states and the peculiarities of the determinant (or determinants) and the pathologic system (or systems) are known. Some questions of the nature and pathogenesis of schizophrenia should be considered in this respect.

It has already been shown (Chapter 12) that the enhanced functional activity of the DA system in the respective forms of schizophrenia may be due not to the enhanced activity of the dopaminergic neurons, as has been proposed earlier (see Snyder et al., 1974a; Meltzer, Stahl, 1976; Snyder, 1976; Carlsson, 1978). New data shift the focus of attention in the study of the mechanisms of enhanced DA effects from the presynaptic structures (terminals of the depaminergic neurons which secrete DA) to the postsynaptic structures which have DA receptors.

There may be two hypotheses of the mechanisms of enhanced post-synaptic effects (see Cros, Johnstone, 1978). According to one of them, such effects may originate when functionally dopamine-antagonistic influences on the mesolimbic neurons drop out. The number of postsynaptic DA receptors on these neurons may not increase, but the DA effect in the form of a change in the activity of these neurons is enhanced. According to the other hypothesis, an enhanced dopamine effect is connected with an increase in the number of DA receptors on the given postsynaptic mesolimbic neurons. This hypothesis accords with the data of the postmortem examination of the brain of patients who suffered from schizophrenia, in the case of which a number of DA receptors was detected (Lee et al., 1978; Owen et al., 1978; Crow, 1979). This increase in the number of receptors cannot be attributed to the action of neuroleptics, since it was observed also when treatment did not involve these drugs (Owen et al., 1978). Among the two types of receptors that were detected in schizophrenia, the number of only the type of receptors which were labelled by the butyrophenone antagonist drugs (D2 receptors) increases (Owen et al., 1980; Cross et al., 1981), while the number of receptors labelled by both antagonist and agonist drugs grows under the influence of neuroleptics (Müller, Seeman, 1977). Thus, an increase in the number of receptors in the brain can be regarded as an essential pathogenic factor of schizophrenic forms that are connected with DA activity (Crow, 1982).

The mechanisms of an increase in the number of DA receptors on the postsynaptic mesolimbic neurons are unclear. At the same time, the disturbances of the influences which normally control the number of DA receptors on the postsynaptic membrane can evidently play an important role in those mechanisms. In other words, the defect of certain antisystems (extraneuronal or intraneuronal ones) is involved in this case, too. It has been discovered that the dissociation constant (K_d) increases with respect to the DA antagonist in schizophrenia (Owen et al., 1978; Cross et al., 1981). This fact indicates that the structure of the receptor which binds the DA antagonist is changed. If it is assumed that there are natural DA ligand-antagonists which, like pharmacological antagonists (e.g. haloperidol), become bound with the respective sites of the membrane and, consequently, reduce the ability of the DA receptors to bind DA, a change in the structure of the receptors of this ligand or an insufficiency of the ligand itself can cause the mechanisms of DA receptor formation to be disinhibited. Since receptor formation occurs via the synthesis of structural proteins, this process should be connected with the derepression of the appropriate genes. Such derepression can be due to causes which change the genetic apparatus, e.g. the disturbance of the controlling exogenous (extraneuronal) or endogenous (intraneuronal) factors (including the influence of possibly the virus on the genome). The main task is to ascertain the primary determinant. The solution of this problem is a prerequisite in elaborating causal pathogenetic therapy.

It follows from the above concepts that although the pathologically enhanced dopamine effects are a very important pathogenic mechanism, they are not the initially pathogenic link. An increase in the number of receptors and enhanced dopamine effects are the secondary determinant which establishes the subsequent chain of events, including the development and characteristics of the clinical syndrome. However, the blockade of DA receptors is an important part of pathogenetic therapy of the dopamine-dependent syndromes in schizophrenia: such a blockade eliminates or, to be more exact, suppresses the secondary determinant of these syndromes.

Since the combined influence on the determinant structures produces an optimum effect, it is expedient to use complex pathogenetic therapy to suppress the secondary dopamine determinant. The initial results in this respect are given in the form of the above data on the use of complex pathogenetic therapy of some forms of schizophrenia. It has already been shown that the set of drugs used

is aimed at reducing the synthesis of DA, at releasing it, at blocking the postsynaptic DA receptors, and at precluding the formation of
new receptors.

Schizophrenia cannot be cured when therapy produces a clinical
effect due to the suppression of the secondary determinant, while the
primary determinant remains. In this case, the disease becomes
latent and can even develop latently, as is seen when sustaining
therapy is ceased. Sometimes, this occurs immediately, while occasionally it occurs even after the state improves at first. For
complete recovery, the main, causal determinant should be eliminated.
Such an outcome is observed in the case of definite forms of schizophrenia. It is known well to clinicians, although the mechanisms of
this process are unclear. In some cases, the prolonged and continuous suppression of the secondary (dopamine) determinant can cause the
suppression of the primary determinant, since these pathogenetic
structures are interconnected and constitute the pathologic system's
basis. Such forms of therapy as electric, insulin, hyperthermic or
chemical (pharmacological) shock, while greatly changing the metabolism and state of neurons, can probably influence also their genome
and the whole pathologic system. Of course, these therapeutic
methods are not being advocated here. The matter in question is the
need to look for forms of complex therapy that would eliminate the
secondary determinant and especially the primary determinant.

In considering the causes of the failure in treating schizophrenia, it should be taken into account that the term 'schizophrenia' covers syndromes which differ in their neurobiological
nature. According to Crow (1982), at least two main syndromes which
represent different pathologic processes are covered by that term.
The first process, i.e. the syndrome of type I, being described as
'acute schizophrenia' and characterized by definite positive symptoms, is connected with both the changes in dopaminergic transmission
and an increase in the number of D2 receptors (Crow et al., 1981).
The second process, i.e. the syndrome of type II, being described as
the 'defect state' and characterized by negative symptoms, does not
pertain to the changes in dopaminergic transmission, but is probably
connected with both intellectual damage and the structural changes in
the brain. Treatment, results and prognoses differ in such cases.
Of course, therapy which is intended to suppress enhanced dopamine
effects can succeed only when dopamine-dependent syndromes are involved.

Returning to our consideration of CSPT in general, it should be
noted that the term specific pathogenetic therapy does not lose any
of its significance when the matter in question is the suppression of
not the primary, but the secondary and subsequent determinants and
the pathologic systems induced by them. A new determinant (or determinants) and a new pathologic system (or systems) can originate at
every stage of the pathologic process. When the drugs used are

selectively directed at definite, pathogenetically connected links of
the determinant and the pathologic system, and they specifically
influence these links, the combined use of such compounds constitute
CSPT. It has already been emphasized that CSPT is not nosologic
therapy, but syndromic therapy.

The appearance of drugs which do not produce pronounced side
effects in the form of extrapyramidal disorders (e.g. closepin and
thioridazine) does not make CSPT less significant. It has been
learned that these drugs simultaneously act cholinolytically. The
fact that they do not produce a side extrapyramidal effect is connected with their ability to block the central muscarine cholino-
receptors (Anden, Stock, 1973; Miller, Hiley, 1974; Snyder et al.,
1974; Kelly, Miller, 1975; Herberg, Wishart, 1981). It has been
shown (Chapter 10) that GPEE formed in CN by cholinergic neurons
which were disinhibited due to the failure of DA influences is the
determinant structure in parkinsonism. Closepin and thioridazine
block DA receptors in both CN and the mesolimbic structures. However, the consequent action of GPEE formed by cholinergic neurons in
CN is blocked by these drugs' cholinolytic action. Thus, the given
drugs act dually, and their effect is in essence an example of
another type of complex therapy whose purpose is to produce a therapeutic effect and suppress the side effect. The purposeful production of such compounds and the combined use of drugs which make it
possible for such a dual effect to occur is another way of elaborating complex pathogenetic therapy.

Each neuropathologic syndrome should have its own CSPT which
corresponds to the neurochemical nature of the determinant's pathogenic links. CSPT may include the electric stimulation (ES) of the
respective antisystems' structures, which should also be specific.
As syndromic therapy, CSPT may be part of general complex etiopathogenetic nosologic therapy.

COMBINATION OF ABLATIVE AND STIMULATIVE INTERVENTIONS
WITH PHARMACOLOGICAL EFFECTS

All the given data show that the elimination of the hyperactive
determinant structure either surgically or electrocoagulatively is a
fundamentally important method of the central neuropathologic syndromes' pathogenetic therapy. The elimination of GPEE in the peripheral nervous system in the case of neuropathologic syndromes of
peripheral origin (e.g. some types of pathologic pain) is also part
of pathogenetic therapy. However, it is far from always that
ablative interventions produce favorable effects. There may be a
recurrence much later, although it may seem at first that a favorable
effect is produced. We have also observed such a result in experiments involving the above-mentioned models of syndromes. Clinicians
have frequently seen it in different forms of the central nervous

system's pathologic state. Recurrences in the pain syndromes' surgical treatment are a vivid example in this respect.

There may be a recurrence due to the restoration of GPEE which was incompletely removed, to its reactivation after it was temporarily weakened as a result of the switch-off of the afferent inputs after the section of the neural pathways, or to the formation of a new GPEE in a given pathologic system.

Therefore, it seems to be expedient to combine ablative interventions with other types of therapy whose purpose is to suppress the pathologic system and eliminate the conditions which are conductive to either the restoration of a former determinant structure or the origination of a new one. Pharmacotherapy, ES of the antisystems' structures and other methods of eliminating the pathologic system which were considered earlier may be used in this respect.

It has been clinically observed that a better therapeutic effect is produced when the cerebral structures' ES, being performed in order to destabilize a resistant pathologic state, is combined with the ablation of the determinant focus than when each of these acts is performed separately (Bekhtereva et al., 1978).

The combination of ES of the antiepileptic system's structures and the employment of medicine is a very important form of complex therapy. Fundamentally, this therapy is based on the fact that anticonvulsants, particularly barbiturates in respective doses, do not suppress the inhibitory mechanisms. Barbiturates do not affect both the synaptic processes which directly participate in the generation of the inhibitory postsynaptic potential and the excitation of inhibitory neurons. Moreover, barbiturates increase the duration of that potential (Serkov, 1977). The resistance of inhibition in cortical neurons has also been observed in experiments involving the direct stimulation of the cortex (Krnjevic et al., 1966). Barbiturates facilitate conductance in the GABAergic synapeses (Ransom, Barker, 1976; Lodge, Curtis, 1978; Nicoll, 1978). When barbiturates are bound with one of the sites of a GABA receptor, they enhance GABA-dependent conductance in the Cl^- channel (Olsen, Leeb-Lundberg, 1981; Johnston, Willow, 1981). This mechanism may also be connected with the elimination of the endogenous inhibitors of GABA binding (e.g. the phosphoethanolamine derivative; Johnston, Willow, 1981). Therefore, the inhibiting effects of the antiepileptic structures' ES, which is performed via the inhibitory mechanisms, not only are unaffected, but also can be enhanced when barbiturate actions are combined. Potentiation can be attained also as a result of the functional changes in the antiepileptic system's structures which are electrically stimulated. For instance, barbiturates suppress the activity of the neurons of the cerebellar cortex (Murphy, Sabah, 1970; Van Gilder, O'Leary, 1971; Gordon et al., 1973; Fanaryan, 1975), as a result of which cerebellar nuclei are disinhibited and, consequently, their inhibiting antiepileptic effects in general and those during ES in particular increase. At the same time, barbiturates weaken the pathologic system's activity and reduce its resistance, owing to which it can be suppressed more rapidly and more considerably, breaking down under the influence of ES.

It has already been shown that the effects of benzodiazepines are also realized by GABA inhibition. Consequently, a potentiating effect is likely to be produced when benzodiazepines are used together with ES. Indeed, our researches (Shandra et al., 1982) have shown that the effect of the suppression of epileptic activity of a separate focus as well as a set of foci is greater and lasts longer when the cortex of the caudal part of the vermis is electrically stimulated as phenazepam is used. When ES is successively performed against the background of phenazepam, the set of foci is reduced more rapidly in the above-mentioned sequence of focal inhibition. Such effects were observed in the case of penicillin- and strychnine-induced foci. A very important fact is that such effects could be produced when relatively small phenazepam doses, which did not noticeably suppress epileptic activity, were used. In some cases, an inhibiting effect was produced when benzodiazepines were used together with ES as the strength of the ES current was reduced. It has already been shown that epileptic activity is restored when the structures whose ES causes this activity to be suppressed are destroyed (i.e. when they coagulate). If epileptic foci were induced by a sufficiently large penicillin dose, epileptic activity which was being restored often became generalized. When epileptic activity of the foci was suppressed by combining ES with the use of benzodiazepines, the coagulation of the structure being sitmulated (in this case, the caudal part of the vermis) was followed by the restoration of epileptic activity to some extent, but the latter never became generalized.

The combination of phenazepam and ES produces another important effect: the arrest of the exciting component that either precedes epileptic activity or originates after its suppression. Such a result was produced also when diazepam (Seduxen) (Kryzhanovsky et al., 1981c) or phenazepam (Shandra et al., 1982) was used together with ES, although diazepam was less effective than phenazepam. This result is explained well by the concepts of the dual functional message which originates when the antiepileptic structures are electrically stimulated (this was discussed earlier): the effect of an exciting message is suppressed by benzodiazepines (or phenobarbiturates), while the effect of an inhibiting message is not only preserved, but also enhanced due to the inhibitory mechanisms' potentiation.

At the same time, it should be borne in mind that not any combination of pharmacological antiepileptics and ES produces a favorable result. Unlike the ES of the fastigial nucleus, the ES of the

dentate nucleus against the background of diazepam (systemic administration) produces neither an inhibiting nor an enhancing effect for
the cortical epileptic focus. These data show that the mechanisms
which realize the effects during the ES of the fastigial and dentate
nuclei are different. Unlike the pathways which are activiated
during the ES of the fastigial nucleus, the pathways which are activated during the ES of the dentate nucleus are sensitive to diazepam's inhibiting action.

It is suggested that the combination of the ES of definite
cerebral structures and the respective antiepileptics can be used in
clinics under the corresponding conditions. ES treatments against
the background of pharmacological effects, while allowing the antiepileptic system to be activated repeatedly without any exciting
proepileptic side effects, can promote the consolidation of tonic
activity and the antiepileptic system's stabilization, i.e. it can be
conducive to a stable antiepileptic effect.

Neurotropic drugs can be used also to preliminarily destabilize
the pathologic system against the background of which the inhibiting
effects of the ES of the cerebral structures are more effective. In
such cases, ES can be replaced by external stimuli. For instance,
light and acoustic stimuli can suppress various types of hyperkinesis
against the background of a nonspecific connector, i.e. etymisole
(bismethylamide-1-ethylmidasole-4,5-dicarboxylic acid) (Smirnov,
Borodkin, 1979). The pathologic system can be preliminarily destabilized also without drugs, e.g. by special muscular loads.

Neurotropic pharmacological agents can be used in complex
therapy in the case of intracerebral or extracerebral therapeutic
stimulations: they can promote the formation and stabilization of
artificial antisystems and the activation of natural antisystems.
Etymisole, which stimulates the processes of learning and memory,
causes artificial stable functional connections, which suppress the
pathologic syndrome, to be formed and consolidated more quickly
(Smirnov, Borodkin, 1979).

METABOLIC THERAPY

The primary pathogenic factor and the secondary pathologic
changes, particularly the pathologic system which originated, promote
the structural and functional disorganization of the nervous system,
its abnormal activity, the disturbance of metabolism, and the exhaustion of the nerve cells. In such cases, the realization of the
sanogenic effects of the antisystems and other defence mechanisms is
hampered, and it becomes more difficult to restore and normalize the
functions which were disturbed. Therefore, metabolic therapy, which
furthers the restoration and maintenance of the trophic and plastic
potentials of the damaged nervous system, should be an indispensable
part of complex pathogenetic therapy.

Metabolic therapy can be divided into two parts: one part is therapy whose purpose is to normalize the changed metabolism with which the activity of the pathologic determinant and the pathologic system is connected, to activate the respective antisystems, and to restore the damaged functions and the reactivity (binding capacity) of the neuronal receptors; the other part is therapy whose purpose is to maintain the general metabolism of the nervous tissue at the required level and to preclude secondary nonspecific tissual disturbances. This division is very conditional, since the same metabolic drug can participate in the realization of both effects.

Present data show that substances which regulate metabolism, e.g. vitamins, can influence the synaptic processes, the transmitters' synthesis and effects, the membranous state, etc. On the other hand, transmitters and their metabolites play a nonmediatory role, regulating the nerve cell's metabolism. All these data make it necessary to take a new approach when using the substances which are included in metabolic therapy and when learning about the mechanisms of their action. It will be shown further that the principle of complex therapy, i.e. the combined use of substances which act on various links of a pathologic process, is likewise very important. This principle is expressed also in the possibility of using metabolic and medicamentous therapies together.

Nicotinamide Effects

Nicotinamide is known to be in the enzymic system of codehydrogenases which participate in redox processes; it is a hydrogen carrier. It also participates in carbohydrate metabolism, and insulin action is enhanced under its influence. Nicotinamide can accelerate the reparation of damaged DNA (Berger, Sikorski, 1980). These properties make nicotinamide an important means of metabolic therapy. According to some authors (Polc et al., 1974; Beaton, 1976; Möhler, Okada, 1977; Polc, Haefely, 1977; Möhler et al., 1979; Kennedi, Leonard, 1980), nicotinamide binds with the benzodiazepine receptors and, in large doses, can produce tranquilizing, hypnotic and anticonvulsive effects, i.e. its action is similar to that of benzodiazepines.

Our researches (Kryzhanovsky et al., 1980d; Kryzhanovsky, Shandra, 1981; Kryzhanovsky et al., 1981f) have shown that nicotinamide administered intravenously in a dose of 100-350 mg/kg suppresses a separate focus as well as a complex of foci of epileptic activity induced by applying strychnine to a cat's cerebral cortex (Figure 119). The sequence of the suppression of foci and the breakdown of the epileptic complex was the same as that which occurred under the anitconvulsant's action and ES of the antiepileptic system's structures (see the previous pages); the determinant focus was the last to disappear. Nicotinamide also suppressed the convulsive focus which

was produced by acetylcholine with proserine. However, it was hardly
effective as regards a complex of foci and a separate focus of epileptic activity induced by applying penicillin (Figure 119 B). Only
relatively weak epileptic foci were suppressed when the nicotinamide
doses were enlarged.

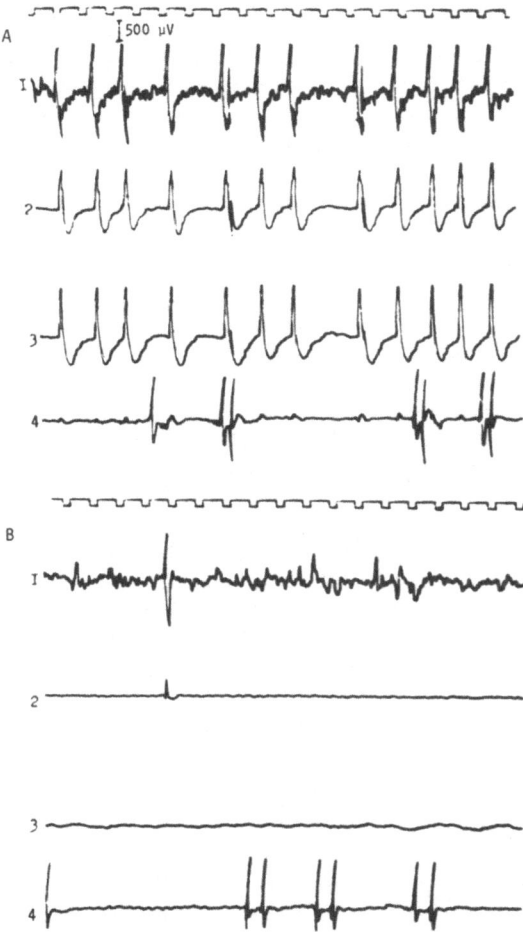

Fig. 119. Influence of nicotinamide on the epileptic complex induced
          in a cat's cerebral cortex by strychnine in various con-
          centrations and on a separate penicillin-induced focus.
          A: Focal complex induced by applying 3% strychnine to zone
          1 and 0.1% strychnine to zones 2 and 3; a separate focus
          induced in zone 4 applying 2% penicillin. B: 4 min after
          i.v. injecting nicotinamide (350 mg/kg). 1: Orbital
          sigmoid gyrus; 2: coronary sigmoid gyrus; 3: posterior
          sigmoid gyrus; 4: anterior sigmoid gyrus. Amplitude:
          500 μV; time: 1 s.

In semichronic experiments with rats which were partly immobilized, nicotinamide in large doses (800-1000 mg/kg intraperitoneally) suppressed penicillin foci in the cerebral cortex (Braslavsky et al., 1982). Under these conditions, nicotinamide caused the disappearance of the ictal discharge, an increase in the number of interictal spikes, and occasionally the appearance of brief, grouped interictal discharges (Figure 120). These effects were like those which were observed when diazepam was used (this was discussed earlier; Chapter 3). Later, interictal spikes also disappeared; they disappeared at first in the 'mirror' focus, remaining in the determinant focus. Epileptic activity was precluded when nicotinamide was administered. In such cases, only separate low-amplitude spikes were recorded.

Those nicotinamide effects can be explained in the following way: if nicotinamide is the benzodiazepine receptors' endogenous ligand, its pharmacological effects are realized by activating GABA-ergic inhibitory control (see above). When strychnine and acetylcholine act, GABA inhibition remains, but it is disturbed under the action of penicillin (see above and Chapter 5). Therefore, the inhibitory effects of nicotinamide are realized well in the case of strychnine and acetylcholine foci, and worse in the case of penicillin foci. To suppress the latter, much larger nicotinamide doses and the appropriate experimental conditions are required.

When the data on the antiepileptic effectiveness of nicotinamide are being discussed, it is necessary to take account of the low level of nicotinamide penetration into the cerebral tissue, the disturbance of the permeability of the hematoencephalic barrier under the influence of certain factors, the doses of a drug and the possible species specificities of the effects of nicotinamide and convulsants.

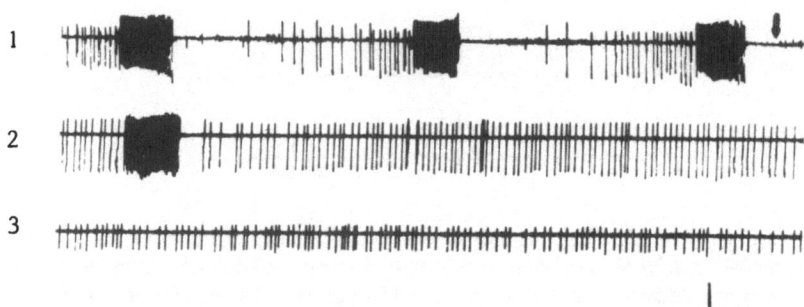

Fig. 120.   Influence of nicotinamide on a penicillin-induced focus in a rat's cerebral cortex. The arrow indicates the moment of nicotinamide administration (1000 mg/kg, i.p.). 1,2,3: 30, 35 and 120 min, respectively, after the application of penicillin. Amplitude: 1 mV; time: 45 s.

Pyridoxal Phosphate Effects

In the organism, pyridoxine is phosphorylated and converted into pyridoxal-5-phosphate (P-5-P), which is in the prosthetic group of enzymes that decarboxylate and transaminate several amino acids. P-5-P stimulates various types of metabolism and participates in the regulation of trophic, particularly neurotrophic, processes. These properties and some others make it an important part of metabolic therapy. At the same time, P-5-P can participate in the realization and activation of the GABAergic mechanisms, since, being a coenzyme of decarboxylase (Roberts, Frenkel, 1951; Tapia, 1974; Taia et al., 1975), it allows GABA to be synthesized from glutamate (Wu et al., 1973). The content of GABA in the brain is connected with the activity of P-5-P (Lott et al., 1978). Although the data on the properties of P-5-P are contradictory, there is evidence which shows that P-5-P, pyridoxine, or pyridoxine together with ATP favorably influences the epileptic process (Hunt et al., 1954; McKhann et al., 1960; Perez de la Mora et al., 1973; Clarke et al., 1979). An important fact is that P-5-P normalizes membranous electrogenesis (Steiner, 1969).

Our researches (Kryzhanovsky, Shandra, 1981; Kryzhanovsky et al., 1981e) have shown that P-5-P suppresses the activity of a separate focus as well as a complex of foci produced in the cat's cerebral cortex by applying penicillin (Figure 121). The sequence of the breakdown of a complex of foci and the suppression of individual foci was like that described earlier when other anticonvulsants were used: the dependent foci were the first, while the determinant focus was the last, to be suppressed. However, P-5-P was less active as regards a complex of foci and individual foci produced by strychnine (Figure 121 B). The inhibiting effects of the compound were expressed when the foci were induced by weak strychnine solutions or when relatively large doses of it (20 mg/kg) were used. It is necessary to specially study why P-5-P is relatively poorly effective as regards strychnine-induced foci. An important fact is that the compound is active differently concerning epileptic foci of a diverse nature. It has been shown (Sakurai et al., 1980) that the content of P-5-P is reduced in epileptic foci induced by penicillin, thiosemicarbazide and semicarbazide. The administration of P-5-P normalizes the compound's content and precludes the development of convulsive activity caused by penicillin and thiosemicarbazide, but not by semicarbazide. These data indicate the role which the insufficiency of P-5-P plays in epileptogenesis and show that the compound's antiepileptic effectiveness differs as regards epileptic foci of a different nature.

Thus, nicotinamide and P-5-P seems to produce inhibiting effects as regards different types of epileptic activity. The dependence of the therapeutic effect on the nature of GPEE is involved in this respect, just as in the case of other forms of the nervous system's

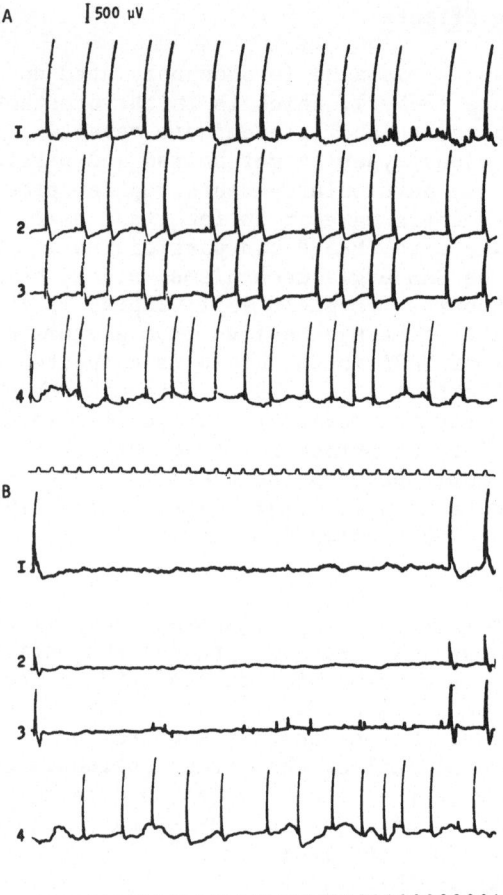

Fig. 121. Influence of pyridoxal phosphate on an epileptic complex
           induced in a cat's cerebral cortex by penicillin in vari-
           ous concentrations, and on a separate strychnine-induced
           focus. A: Complex induced by applying 2% penicillin to
           zone 1, and 0.5% penicillin to zones 2 and 3; a separate
           focus induced in zone 4 by applying 0.1% strychnine.
           B: 7 min after i.v. injecting pyridoxal phosphate
           (10 mg/kg). The notation is the same as in Figure 119.

pathologic state (e.g. central pain syndromes, spinal myoclonia; this
was discussed earlier).

      The above features of the antiepileptic effects of nicotinamide
and P-5-P have naturally given rise to the idea of using these compounds together: in this case, both links of GABAergic control, i.e.
the GABA receptor (the nicotinamide effect) and GABA synthesis (the
P-5-P effect), would be activated.

Researches have shown that the complexes of epileptic activity, which consist of strychnine- and penicillin-induced foci, are completely suppressed when nicotinamide and pyridoxal phosphate are used together in relatively small doses (250 mg/kg and 10 mg/kg, respectively). Another important feature has been revealed. When generalized epileptic activity, caused by applying epileptogens in a high concentration, was not suppressed by either nicotinamide or P-5-P used separately, it was suppressed when they were used together (Figure 122), and the desired effect could be produced even when their doses were smaller (Kryzhanovsky, Shandra, 1981). Generalized epileptic activity induced in various areas of the cerebral cortex was the first to disappear, and then the primary focus, i.e. the source of generalized epileptic activity, was suppressed.

Antioxidants' Effects

Enhanced lipoperoxidation of the neuronal membranes plays an important pathogenic role in the formation and maintenance of GPEE activity (Kryzhanovsky et al., 1980f; Nikushkin et al., 1981) (Figure 123). Epileptic activity in both the primary focus and the 'mirror' focus as well as in the system of two primarily induced synchronized foci was considerably suppressed when a natural antioxidant, α-tocopherol, was administered at first. This inhibiting dose-dependent effect correlated with the degree of diminution of lipoperoxidation (Kryzhanovsky et al., 1980f; Nikushkin et al., 1981). Other antioxidants, such as Ionol (2,6-di-tert-butyl-4-methylphenol, a synthetic fat-soluble compound) (Nikushkin et al., 1980) and a synthetic water-soluble antioxidant acted similarly. It was found that nicotinamide is able also to suppress the enhanced lipoperoxidation of the neuronal membranes (Braslavsky et al., 1982). Hence, its antiepileptic effects may be connected also with the antioxidant properties. The fact that epileptic activity could be suppressed by using antioxidants of a different nature shows the universal significance of enhanced lipoperoxidation in the mechanisms of epileptic activity. The antioxidants' effects can also be connected with the fact that the receptors' disturbed reactivity, i.e. their ability to become bound with drugs, including endogenous ligands which regulate the neuron's state, is restored as the neuronal membranes' lipid layer is normalized under the antioxidants' influence.

Effects of the Combined Use of Vitamin and Anticonvulsant Compounds

Researches were carried out with the model of generalized clonic-tonic pentylene tetrazole convulsions in mice (Kryzhanovsky et al., 1982d). Such a model of epileptic activity is known to be very severe. Besides the above vitamin compounds, which can produce an antiepileptic effect (nicotinamide, P-5-P, α-tocopherol), use was

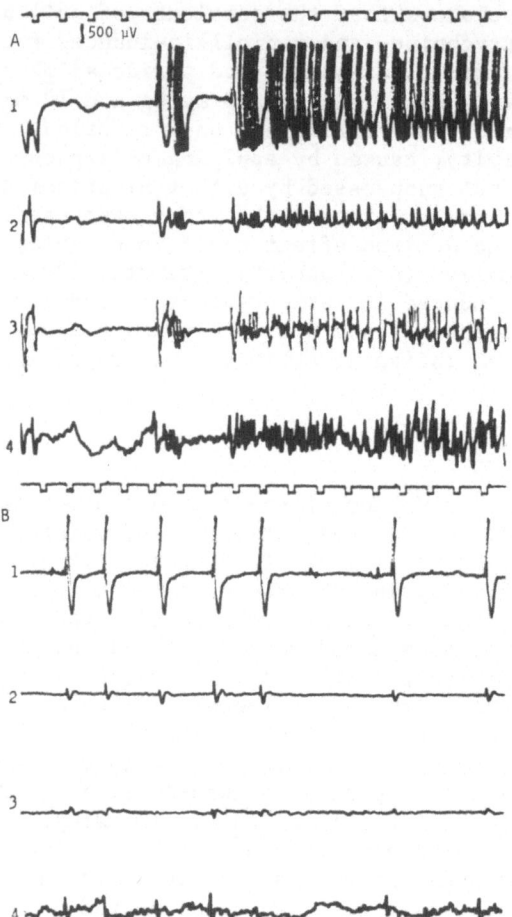

Fig. 122. Influence of the combined administration of nicotinamide
           and pyridoxal-5-phosphate on an epileptic complex in a
           cat's cerebral cortex. A: Complex produced by applying
           crystal strychnine to zone 1 and 0.5% strychnine to zones
           2 and 3. B: 3 min after i.v. injecting nicotinamide (250
           mg/kg) and pyridoxal phosphate (5 mg/kg). The notation is
           the same as in Figure 119.

made of pantogam [the calcium salt of D-homopantothenic acid, or
D/+/-(α,γ-dioxy-β,β-di-methylbutyryl)-γ-aminobutyric acid], which, on
the one hand, produces the effects of pantothenic acid, and, on the
other, passes well through the hematoencephalic barrier and causes
the anticonvulsive effects of GABA (Kovler et al., 1980).

     The combined use of nicotinamide, P-5-P and α-tocopherol clearly
potentiated the antiepileptic effect of every compound, i.e. the

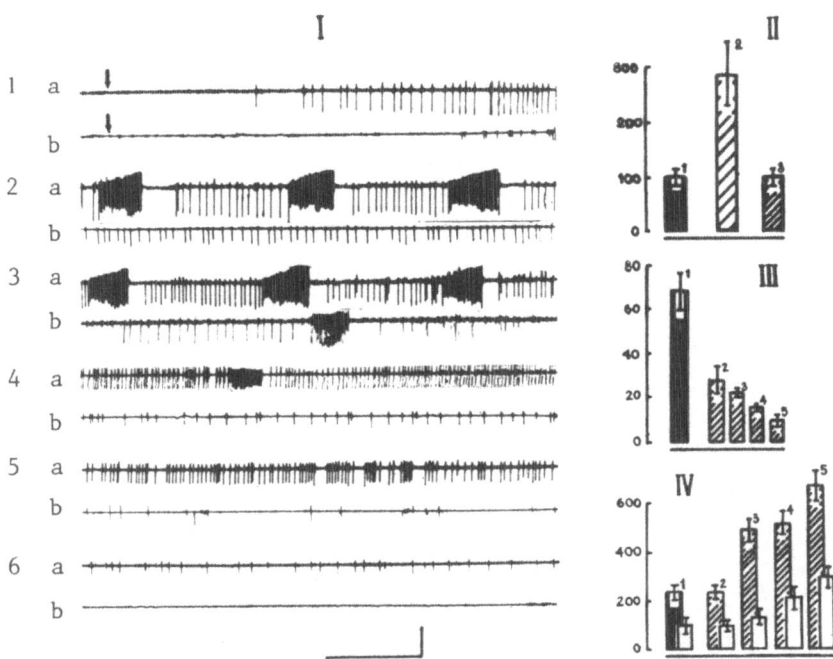

Fig. 123. Electric activity and lipoperoxidation in the cortical
           penicillin-induced epileptic focus: α-tocopherol effects.
           I: EA in the epileptic focus of a rat's cortex not treated
           (a) and treated (b) with α-tocopherol (100 mg/kg, before
           48 h). The figures stand for ECoGs at different intervals
           after penicillin application: 0-5 min (1), 5-10 min (2),
           10-15 min (3), 60-65 min (4), 90-95 min (5), and 120-125
           min (6). The arrow indicates penicillin application.
           Amplitude: 1 mV; time: 1 min. II: Content of lipoperoxi-
           dation products in crude synaptosomal fractions isolated
           from the cerebral cortex of healthy rats (1) and animals
           not treated (2) and treated (3) with α-tocopherol (100
           mg/kg, before 48 h). Ordinate: content of malonic
           dialdehyde (%). III: Number of seizures recorded during
           the existance of an epileptic focus in animals not treated
           (1) and treated with α-tocopherol in different doses: 0.1
           mg/kg (2); 1.0 mg/kg (3); 10 mg/kg (4); and 100 mg/kg (5)
           before 48 h. Ordinate: number of seizures. IV: Latent
           period of the origin of the first interictal discharges
           (lower columns) and the first seizure (upper columns) in
           animals not treated (1) and treated with α-tocopherol in
           different doses: 0.1 mg/kg (2); 1.0 mg/kg (3); 10 mg/kg
           (4); and 100 mg/kg (5) before 48 h. Ordinate: time, s.

latent period of the initial convulsive manifestations increased and
the severity of clonic-tonic convulsions decreased. This effect
could be achieved when the dose of every component was reduced (to
one-half or one-third). The antiepileptic effect was enhanced when
those drugs were administered together for several days. The antiepileptic effectiveness of the compounds as a whole and each anticonvulsant was mutually enhanced when phenobarbital or phenazepam was
added them. When pantogam and nicotinamide (as a possible endogenous
ligand) or phenazepam (as the benzodiazepine receptors' exogenous
ligand) were combined, each drug's effectiveness was potentiated. An
important fact is that the doses of phenobarbital and phenazepam
could be reduced against the background of vitamin anticonvulsive
therapy and that the repeated administration of medicamentous anticonvulsants together with vitamins did not reduce their activity.
All these data create prerequisites for practically using complex
vitamin and pharmacological antiepileptic therapy.

When those compounds are used, it can be assumed that the
trophic and plastic potentials of the brain will increase, the conditions for compensatory reorganizations and recovery processes will
be optimized, and further structural and functional damages to the
brain will be precluded. The use of those compounds and other meta-
bolically active preparations would become an effective link in
complex therapy whose purpose is to maintain the antiepileptic
status.

In that therapy, account is taken of the significance of therapeutic influences, including surgical intervention in order to remove
the irreversibly changed structures, which act as pathologic determinants that sustain and stabilize the pathologic system, and the
electric stimulation of the cerebral structures that is intended to
activate the physiological antiepileptic mechanisms. It can be
assumed that these therapeutic methods would be far more effective
when they are used in combination with pharmacological metabolic
therapy.

All that is important in treating not only epilepsy. The above
considerations can be used also for other neuropathologic syndromes
which are based on hyperactive pathologic systems. This can be seen
from the effectiveness of the above complex therapy of dopamine
psychoses, the stereotyped behavior syndrome, and some forms of
schizophrenia. Our investigations have revealed the effectiveness of
complex pathogenetic therapy of severe pain syndromes of central
origin. This therapy consists in jointly influencing the hyperactive
nociceptive system's various neurochemical structures. Clinicians
working on the problem of pain have also drawn the conclusion that it
is necessary to use complex therapy of the pain syndromes which
covers a very wide range of influences, including pharmacological,
psychological and social factors.

All that equally applies to the metabolic therapy of neuropathologic syndromes. For instance, combat with enhanced lipoperoxidation, which is known to occur in all the forms of the nervous
system's pathologic state, should be part of all the neuropathologic
syndromes' treatment. This relates also to other drugs which produce
a pharmacological metabolic effect. Our researches involving nonspecific encephalopathy have shown that the progressive course of the
neurodystrophic and auto-immune processes, which cause the brain to
be destroyed, can be arrested by a set of specially selected drugs.

SUMMARY

Pathogenetic therapy is the center of gravity in the treatment
of neuropathologic syndromes which are characterized by the systems'
hyperactivity. However, this does not belittle the role of etiologic
therapy, which is intended to eliminate the influences of the pathogenic factors that either cause or promote the pathologic systems'
formation. Therefore, combined etiopathogenetic therapy is the
optimum type of treatment.

The main task in the neuropathologic syndromes' pathogenetic
therapy is the elimination of the pathologic systems on which the
syndromes are based. The elimination of the pathologic system's key
link, i.e. its determinant, is a radical way of achieving such a
result. In this case, not only is the pathologic system eliminated,
but also a recurrence is ruled out. Such a result can be obtained in
the case of monodeterminant syndromes also when other parts of the
pathologic system do not act as secondary determinants. However, it
is expedient to eliminate the determinant structure under these
conditions, since it is conducive to the effect of other therapeutic
influences and to the pathologic system's breakdown. The suppression
of the pathologic system's intermediate links causes also the
effector links, which are deprived of the determinant's influence, to
be simultaneously eliminated from the system, substantially reducing
the pathologic system. In this case, the determinant structure loses
its guiding role and becomes a generator of excitation that is of
local functional significance. Such a weakened generator is less
resistant to therapeutic influences than the determinant structure in
the pathologic system. However, it is very difficult to completely
suppress it. Its presence even in reduced form makes a recurrence
possible. This recurrence may originate when the generator's
activity is enhanced. Although the elimination of only the effector
link causes the syndrome's clinical manifestation to disappear, it
does not cause the pathologic system's basis, i.e. the determinant
structure and the intermediate links connected with it, to be
removed. In such cases, the possibility of a recurrence is a constant threat. Many presently used therapeutic methods are of such a
nature: they are aimed at normalizing only the effector link.

The pathologic system's resistance is determined by its power,
which depends on the power of the determinant structure and other
parts of the system that are connected with it, and also on the
amount of its constitutents and the stability of the connections
between them. Therefore, the effects intended to reduce the pathologic system can be useful therapeutically even if they do not eliminate it. The pathologic system is reduced as its parts which depend
less on the determinant structure are 'cut off'. The determinant
structure, which is the most powerful link and is based on GPEE, is
the last to be suppressed. It is the last to die. This regularity
is expressed in the case of various therapeutic influences which are
intended to suppress the whole pathologic system. The fact that the
determinant is the last to be suppressed can be used for diagnostic
purposes, i.e. for ascertaining the determinant structure when drugs
which generally suppress CNS are used.

The pathologic system's destabilization and the weakening of the
connections between its parts and especially the connections with the
determinant are an important prerequisite of the pathologic system's
elimination. Under these conditions, the system is reduced and
breaks down more rapidly, causing the system's resistance to weaken
further.

The antisystems' activation is an essentially important method
of suppressing the pathologic system. This method is based on the
existence of natural functional antagonistic relations between definite formations and systems of CNS, being a form of the physiological
regulation of the functions. The appropriate pathologic system and
its clinical expression, i.e. the neuropathologic syndrome, can be
suppressed by activating the antisystem as definite CNS areas are
specially stimulated or naturally activated. The pathologic system
can be suppressed for a long time by producing long-term generators
of excitation in the antisystems' definite structures. An analysis
has shown that, besides a general effect, many pharmacological agents
produce their characteristic effect indirectly, i.e. by activating
the appropriate antisystems. It is hypothesized that special substances, which can specifically activate the appropriate antisystems
and possibly directly suppress the pathologic system, originate in
the organism as the pathologic process develops.

Rational pathogenetic therapy should be complex: it should
consist in combined action on various links of the pathologic system.
Such an approach makes it possible to eliminate the pathologic system
more effectively and rapidly. Complex pathogenetic therapy can
consist in the combined use of various methods, i.e. the ablation of
the determinant structure, the electric stimulation of the antisystems' structures, and the use of pharmacological agents.

Complex specific pathogenetic therapy (CSPT) is a special type
of pathogenetic therapy. It consists in the combined use of com-

pounds which specifically act on the determinant structure's pathogenic links, as a result of which an additive effect is produced and,
what is more, the compounds' action is really potentiated. Consequently, the general ultimate effect is considerably enhanced. CSPT
is based on the fact that the action of the drugs used is specific
and is selectively aimed at the determinant structure's interconnected pathogenic links. Therefore, their influence is focused
there, and an enhanced general effect originates. This feature makes
it possible to reduce the dose of every compound and to minimize
their side effects. The ratio of the compounds in the complex can
vary.

Metabolic therapy, whose purpose is to restore and sustain the
damaged nervous system's trophic and metabolic potentials, is an
indispensable part of complex pathogenetic therapy. The antioxidants
suppress the neuronal membranes' enhanced lioperoxidation, which is a
mechanism of GPEE formation, inhibit GPEE activity, and normalize the
receptors' state on the neuronal membranes. As a result, pharmacological agents become more effective. Vitamins and their derivatives
(nicotinamide, pyridoxal-5-phosphate, α-tocopherol, pantogam), which
normalize the nervous tissue's metabolism, also produce a direct
antiepileptic effect when they are given in the appropriate doses.
The combined use of the corresponding vitamins enhances their antiepileptic effect and makes it possible to suppress various types of
epileptic activity that are resistant to separate vitamins. Vitamins
can be used also together with antiepileptics. This produces a
potentiating effect and makes it possible to reduce the doses of
antiepileptics. Metabolic therapy is an important part of complex
pathogenetic therapy concerning not only epilepsy, but also various
neuropathologic syndromes which are based on hyperactive pathologic
systems.

14
Recovery mechanisms

Recovery is a physiological process of the elimination of a disease that develops according to definite laws. These laws are the same regardless of whether treatment is given or not. If treatment is not special urgent therapy or substitution therapy (including therapy of the organ engineering type), it only promotes the development of the natural recovery mechanisms, furthering and consolidating the recovery process as a whole, but it does not replace the mechanisms of this process.

In the case of any neuropathologic syndrome, recovery is effected by specific mechanisms, and definite therapeutic methods are used in accordance with these mechanisms. At the same time, there are general regularities of the development of recovery as an intricate chain process which occurs in stages. The recovery states are determined by the essence of the physiological mechanisms and have a definite clinical expression.

It follows form the preceding chapters that the recovery process involving neuropathologic syndromes consists in the elimination of the pathologic systems on which the syndromes are based and in the normalization of the functions which were disturbed due to the activities of both the pathogenic agent and the pathologic systems, and to the brain's reaction to a pathologic system. The elimination of the pathologic system is a complex process which has definite stages and which depends on several factors. In general, it consists in the elimination of the hyperactive determinant structure (DS) and the suppression of the activity of other parts of the pathologic system.

Researches have shown that, in the case of natural recovery, the pathologic system is so eliminated that at first it is deprived of

its connections which realize the determinant's influence. The first
to 'drop out' of the system are its parts which are influenced least
of all by the determinant; then, other parts which are more dependent
on DS 'drop out'. The last to remain is hyperactive DS. The determinant is the last to die. However, it now loses its significance as
a controlling structure, since the pathologic system no longer exists
as a whole. Instead of DS, a hyperactive structure of local significance may remain. Such a structure may be a weakened GPEE, a disinhibited apparatus of transmitter secretion, or the supersensitivity
(an increased number) of postsynaptic receptors. When GPEE remains,
its power gradually decreases, its dimensions diminish, and it disappears, as can be seen from the respective clinical, electrographic
and pathophysiological phenomena. We have observed this process in
models of various pathologic systems produced in different regions of
the central nervous system.

Figure 8 (see Chapter 1) shows an example of the breakdown of
the epileptic complex, which was produced in the cerebral cortex by
applying strychnine to several regions, after strychnine was washed
away. This example can be regarded as a model of the natural restoration of the structures' normal function when the pathogenic agent
is removed and there are no additional pharmacological influences.
In this case, the activity of the whole complex diminishes. This
diminution is especially clearly expressed in the zone which is far
away from the determinant focus (fragment C). After a definite
length of time, the complex disappears and epileptic activity in the
determinant focus (fragment D) ceases. However, the pathologic
process is not completely eliminated yet. If bemegride, which
enhances the brain's general excitability, is administered at this
stage, the whole epileptic complex is also restored. This restoration begins with the generator's reactivation in the former determinant focus (fragments E, F), which becomes the determinant again; a
complex of foci of epileptic activity is restored under its influence
(fragment G). The repeated breakdown of the complex under the action
of an anesthetic (halothane) occurs in the same way as when the
complex naturally disappears: at first, activity in the dependent
foci is suppressed, while the determinant focus remains, and so forth
(fragments H, I). When bemegride is given, the complex may originate
again, and the determinant focus is the first to be activated. Such
a procedure can occasionally be repeated several times before the
natural activity of the system's parts ceases. In this case, activity in the former determinant focus may remain. This experiment
simulates the behavior of the latent focus, which at first becomes
a local generator when it is activated, and then it acquires DS
properties.

The peculiarities of the natural elimination of GPEE can be
clearly seen by taking the example of GPEE produced by TT in the
efferent output system in the lumbosacral segments of the spinal cord
(see Chapter 2). GPEE becomes larger as TT spreads in the ventral

horn. This is expressed clinically in the origination of rigidity at
first in one muscle which is directly connected with the segments
where GPEE originated, and then in the neighboring muscles, and
finally in all the muscles of a leg. Electrographically, this process is expressed in the appearance of increasing, prolonged asynchronous electric activity (EA) in those muscles (Figure 40). In
recovery, EA is reduced in the opposite sequence: at first, it is
reduced in distant muscles, then in the adjacent muscles, etc.
Rigidity in the same muscles weakens and the respective joints can
move. In some cases, bursts of enhanced EA can be provoked for a
long time, indicating that reduced GPEE remains. In one patient,
such activity was observed for several years after recovery from
local tetanus (Kryzhanovsky, 1966).

The dynamics of the changes in the same muscle during recovery
is noteworthy (Figure 124). At first, spontaneous (background) and
enhanced after-discharge activities diminish, while provoked burst
activity is reduced less. The latter remains augmented even at the
stage when background and trace electric activities disappear.

Such disappearance of EA in natural recovery coincides with the
dynamics of the reduction of EA under the influence of pharmacological drugs, particularly antiepileptics. It has already been shown

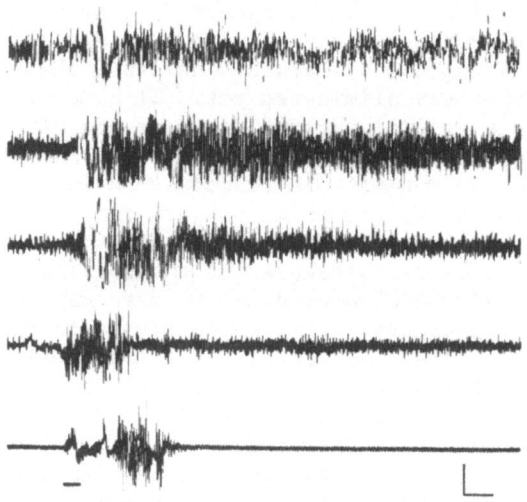

Fig. 124. Electric activity in the gastrocnemial muscle 3 (A) and 7
           (B) days and 1.5 (C), 2.5 (D) and 3 (E) months after TT
           injection into it. The illustration shows background EA,
           bursts of evoked EA caused by stimulating the left hind
           leg (squeezing the left foot: marked by a short line), and
           after discharges being transformed to background EA.
           Amplitude: 0.3 mV; time: 0.2 s.

(Chapter 13) that after-discharge and background activities are the
first to disappear under the influence of those drugs. Then, the
duration of tonic activity decreases and its 'tail' disappears.
Provoked burst activity can be suppressed under the action of only
large doses of anticonvulsants. Hence, the dynamics of a change in
EA is the same in natural recovery as during the action of anticonvulsants.

That course of the involution of enhanced EA during both
recovery and the action of anticonvulsants is due to the peculiarities of the same process of GPEE reduction. Those signs of the
disappearance of different constituents of enhanced EA in the muscles
show that the inhibitory mechanisms in the GPEE neuronal population
are being gradually restored, the amount of the GPEE neurons and
their excitability have decreased, and that positive connections and
other mechanisms which allow GPEE to function have weakened (see
Chapter 3). At the same time, they are in accord with the above data
which show that the GPEE neuronal population is functionally heterogeneous and that neurons which are less excitable and are under
greater inhibitory control (neurons of group III) are the first to be
relatively normalized and to 'drop out' of the neuronal aggregate of
GPEE. Probably neurons of group II escape the influence of the
trigger neurons of group I during recovery, becoming neurons of group
III, and then they lose their hyperactivity, as a result of which the
dimensions and power of GPEE diminish. The GPEE 'nucleus', which
consists of neurons of group I, also diminishes in this way. That is
the main process of the elimination of GPEE during recovery. A
population of neurons with enhanced activity and excitability remains
even at the late stages of recovery. Normally, these neurons' activity is not exhibited clinically. However, they can react with
enhanced provoked activity under the action of the corresponding
stimuli.

The ordinary division of the recovery process into two periods,
i.e. the clinical recovery period and the so-called complete recovery
period, does not reflect all its peculiarities. An analysis has
shown that the recovery process should be divided into more parts in
accordance with the clinical and pathophysiological specifics of its
stages.

Even inconsiderable signs of the neuropathologic syndrome shows
that a functionally significant and active GPEE still remains,
although it is substantially reduced. At this stage of clinically
incomplete recovery, not only pathogenically specific agents, but
also some nonspecific agents are capable of activating the generator
and, consequently, augmenting the syndrome's manifestation. Nonspecific influences produce an especially strong effect if they are
capable of not only activating the generator, but also simultaneously
disturbing the mechanisms of compensation and general integrative
control by the brain. Clinicians know the danger of this stage, when

a patient seems to be well, but then his condition exacerbates, and
sometimes this may be tragic.

At stage I of clinically complete recovery, the syndrome's
clinical signs are not exhibited, and they are not observed under
ordinary conditions without any additional influences, i.e. without
any tests. At this stage, nonspecific influences are hardly effective. However, they can to a certain extent cause the former syndrome to be exhibited. Pathogenically specific influences can
enhance the generator's activity, restore its power and even turn it
into an active determinant structure, causing both the pathologic
system's reactivation and a recurrence. As regards a model, stage I
of clinically complete recovery can be illustrated by taking the
example of the natural disappearance of the determinant focus and the
whole epileptic complex in the cerebral cortex and their restoration
under the influence of bemegride, as has been described earlier
(Figure 8). A similar effect of the reactivation of epileptic foci,
which disappeared, by convulsants has been described. They were
induced by applying antiserum to ganglioside and onto the cerebral
cortex (Karpiak et al., 1976), and to actomyosin-like brain protein
(Bowen et al., 1976). The epileptic focus can be activated by
pentylenetetrazole even four weeks after it disappears (Karpiak et
al., 1976).

The phenomena of a recurrence can be clearly seen by taking the
example of the syndrome of muscular rigidity induced by GPEE in the
efferent output system in the spinal cord. If a small TT dose is
injected intravenously at the stage of clinical recovery when the
signs of muscular hypertonicity disappear and the animal freely use
the leg, hypertonicity and rigidity are observed in the leg against
the background of general rigidity under the influence of TT.

The phenomenon of the origination of the signs of the former
pathologic process can be clearly seen when phenol is used for
repeated pathogenic action. This agent acts also on the spinal mc
apparatus, particularly on the efferent output system in the spina
cord. Phenol usually causes generalized muscular jerks of the sli
myoclonus type when it is injected subcutaneously in small doses.
Such general myoclonic jerks originate also in animals which previously experienced hypertonicity and rigidity of the hind-leg
muscles. Against this background, however, the leg which was for
merly rigid soon extends (Figure 125 A).

At stage II of complete clinical recovery, no pathogenic inf
ences, neither specific nor nonspecific ones, can reproduce the
symptoms of the former syndrome regardless of the dose. This sta
can be called the stage of full recovery. However, it is still
unclear whether the generator is completely eliminated at this s
or it is so repressed and disorganized as a functional structure
its constituents cannot form a functionally significant aggregat

Fig. 125. Extension of the hind leg after phenol administration at
           the stage of clinical recovery from local rigidity of
           spinal origin. A: A rat to which phenol (2 ml of 2%
           solution) was injected subcutaneously at the moment when
           signs of muscular rigidity of the left hind leg dis-
           appeared. B: A healthy rat to which the same dose of
           phenol was injected.

under the action of pathogenic agents. Probably trace pathologic
changes in the appropriate structural processes may remain indefinitely and even throughout life, and they can reveal themselves
somehow when conditions are favorable, adding certain features to the
clinical picture of a new pathologic process. The mechanisms of
so-called locus minoris resistentiae may be connected with this fact.
That is also one of the causes of the individual specifics of the
disease's manifestations.

     Hence, the following stages in the last steps of recovery can be
singled out: incomplete recovery and clinically complete recovery of
stages I and II. The latter can also be regarded as full clinical
recovery. Such a division of the process corresponds to its clinical
manifestations and pathophysiological mechanisms. Of course, the
boundaries of these stages are nominal, and it cannot be seen where
one stage ends and another begins. All the same, they objectively
exist, and it is very important to establish a definite stage in
order to predict the possibilities of an aggravation of a process, a
recurrence, and the specifics of the further course of a process, and
also to select effective rational therapy which corresponds to a
certain stage of recovery.

     One of the peculiarities of the phenomenon of the reproduction
of trace pathologic effects is noteworthy. Using phenol to extend
the leg which was formerly affected by extensor rigidity, we could
not reproduce the same rigidity or hypertonicity which occurred
earlier even at stage I of clinical recovery. At this stage, TT
alone could induce rigidity, i.e. in essence, it induced a pathologic
process with the same features as the former one not simply by exciting the generator, but by additionally forming it. On the other
hand, experiments involving the natural extinction of epileptic foci
in the cerebral cortex have shown that bemegride can cause the mani-

festation of latent epileptic foci which were induced by strychnine.
Apparently, a general rule of the reproduction of the extinguished
syndromes can be formulated in the following way: the more complex a
neuropathologic syndrome and the longer the period since the beginning of recovery, the more difficult it is to completely reproduce
the syndrome by additional pathogenic influences. Obviously, nonspecific pathogenic agents cannot, in principle, reproduce the clinically extinguished neuropathologic syndrome. At best, they can
reproduce its separate constituents. The specifics of a pathogenic
agent with respect to a syndrome especially apply to GPEE owing to
its neurochemical organization.

It has already been shown (Chapter 3) that specific trigger
stimulation plays an important role in starting GPEE and, consequently, activating the pathologic system. Of course, the activation
of GPEE by modally specific trigger stimulation is not equivalent to
its formation. If a syndrome or its constituents can be reproduced
by specific trigger stimulation, this does not mean yet that a new
GPEE is produced; it may mean that a latent GPEE is reactivated.
GPEE is reproduced by modally specific trigger stimulation only when
it was produced earlier by the same stimulation.

An exception in this case are the syndromes in which GPEE originated in connection with the nonspecific pathogenic conditions, e.g.
in the case of a vascular pathologic state. Then, new pathogenic
influences, even if they are nonspecific, can cause the whole neuropathologic syndrome to recur when they either aggravate the vascular
pathologic state or induce its recurrence.

In his day, Speransky (1937) formulated the 'second blow' concept on the basic of data concerning the reproduction of a pathologic
process by secondary pathogenic influences. This concept was very
important in pathophysiologically clarifying recurrences and mnestic
pathologic reactions of various systems, in assessing the significance of anamnesis, etc. At present, we are fully aware of the
significance of the concept as regards the nervous system's pathologic state. At the same time, the boundaries of the extinguished
pathologic processes and the regularities of their reproducibility,
and the pathogenic role which specific and nonspecific pathogenic
agents play in such a process should be clearly understood. This is
especially important in neuropathology, since neuropathologic syndromes, particularly those which were discussed, are based on structural hyperactivity, while hyperactivity in itself is not specific
and can be induced by various agents, as has been shown when the
mechanisms of the origination and activity of GPEE were analyzed
(Chapter 3).

The role which neuropeptides may play in recovery and in the
pathologic system's stabilization and activation is especially
interesting.

Figure 126 illustrates experiments involving sectapeptide
(Tir-Gli-Gli-Fen-Lei-Arg)*, which, under normal conditions, engenders
posture asymmetry ('right-handed' and 'left-handed' rats). It has
been observed that the animal's hind leg, whose muscles were rigid

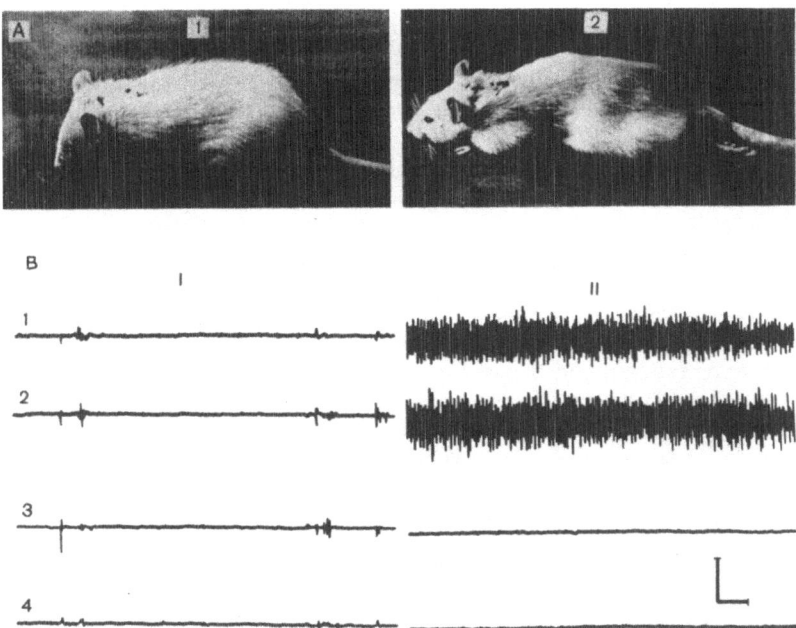

Fig. 126.   Reproduction of the signs of the former neuropathologic
            syndrome. A: Hypertonicity in the formerly rigid muscles
            of the left hind leg as peptide was administered sub-
            occipitally at the moment when the clinical signs of
            rigidity disappeared: animals before (1) and after (2)
            peptide was administered suboccipitally (100 µg/kg); 30
            days after TT was injected (1/30 DLM for the rat) into the
            left gastrocnemial muscle. B: Enhanced EA in the formerly
            rigid muscles of the left lind leg as peptide was admin-
            istered suboccipitally at the moment when signs of rigid-
            ity disappeared. EA in the muscles of both hind legs
            before (I) and after (II) peptide was administered sub-
            occipitally (100 µg/kg). Recordings of EA: 1: Left
            gastrocnemial muscle; 2: left group of femoral muscles;
            3: right gastrocnemial muscle; 4: right group of femoral
            muscles; TT was injected (1/30 DLM for the rat) 30 days
            earlier. Amplitude: 250 µV; time: 1 s.

---

* This peptide was synthesized by M. I. Titov and his associates in
  the Cardiology Center of the USSR Academy of Medical Sciences.

(due to GPEE created in the lumbosacral segments by TT) but whose
rigidity disappeared by the time when the peptide was administered
(stage I of complete clinical recovery), clearly extended when the
peptide was administered suboccipitally (Figure 126 A) (Kryzhanovsky
et al., 1981a). This effect is relatively specific: it could not
be produced by other peptides (three compounds were tested), and it
was exhibited only in the formerly 'tetanus' leg, whether left or
right, regardless of the congenital functional locomotor asymmetry.
The effect was connected with the activation of extinguished GPEE in
the ventral horns of the lumbosacral segments. After peptide administration, not only a clinical effect of enhanced muscular tonicity,
but also an increase in electric activity in the muscles of a given
leg was observed (Figure 126 B), showing that the spinal motoneurons
of GPEE are hyperactive. This effect originated when the animals
were no longer in anesthesia and lasted 10-15 minutes, i.e. apparently until the peptide's action ended.

A noteworthy fact is that the given peptide can 'exhibit' a
pathologic process at the latent stage: muscular hypertonicity is
clearly revealed when the peptide is administered suboccipitally to
an animal which still has no signs of such hypertonicity after TT is
injected. In this case, enhanced electric activity originates in the
muscles (Figure 127), showing that GPEE in the ventral horns of the
spinal cord is being activated.

Hence, a comparison of the experiments involving the 'exhibition' of the pathologic syndrome at the clinical recovery stage,
when the syndrome's symptom no longer exists, and during the latent
period, when such symptoms still do not exist, suggests that the same
pathogenetic mechanism, i.e. the activation of GPEE being either
formed (latent period) or eliminated (clinical recovery), is concerned in both cases.

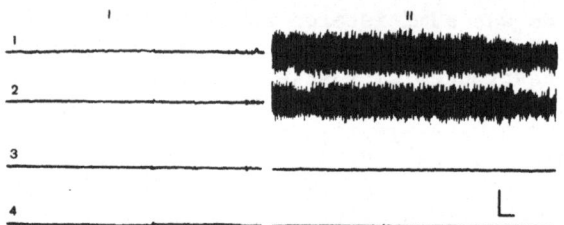

Fig. 127. Enhanced EA in the left hind leg muscles after
           suboccipital peptide administration during the latent
           period following TT injection. Electric activity in the
           hind leg muscles before (I) and after (II) suboccipital
           peptide administration (100 µg/kg); 48 h after TT (1/40
           DLM for the rat) injection into the left gastrocnemial
           muscle. Recordings of EA: same as in Figure 126.
           Amplitude: 250 µV; time: 1 s.

The role of peptides is apparently not confined to the activation of GPEE at the stage of its formation and elimination. Probably peptides participate in pathologic processes, influencing the
pathologic system in various ways, e.g. influencing its stabilization
or destabilization. In this respect, it was interesting to study the
effects of peptides which promoted a stimulating and stabilizing
effect on memory and, therefore, could influence the course of the
processes underlying neuropathologic syndromes. Use was made of the
same model of the relatively simple neuropathologic syndrome, namely,
local muscular rigidity of the hind leg that originates when GPEE is
produced by TT in the ventral horns of the spinal lumbosacral segments. The agent used was tetrapeptide $AKTG_{4-7}$*, which produces a
stimulating and stabilizing effect on the mechanisms of memory (De
Wied et al., 1975; Ashmarin et al., 1978: De Wied, Jolles, 1982).
Researches have shown that enhanced electric activity remains much
longer in the hind leg muscles of an animal which was repeatedly
treated with peptide at the height of muscular rigidity as compared
with the control animals (Figure 128) (Kryzhanovsky et al., 1981b).
This fact indicates that GPEE is preserved much longer under the
peptide's influence. The same result (prolongation of GPEE activity)
was produced when use was made of another peptide, lysyl vasopressin,
which also has a stabilizing effect on memory. These experiments
suggest that physiological neuropeptides which participate in consolidating physiological memory can also participate in stabilizing
GPEE and the pathologic system sustained by it.

The data obtained give us a better understanding of the factors
which promote the stabilization and preservation of the pathologic
systems and the GPEEs on which the systems are based. Those data are
all the more significant when account is taken of the fact that GPEE
was produced on the spinal level, where the effect of also $AKTG_{4-7}$
and lysyl vasopressin was exhibited. This concerns not the physiological processes of memory and training, but the pathologic functional organization in the form of GPEE. Consequently, such an
organization also has a peculiar form of memory. Long-term GPEE is a
type of such 'pathologic' memory. The latter may play an important
role in the pathologic system's stabilization and a recurrence. The
above experiments involving the reproduction of the symptoms of the
'extinguished' neuropathologic syndromes and epileptic complexes
under the action of the respective pathogenic agents show that this
type of memory plays such a role.

Afferent stimulation may produce various effects during a pathologic process. When such a process reaches its peak, additional
afferent influences nonspecifically activate GPEE and, consequently,

---

* The peptide was synthesized by M. A. Ponomareva-Stepanova and V. N.
  Nezobatko in the Institute of Molecular Genetics of the USSR
  Academy of Sciences.

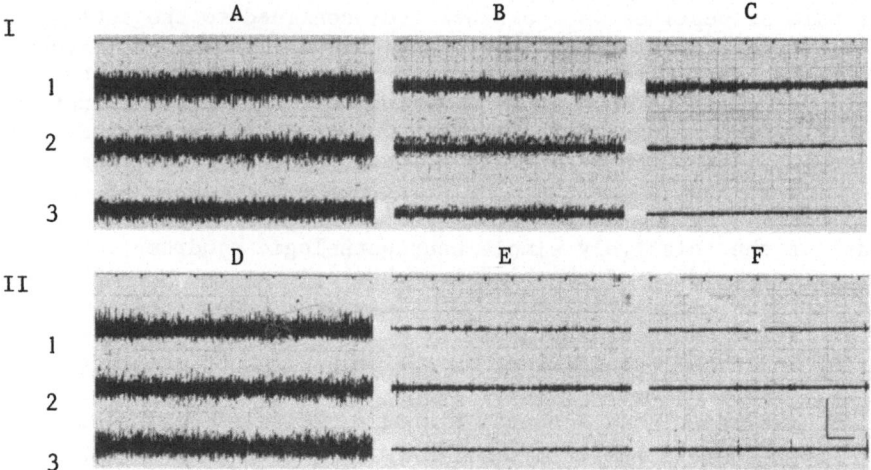

Fig. 128. Electric activity in the hind leg muscles of rats treated
with $ACTH_{4-7}$ (I) and control animals (II) after GPEE
formation in the ventral horns of the spinal lumbosacral
segments. (A) Electric activity 3 days, (B) 28 days, and
(C) 34 days after TT injection into the left gastrocnemial
muscle. 1: Gastrocnemial muscle; 2: anterior crural
muscles; 3: posterior group of femoral muscles of the left
hind leg. Amplitude: 250 µV; time: 1 s.

augment the pathologic syndrome's exhibitions (Chapter 3). This
conclusion is underpinned by data obtained from a study of the hypertonicity of the hind leg that was caused by GPEE produced by TT in
the ventral horns of the lumbosacral segments: at the height of the
process, various afferent stimuli intensify electric activity which
is already enhanced in a given leg's muscles. As regards GPEE
induced by TT in small doses in the same segments on the side of
their deafferentation (ipsilateral section of the posterior lumbosacral roots), it is formed more slowly and its excitation is less
powerful, being exhibited in electric activity which is less enhanced
in the muscles. However, rigidity and contracture disappear more
slowly in such animals, although GPEE which is less powerful originates in them, and it may seem that the process should be eliminated
more quickly. In the deafferentation of the segments where GPEE is
localized, secondary contractures are observed for a long time (Figure 129), and sometimes they can remain for an infinitely long time
(Kryzhanovsky, 1966).

Research has shown that passive flexion and extension of the
rigid leg with definite frequency and amplitude (an imitation, as it
were, of the leg's natural movements) cause rigidity to diminish so
that later the animal can bend its leg (Figure 130) and can use it
again in walking as it makes the necessary movements. This effect

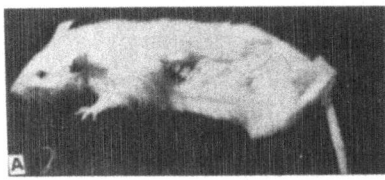

Fig. 129. Maintenance of contracture after muscular rigidity of the
           leg on the side of the deafferentation of the lumbosacral
           segments of the spinal cord. A: A rat in which the dorsal
           roots (L_2–S_1) were cut a week before TT was injected (1/15
           DLM) into the left gastrocnemial muscle. B: A rat into
           the left gastrocnemial muscle of which the same dose of TT
           was injected during the same period. The photograph was
           taken 4 months after TT was injected. In the control rat,
           muscular rigidity in the left hind leg disappeared within
           a month.

can be produced not only during recovery, but also at the height of
muscular rigidity. Gradually, extensor tonicity increases, rigidity
grows, articular mobility decreases, and the leg is back again in a
state of extensor rigidity within a few minutes.

Hence, natural afferentation, including probably above all the
feedback from a system's efferent connections, is needed additionally
to suppress GPEE's activity. That factor is important even at the
stage when the pathologic process is expressed to the utmost,
although its effect is brief under the given experimental conditions.
At the stage of recovery, its role is especially important. Apparently, it promotes the normalization of intrasystemic relations and
processes in the GPEE neuronal population.

In analyzing the mechanisms of those effects, it should be taken
into account that neuronal deafferentation may be a prerequisite of
neuronal epileptization and the formation of GPEE, and that epileptic
activity can be suppressed by afferently activating the neurons
(Chapter 3). These mechanisms probably participate also in the given
phenomenon. Natural afferentation is important also in normalizing
intersystemic relations, in enhancing integrative control, and in
rearranging the intracentral relations in the brain.

A comparison of the dynamics and peculiarities of the elimination of the pathologic system in natural recovery and during pathogenetic therapy (Chapter 13) shows that the main way of eliminating
the pathologic system is the same in both cases. In natural recovery
and during therapeutic interventions, the pathologic system is
reduced in the same sequence: the first to be eliminated are the
system's parts which depend least of all on the determinant structure; the next are the parts which depend more on the determinant

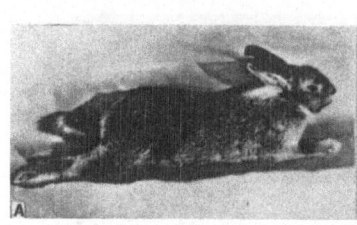 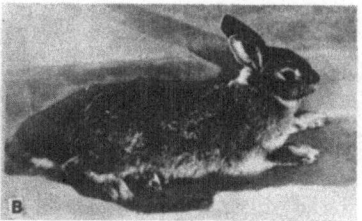

Fig. 130.   Effect of the passive flexion and extension of the leg on
            the muscular tonicity of spinal origin. A: Extensor
            hypertonicity of the right hind leg of a rabbit caused by
            the production of GPEE in the ventral horn of the spinal
            lumbosacral segment; three days after injecting TT into
            the gastrocnemial muscle. B: Position of the same leg
            after three minutes of passive flexion and extension.

structure, etc. The last to disappear is the former determinant as a
reducing generator. Apparently, treatment stimulates and accelerates
recovery. It means that the effect of therapeutic factors is
realized through recovery mechanisms.

It has already been shown that the activation of the existing
antisystems and the formation of new ones are important as a universal physiological mechanism of recovery. This mechanism has its
specifics in every physiological process. Tonic activity or readiness for the rapid activation of the respective antisystems, which
remain after a pathologic process, is a mechanism of acquired resistance. Besides the generators of excitation, which maintain the
sanogenic antisystems' tonic activity, probably specific agents,
which are witnesses and inductors of that stable state of health, are
produced in the nervous system, causing resistance to a given pathologic process. It is important to ascertain the neurobiological
factors which maintain this state. Moreover, it is just as important
to reveal the obstacles of those mechanisms and the causes of the
chronicity of a process, the unsteady state of health, and the predisposition to a disease.

SUMMARY

Recovery in the case of neuropathologic syndromes, which is
characterized by the systems' hyperactivity, is a process involving
the elimination of both the pathologic systems on which the syndromes
are based and the hyperactive DSs which induce them.

The following concluding stages of recovery can be singled out
in accordance with pathophysiological and clinical data: clinically
incomplete and clinically complete recovery. The latter is sub-
divided into stages I and II. At the incomplete recovery stage,

various pathogenic factors can activate the hyperactive determinant
and the pathologic system, aggravating the process and causing its
recurrence. At stage I of clinically complete recovery, a recurrence
can be caused only by specific pathogenic factors, while nonspecific
pathogenic agents can reproduce only separate signs of a syndrome.
At stage II of clinically complete recovery, neither specific nor
nonspecific pathogenic agents can reproduce a former syndrome or its
constituents. It is still unknown whether the former GPEE is completely eliminated at this stage. Its pathobiological traces probably remain, as a result of which the clinical picture of the subsequent pathologic process may somewhat change.

In natural recovery, the pathologic system and the determinant
are eliminated by the same regularities according to which the processes occur during treatment, namely, the system's destabilization
and breakdown, its reduction at the expense of its parts, first those
which are influenced least of all by the determinant, and then those
which depend more on it, etc. Afterwards, the determinant, which is
deprived of the system's parts, loses its leading role. Instead, a
local active structure, e.g. a weakening generator, remains.
Finally, the generator is also eliminated. That sequence of the
pathologic system's elimination illustrates only the main stages.
However, the process as a whole simultaneously covers various parts
of the system, including the determinant structure.

It has been shown on the basis of the model of the muscular
rigidity syndrome caused by producing GPEE in the efferent output
system in the spinal lumbosacral segments that natural afferent
stimulation, including feedback from the effector links, is needed
for optimally realizing the recovery mechanisms, i.e. for eliminating
GPEE and the pathologic system and restoring the disturbed functions.

Some peptides can modulate the course of a pathologic process
and activate ('reveal') GPEE at stage I of clinical recovery and
during the latent period. Peptides, which stimulate memory processes, (e.g. $AKTG_{4-7}$, lysyl vasopression), can maintain GPEE's
activity longer.

The action of therapeutic factors is realized through physiological mechanisms, stimulating those processes and accelerating
their development. Rational pathogenetic therapy stimulates the
recovery mechanisms.

The activation of antisystems, i.e. the functional antagonists
of the physiological systems which became pathologic ones, is a
universal sanogenic mechanism. Hence, a stable state of health
should be maintained after a syndrome by special long-term factors,
i.e. by the antisystems' tonic activity which is sustained by physiological mechanisms, and probably specific compounds which originate
during a pathologic process.

In lieu of a conclusion

The data of all the research presented in this book and the general theory worked out on their basis can be summarized in the following way.

The <u>pathologic system</u> (PS) is a universal pathogenetic basis of various neuropathologic syndromes. It is a new functional organization formed by the elements of damaged physiological systems under the conditions of pathogenic influences, and is characterized by new regularities and mechanisms. An important feature of PS which distinguishes it from the physiological system is that the result of its activity is of not adaptive, but disadaptive and even pathogenic significance to the organism.

Unlike the physiological system, which is formed at every given moment as a dynamic organization for a definite functional task in connection with the influences of the environment or an organism's requirements, PS is induced by a hyperactive structure which originates in CNS due to the effect of the exogenous or endogenous damaging factors or to a genetic defect. Such a structure subordinates to itself other parts of PS connected with it and determines the peculiarities of their activity and, consequently, the peculiarities of the pathologic system's behavior and the result of its activity. Owing to these features, this structure is a pathologic <u>determinant</u>.

The pathologic determinant escapes both the control of the system in which it originates and that of the brain's homeostatic mechanisms. Consequently, PS induced by it also gets out of integrative control. The functional effects which are realized by the pathologic determinant do not correspond to the acting stimulant,

326

situational afferentation, and the organism's requirements. Their
peculiarities are due only to the peculiarities of the determinant's
activity.

A hyperactive determinant structure makes the functional organization and activity of PS rigid, as a result of which the modulating
influences of other systems and of the CNS's regulating mechanisms
are ineffective. The extent of this rigidity depends on the intensity of the functional message produced by the hyperactive determinant structure and on several peculiarities of the determinant.
Owing to their intensity, the pathologic determinant's functional
influences overcome the regional mechanisms of control and autoregulation in the CNS formations which receive these influences and,
consequently, become parts of PS.

Unlike the physiologic system's activity, which is modulated
during the achievement of a programmed result due to the feedbacks of
the system's efferent and intermediate links, PS activity is modulated either insufficiently or not at all, although feedbacks remain.
Moreover, feedbacks from the intermediate or effector links can even
be functionally enhanced, but they are ineffective, since their
influences cannot be realized in either the determinant's region or
the system's intermediate links due to the inhibitory mechanisms'
insufficiency.

Unlike the physiological system, which disappears as a dynamic
organization after its programmed result is produced, PS can remain
and act for an indefinitely long time. The duration of PS activity
depends not on the achievement of a 'result', but on the determinant
structure's activity.

At the early stages of a process, all the peculiarities of PS
behavior and its main features were established by the determinant's
activity. During this period, other links of the system may remain
unchanged as individual functional formations: being parts of other
physiological systems, they can adequately perform their function.
At the late stages, however, these links can undergo secondary
changes; some of them begin to act as secondary determinant structures, promoting the stabilization of a given PS, enhancing its
resistance, or inducing new pathologic systems.

The determinant is the main factor of the stabilization and
resistance of PS: its elimination causes the breakdown of PS. At the
same time, the system's subordinate parts make a definite contribution to the stabilization and resistance of PS due to the establishment of mutually sustaining and mutually potentiating positive links
between them. This form of relations between the PS parts is well-
known practically as the classical circulus vitiosus. The dimensions
and structure of PS are important in establishing its resistance.
Researches involving a model of an epileptic complex in the cerebral

cortex have shown that the larger the number of epileptic foci in the complex, the greater is the latter's resistance to inhibiting effects. The complex does not break down when its separate subordinate foci are suppressed. Nevertheless, this suppression reduces the general resistance of the epileptic complex. As PS develops and becomes stabilized, it acquires new mechanisms and becomes more resistant.

Like any biological system, PS can become adapted to the influences directed at it. The adaptation of PS is another mechanism of its resistance. It is especially important when therapeutic influences are either insufficiently effective or inadequate. In such cases, adapted PS can engender a state of enhanced resistance. All these phenomena act as the pathogenetic mechanisms of the resistance of neuropathologic syndromes and their passage into a chronic state.

The neuropathologic syndromes are a clinical expression of the corresponding pathologic systems' activity. Every syndrome has its PS and its determinant. The specifics of a neuropathologic syndrome are established by PS, since the latter is a changed physiological system whose activity has its own functional specificity and modality. By establishing PS behavior, the determinant gives rise to the peculiarities of a neuropathologic syndrome's manifestation and development.

The pathologic system develops by involving new nervous structures in a given process due to the pathologic determinant's growing influence with an increase in its power and to the attenuation of the regional mechanisms of control and autoregulation of the structures which are influenced by the determinant. A pathologic process can develop also due to the formation of secondary determinant structures and new PS, being exhibited in either a complex of new pathologic syndromes or an intricate multisystemic syndrome. Such intricate forms of a pathologic state, i.e. complexes of pathogenetically connected syndromes, are usually observed. In natural recovery, intricate complexes are eliminated (expressed in the disappearance of separate syndromes) usually in the reverse order of their origination. In treatment with special drugs, the elimination of separate syndromes depends on the selective influence of the drugs on the appropriate PSs.

The development of PS, being connected with the involvement of new cerebral structures in a process, causes the physiological systems' disorganization. The formation of a powerful PS is accompanied also by the suppression of the activity of antagonistic physiological systems and defensive compensatory mechanisms. All these processes disturb and disorganize the brain's integrative activity. Therefore, interventions which can preclude the development of PS are very important both therapeutically and prophylactically.

The data in this book show that the following can be the neurobiological, i.e. neurophysiological and neurochemical, basis of the
hyperactive determinant structures: (1) generators of pathologically
enhanced excitation; (2) pathologically augmented transmitter
release, and (3) supersensitivity (increase in the number) of postsynaptic receptors.

Generators of pathologically enhanced excitation (GPEEs) are
populations of neurons with an insufficiency of inhibitory mechanisms
and low thresholds of excitability. The insufficiency of inhibitory
mechanisms may be either absolute (primary) or relative (secondary).
Primary inhibitory insufficiency originates due to the pathogenic
agents' directly damaging effect on inhibitory synapses; in such
cases, GPEE is formed as a result of neuronal disinhibition. Secondary inhibitory insufficiency originates due to neuronal hyperactivation under the influence of pathogenic conditions which cause
depolarization; in such cases, the existing inhibitory mechanisms
remain, but they are functionally ineffective.

The functional organization of GPEE is connected with two processes. On the one hand, the neuronal populations are functionally
'homogenized'; neurons lose their basic distinctions to a certain
extent due to the failure of inhibitory control. On the other, the
GPEE neurons are 'differentiated' owing to the different degree of
their pathologic changes. There are neurons with stable signs of
epileptic activity; they act as leading, or trigger, neurons (group
I) and can be regarded as primary microgenerators in GPEE. Other
neurons become involved in general electrogenesis under their influence, and some of them acquire the properties of epileptic neurons
(group II). The neurons of group III do not acquire these properties, participating only in the general production of pathologically
enhanced excitation. The critical mass of the neurons of group I is
needed to activate the whole neuronal population which constitutes
GPEE. The critical mass of the GPEE operant neurons is needed for
GPEE to operate steadily and produce a pathologic effect. The power
of GPEE is established by the number of the operant neurons and their
ability to generate action potentials with increased frequency.

The peculiarities of the activity of GPEE and its ability to
become activated depend on the extent of the processes involving the
functional reorganization of the GPEE neuronal populations. When the
inhibitory mechanisms still remain, the thresholds of excitability of
the operant neurons are not greatly lowered, and a large number of
neurons of group I is not formed yet (e.g. at the early stages of the
formation of GPEE); to activate GPEE, there must be rather intense
trigger stimulation, which should, in addition, be modally specific
as regards the system in which GPEE originated. At the height of the
development of GPEE, when the given processes of the functional
changes in the operant neurons are exhibited to a considerable
extent, GPEE can be activated even by weak specific stimulation as

well as by stimulation of different modality. At this stage, GPEE
can start operating as if according to the 'all-or-nothing' law; it
can be activated by casual stimulation from different sources and
also spontaneously, i.e. without any visible provoking stimuli. The
GPEE which is complexly formed can be active for different lengths of
time, and even indefinitely in some cases.

Since GPEE constitutes the determinant structure's neurophysiological basis, the peculiarities of the manifestation of the neuropathologic syndrome ultimately depend on the peculiarities of the
exhibition of the generator's activity: the paroxysmal, clonic,
intermittent or tonic character of seizures, their intensity and
duration, the effectiveness of stimuli as a provoking factor, the
spontaneous origin of seizures, their frequency, etc.

It has been shown that GPEE is a universal neuropathophysiological mechanism. GPEEs can originate in various parts of CNS,
nerve centers and nuclei of different organization (linear, reticular, cyclic, etc.).

The conditions under which GPEEs originate and the agents which
engender them can be different. Therefore, neuropathologic syndromes, whose pathogenetic basis consists of GPEEs, are in essence
polyetiologic. At the same time, these syndromes are monopathogenetic: they are caused by a single pathogenetic mechanism, namely, the
formation of GPEE.

Enhanced neurotransmitter release as a mechanism of the determination of pathologic syndromes can occur in the case of neuronal
hyperactivity, or it can be caused by the disinhibition of the terminal apparatus of transmitter secretion. The example of DA release
has shown that a characteristic stereotyped behavior syndrome originates when GABAergic control of DA release by the nigrostriatal
neuronal terminals in caudate nuclei is inadequate. Similar disinhibition of DA released in the mesolimbic structures causes a
psychosis-like state in animals. Such syndromes are suppressed and
behavior is normalized when GABAergic control is restored. There are
grounds to assume that these mechanisms constitute the basis of some
forms of DA-dependent psychosis in man. The hyperactivation of DA
due to the disinhibition of the terminal secretion apparatus may be
connected also with a defect of the feedback from the postsynaptic
structures to the presynaptic apparatus and with a change in the
presynaptic receptors. Such a mechanism of the local disinhibitory
and uncontrollable transmitter release may be found in the systems of
also other neurotransmitters, causing the appropriate neuropathologic
syndromes.

The phenomena of primary supersensitivity (an increase in the
number) of postsynaptic receptors and a change in their structure
that is not connected with medicamentous therapy are especially

noteworthy. As recent works show, a system's DA state in some forms
of schizophrenia is an example of the role played by this pathologic
mechanism. Since a receptor is formed by the synthesis of structural
proteins, this phenomenon of an enhanced formation of receptors
should be connected with the appropriate gene's desuppression. In
this case, the changes in the genome of postsynaptic neurons are the
primary pathologic determinant of that form of pathologic state. An
increased number of membrane receptors and enhanced dopamine effects
are the second determinant which establishes the subsequent chain of
events and the clinical syndrome's development and peculiarities.
The ascertainment of the primary and secondary determinants is of
great importance in understanding the mechanisms of such forms of the
pathologic state of CNS and in elaborating their pathogenetic
therapy.

Experimental models of several neuropathologic syndromes produced on the basis of the above theoretical principles were described
in the second part of this book. Numerous neuropathologic syndromes,
which relate to different spheres of CNS activity, could be reproduced by forming hyperactive determinant structures in definite areas
of CNS and, consequently, turning the appropriate physiological
systems into pathologic ones. The possibility in itself of producing
models of diverse pathologic states of CNS on the basis of a unitary
theory holds out the prospects of such a theory and is a form of
verifying its fundamental principles in practice.

Experimental models cannot thoroughly reproduce the pathologic
states of nervous activity in man and cannot be identical to these
forms. Moreover, a dog's 'psychosis' cannot be reproduced in a cat,
and vice versa, when the matter in question is the disturbance of
higher nervous activity and behavior in different animals. The
stereotyped behavior syndrome described above is manifested in a
characteristic way in different animals in accordance with the
species specificity and the ethological peculiarities of their behavior. However, this syndrome has a single neurobiological basis, i.e.
it has the same neurophysiological and neurochemical mechanisms.
When the neurobiological basis of certain pathologic states of higher
nervous activity is the same, they can be experimentally modeled by
producing their pathogenetic equivalents with regard to the species
specificity and other peculiarities of animal behavior. If the
neurobiological mechanisms of the experimental model and the pathologic state which is being modeled were not the same, drugs could not
be screened by using animals in order to find the most effective
compounds for man. It is necessary to emphasize the importance of
using pathophysiological experimental models in screening researches,
since the effects of drugs on healthy animals may differ from those
on animals with a damaged nervous system.

Various neuropathologic syndromes were experimentally reproduced
on the basis of a unitary principle which issues from the theory

being set forth, i.e. the creation of a hyperactive determinant
structure in order to transform a physiological system into a pathological one. Technically, that principle was realized by either
producing GPEE or disinhibiting neurotransmitter release in the
appropriate CNS areas. These techniques were used to create models
of diverse pathologic states: local muscular rigidity, generalized
spinal myoclonia, various forms of epilepsy, pain syndromes of central origin (spinal, trigeminal, and thalamic origin, and the phantom
pain syndrome), vestibulopathy, parkinsonism, stereotyped behavior,
catatonia, chorea-like hyperkinesis, pathologically prolonged sleep,
neurosis- and psychosis-like states, etc.

     Fundamentally the same regularities of the origin of neuropatho-
logic syndromes as those described above can be observed in the field
of the vegetative nervous system. The peculiarities of these syndromes are connected with the fact that the visceral organs are the
effector of the pathologic systems. That is the basis of a large
group of the so-called neuroses of the visceral organs, various
vegetative crises, dyskinesias, disturbances of secretion, etc. Such
disturbances occur also in the field of neuroendocrinal regulation,
whose changes may result in the corresponding endocrine hyperfunction
or hypofunction. Our recent researches, which are not covered in
this book, have shown that the heart rate (Kryzhanovsky, Pivovarov,
1982; Pivovarov, Kryzhanovsky, 1982), intraocular pressure
(Kryzhanovsky et al., 1983b), and the functions of several visceral
organs can be disturbed when hyperactive determinant structures are
produced in the form of GPEE in definite cerebral areas (hippocampus,
amygdala, some hypothalamic nuclei, etc.). Conversely, the appropriate syndromes can be arrested by suppressing hyperactivity, which
changes the regulation of the visceral organs (e.g. female functional
infertility) (Persianminov et al., 1978). Other authors have
obtained similar data on the disturbances of the visceral organs'
regulation in both experiments and clinical practice when the hyperactivity of certain CNS structures was concerned. All similar forms
of the vegetative nervous system's pathologic state are termed
regulation diseases (Kryzhanovsky, 1978a, 1978b, 1981c, 1981d). This
term is an abbreviation of the definition 'diseases which originate
as a result of the disturbance of regulation'. They cover a considerable part of diverse disturbances, including syndromes caused by
a change in not neural regulation alone. The diseases of neural
regulation constitute the so-called boundary field, which should be
the subject of investigation of not only somatic physicians, but also
neuropathologists and psychiatrists. At the same time, the given
data show that many neuropathologic syndromes relating to different
spheres of CNS activity are also pathogenetically based on regulatory
disturbances.

     The principles of the therapy of the above forms of the patho-
logic state naturally issue from the basis of the. theory being
evolved. The elimination of PSs which underlie syndromes of the

class under consideration (system's hyperactivity syndrome) is the
main task in the pathogenetic therapy of these syndromes. The elimination of the determinant of such a PS is a radical solution. This
way is especially effective at the early stages of a process, when
the stabilization and resistance of PS completely depend on the
determinant. However, such an intervention can be effective also at
the late stages. The importance of diagnosing a determinant structure should be emphasized in this respect. It is necessary to ascertain not only the determinant structure's localization, but also its
nature. For instance, if a leading epileptic focus which acts as a
dominant one is mistaken for a determinant focus, its ablation can
result in not the breakdown of PS (as is the case when the determinant focus is eliminated), but in the activation of the dependent
foci which were suppressed under the influence of the dominant focus.
Such cases are known to occur when epilepsy is being surgically
treated.

It is very important to establish the primary pathologic determinant which acts as the initial pathogenic mechanism. Therapy
intended for eliminating the determinant is described as <u>specific
pathogenetic therapy</u>. Since every stage of the process is characterized by the development of a new determinant (or determinants) and a
new pathologic system (or systems), pathogenetic therapy whose purpose is to eliminate a new determinant (or determinants) and a new
pathologic system (or systems) should correspond to every stage. In
other words, pathogenetic therapy should have its own specific nature
at every stage.

Complex therapy directed at the determinant and various links of
PS is a rational form of pathogenetic therapy. Since these links are
pathogenetically connected with one another, complex therapy can
produce a potentiating effect even when drugs are used in reduced
doses. Therapy which consists in complex selective action on a
determinant's structure is termed <u>complex specific pathogenetic
therapy (CSPT)</u>. It should be noted that CSPT is syndromic and not
nosologic therapy, since it is intended to eliminate a syndrome by
specific action on its determinant and not on a disease as a whole.
Researches involving a model of experimental stereotyped behavior
have shown that CSPT as a complex made up of phenazepam, lithium and
haloperidol is more effective than therapy which involves each of
these drugs used separately; the doses of each of these drugs used in
combination can be greatly reduced, making the origin of side effects
less likely. The first clinical trial has confirmed that this type
of CSPT is effective in some forms of schizophrenia and have shown
that the probability of side effects (extrapyramidal disorders)
sharply lessens.

Complex pathogenetic therapy may be a combination of the determinant structure's ablation and pharmacological influences which
accelerate and maximalize the suppression of PS and the restoration

of the disturbed functions. Metabolic therapy, whose purpose is to restore the trophic and plastic capacities of nervous tissue and to normalize the membranous processes and the receptors' state (their ability to bind drugs), is an important part of complex therapy.

Besides the above mechanisms of the resistance of neuropathologic syndromes due to the pathologic system's biological peculiarities, some factors of ineffectiveness of therapy should also be taken into account. One such factor, being very important, is the nature of the pathologic determinant and the neurochemical organization of GPEE. The neurochemical organization of GPEE differs when GPEE is induced by agents with different mechanisms. At the same time, the PSs induced in such cases are functionally identical, and the corresponding neuropathologic syndromes are clinically the same. It has been shown by taking the pain syndrome of spinal origin as an example that the therapeutic effectiveness of drugs depends on the type of inhibitory control which was disturbed and with which the formation of GPEE is connected. Drugs are ineffective if they act via the activation of the inhibitory mechanisms which were damaged (e.g. when glycine-induced inhibition is damaged). In such cases, compounds with a different action mechanism (e.g. those which activate GABA-induced inhibition) may be effective. These data may help clarify the long-standing question of why the same compound is ineffective in the case of syndromes with a similar clinical picture. On the other hand, they explain the seemingly puzzling question of why different syndromes can be suppressed by the same compounds (e.g. certain anticonvulsants). In the latter case, this applied to the suppression of GPEEs which are sited in diverse systems but which have the same neurochemical organization.

The antisystems' activation is a special form of pathogenetic therapy. It is directly connected with the concepts of the role of PS as the neuropathologic syndromes' pathogenetic basis. Antisystems are functional antipodes, i.e. they are antagonistic systems with respect to physiological systems which became pathologic ones. Pain syndromes of central origin and cortical epilepsy were taken as an example to show the great potentialities of this method in rationally selecting the conditions under which the cerebral structures in an appropriate antisystem are electrically stimulated. Researches have shown that the electric stimulation (ES) of the antisystems' structures can be combined with pharmacological influences which arrest the side effects of ES. Antisystems which inhibit PS, i.e. antisystems which act as a sanogenic mechanism, can also be produced artificially by the coupled activation of definite cerebral structures. An analysis of drug action has shown that many drugs produce their inhibiting effect by activating the appropriate antisystems.

Many presently used methods of pathogenetic therapy are intended to normalize only the target organ. Therefore, they are not pathogenetic therapy in the strict sense of this concept. They are rather

a method of symptomatic therapy. In this case, the desired result,
i.e. the cessation of the manifestations of the neuropathologic
syndrome, can also be produced. But the pathologic determinant and
the PS induced by it remain. Using such a method is the same as
using curare-like drugs to treat epilepsy. A physician who tries to
combat neuropathologic syndromes by influencing only the system's
effector link or the target organ without trying at the same time to
normalize the changed regulation system and to eliminate the pathologic determinant is like a man who uses water-resistant material to
repair the floor instead of plugging the hole in the roof through
which the rain falls. Some routine methods used to treat regulation
diseases are in essence like that. However, if a steady favorable
effect is produced in some cases when the effector links are influenced, it means that this influence helped eliminate PS due to its
destabilization and to a decrease in its resistance, as has been
observed when PS was reduced.

Researches have shown that PS is reduced, breaks down, and is
eliminated according to the same regularities and in the same
sequence as either under the conditions of natural recovery or when
therapeutic effects are produced to eliminate PS. The first to 'drop
out' of PS are its links which are influenced by the determinant
least of all, followed by links which are influenced more by it, etc.
The determinant structure is the most stable structure: it remains
even when all the other parts of PS either are already suppressed (in
the case of inhibiting effects) or have become more or less normal
(in recovery). The determinant is the last to die. However, even
when the determinant structure remains under these conditions, it
loses its decisive significance as a determinant, since the system no
longer exists as such. Hyperactive structures in the form of GPEE,
enhanced transmitter output, or the supersensitivity of the postsynaptic receptors, which are the basis of the determinant, may
remain instead of the latter. The activity of those structures is
more or less reduced; they can be of local functional significance,
and may not be exhibited clinically. But even then they are risk
factors. Researches have shown that those structures can again
become determinants which are capable of restoring PS as they are
activated under certain influence, or when the general mechanisms of
the CNS's homeostatic regulation are disturbed or its general excitability is enhanced. This is how the syndrome recurs.

It follows from the above data that pathogenetic therapy should
be combined with etiologic therapy, whose purpose is to eliminate the
causes of the formation of PS and the conditions that induce its
formation and promote its development. Only complex etiopathogenetic
therapy can be rational.

Frequently, recovery is considered to be a result of the disappearance of the pathologic process. Such a view is wrong. Recovery is not the result, but the very process of the elimination of PS

and the restoration of the disturbed functions. Pathogenetic therapy
(except substitutive therapy) does not introduce any fundamentally
new regularities into this process. Its effects are realized by
natural recovery mechanisms. Rational pathogenetic therapy is a
catalyst of the recovery processes. Recovery mechanisms constitute a
complex multilink sanogenic system. Like any biological system, the
sanogenic system has its own determinant, and it is formed and operates according to definite regularities. Antisystems, the activation
of which causes the suppression of PS, are in the general system of
sanogenic antisystems' tonic activity. The disturbance of their
activity can be the initial mechanism of a pathologic process.

The theory of the generator mechanisms of neuropathologic syndromes, which issues from the general concept of the role played by
the determinant structures in the nervous system's pathologic state,
covers a definite range of neuropathologic syndromes that are characterized by the systems' hyperactivity. At the same time, there are
apparently no syndromes without a component of hyperactivity in the
nervous system's pathologic state. This component plays an important
role even in syndromes which are caused by the nervous tissue's
anatomic defect, since the functional organization of the systems and
parts of CNS is such that damage to any structure inevitably causes
the disinhibition of another structure which is controlled by the
first one. Therefore, the prolapse syndromes are accompanied by the
syndromes of disinhibition, i.e. by the formation of hyperactive
structures, such as GPEEs, which can become pathologic determinants.

Neuropathologic syndromes based on hyperactive determinant
structures are not the nervous system's focal pathology. The syndromes which were described above and similar ones are an expression
of a systemic pathologic state, although the leading pathogenetic
role in this respect can be played by the hyperactive structure or,
to be more exact, by a complex of pathogenetically interconnected
structural, functional and neurochemical changes in a certain area
of CNS.

In the currently popular systemic approach to an analysis of an
investigated phenomenon concerning the processes under consideration,
it is necessary to define the peculiarities of all the connections of
PS and the degree of their participation in the given processes, and
also in the interaction between these processes, as well as under the
conditions of the system's activity, since a structure's properties
change as soon as it becomes part of the system. However, it would
be wrong to confine oneself only to this task, as is frequently done.
In making this systemic approach, another task must be tackled, i.e.
it is necessary to establish the functional hierarchy of systems and
a system's links and to ascertain the determinant of a process. The
first task loses its significance if the second task is not fulfil-
led, because the pathologic mechanisms of a process cannot be established and rational pathogenetic therapy cannot be elaborated when
the most important connection and the key link are unknown.

It should also be noted that the functional plane on which the general concept of the role of the determinant structures in the nervous system's pathologic state, the theory of the generator mechanisms of neuropathologic syndromes, and the conceptions of the pathologic system were elaborated does not rule out the need to make a neurobiological and neurochemical analysis of pathologic processes. On the contrary, the functional approach presupposes the establishment of neurochemical systems and morphological structures as a definite substrate of the pathologic process. Specific pathogenetic therapy cannot be elaborated when the neurochemical mechanisms which constitute the determinant structure are unknown.

The role of the determinant structures in the nervous system's pathologic state is the main subject of this book. Nevertheless, the conception of the determinant has its physiological meaning, too. The determinant's properties are exhibited most clearly when it is hyperactive. It was learned that the determinant and its functional significance were revealed precisely under pathologic conditions. There are many examples in neurophysiology when physiological regularities were discovered as a pathologic situation was created. Sechenov discovered inhibition in CNS in an experiment which represented a real form of a pathologic state: he put a small salt crystal on a section of a frog's thalamus. One of Pavlov's most widely used methods of studying the brain was to produce a 'breakdown', as he put it. The regularities of the nervous system's integrative activity described by Sherrington were discovered in experiments, many of which were of a pathologic nature (it is enough to recall classical decerebration that vividly illustrated the phenomenon of disinhibition, which, as has been shown, is a mechanism of the formation of GPEEs in other hyperactive structures). A structure's hyperactivation has become a routine method of experimentally analyzing its functions.

An artificial PS is in essence simulated and created when a reaction is produced in a classical physiological experiment by electrically stimulating either a nerve or cerebral structures. The reaction thus produced is of no biological significance; it is not corrected by a feedback, since such a feedback cannot change the mode of stimulation under the given conditions. Electric stimulation used engenders a hyperactive determinant structure, the stimulator is an artificial GPEE, the system being stimulated is hyperactive, and the reaction which it realizes is a model of the neuropathologic syndrome. Nevertheless, the results of such an experiment allow us to draw some conclusions on the physiological function of the structure being stimulated, on the system which is activated by it, and on the reaction's peculiarities.

Normally, the determinant structure is the operant part of the program apparatus of a physiological system which originated as a dynamic organization for fulfilling a definite task. This structure

forms a functional message which corresponds to the requirements at a
given moment, the peculiarities of the acting stimulant, the experience gained in encounters with it, and the organism's needs. A
physiological determinant's activity is corrected in the course of a
reaction, and it can be modulated according to changes in a situation. Owing to these properties of the determinant, a favorable
result is produced in the course of a reaction induced by it, while
the physiological system's activity is of adaptive significance.
Every physiological reaction has its own program. Therefore, a
physiological reaction cannot occur without a determinant. The
physiological determinant is a system-organizing factor.

The <u>principle of the determinant</u> as an operant mechanism of
nervous activity is a principle of <u>intrasystemic</u> relations. It
maintains the necessary functional subordination and hierarchy of the
system's parts in order to produce a favorable functional result.
These peculiarities distinguish the principle of the determinant from
another operating principle of the nervous system's activity that was
discovered by Ukhtomsky, i.e. the <u>principle of the dominant</u>. The
latter is a principle of <u>intersystemic</u> relations: a physiological
system which is functionally active at a given moment dominates other
systems due to their coupled inhibition, particularly the systems
which can interfere in the operation of a given active system. Thus,
the principle of the dominant allows a certain reaction to be
realized and optimizes its realization, lessening physiological
interferences (noises) and making the production of a result
rational. Both principles, i.e. those of the dominant and the determinant, indispensably supplement one another: the dominant makes it
possible for a given system to carry on its activity, while the
determinant establishes the pattern of this activity. The combined
action of the two principles is a prerequisite of an adequate physiological reaction. A pathologic state originates when this combined
action is upset.

Both principles, particularly the principle of the determinant,
can be encountered in not only biological, but also other systems
whose activity is intended to produce a definite result. In complex
computer systems which are intended for a definite process, there is
a leading computer which plays the key, regulating role, i.e. it is
the system's determinant. Supervisors may play a similar role of the
determinant. In the system of zoosocial relations, for instance, the
animal herd has its own leader which acts as the determinant, largely
establishing this population's behavior and its inner functional
structure. The role of this leader as the determinant in some zoosocial systems may be so great that when the leader is irretrievably
lost, the system breaks up and its individuals die.

If the determinant becomes hyperactive and escapes systemic and
intersystemic control, it makes the system extremely rigid, and then
the system is not corrected by modulating influences. This is a

pathologic state. When a determinant does not originate, systems
also are not formed. In this case, separate functional structures
which should form a system operate autonomously, disorderly and
independently of one another. This is also a pathologic state.
Ultimately, all the forms of the nervous system's pathologic state
are the hyperbolization of the nervous system's two main forms of
normal activity: powerful programmed activity and stochastic activity.

The principle of the determinant is found also in the sphere of
man's higher nervous activity and behavior, i.e. in all the acts
which are in accordance with a definite program for achieving a
definite result. Upbringing, learning, the gaining of environmental
and life experience, and other processes connected with the establishment of the peculiarities of the systems' behavior in the time
micro- and macrointervals are realized via and are due to the determinant's formation. The central nervous system's activity consists
in the perpetual formation and the constant elimination of the situational determinants. Whenever such a determinant originates, it
must break down (disappear as a functional structure) when the system
activated by it achieves the programmed result. Otherwise, a pathologic state originates.

References

Ambramsky, O., Carmon, A., and Lavy, S., 1971, Combined treatment of Parkinsonian tremor with propanolol and levodopa, J.Neurol. Sci., 14, No. 3, 491–494.

Achar, V. S., Welch, K. M. A., Chabi, E., and Bartosh, K., 1976, Cerebrospinal fluid gamma–aminobutyric acid in neurologic disease, Neurology, 26, No. 8, 777–780.

Agid, Y., Guyenet, P., Glowinsky, J., Beoujonah, J. C., and Javoy, F., 1975, Inhibitory influence of nigro–striatal dopamine system on the striatal cholinergic neurons in the rat, Brain Res., 86, No. 3, 488–492.

Ahlawalia, P., and Singhal, R. L., 1981, Monoamine uptake into synaptosomes from various regions of rat brain following lithium administration and withdrawal, Neuropsychopharmocology, 20, No. 5, 483–487.

Ahtee, L., and Kaariainen, I., 1974, The role of dopamine in pilokarpine induced catalepsy, Naunyn–Schmiedeberg's Arch. Pharmacol., 284, No. 1, 25–38.

Ajmone–Marsan, C., 1961, Electrographic aspects of 'epileptic' neuronal aggregates, Epilepsia, 2, 22–38.

Akaike, A., Shibata, T., Satoh, M., and Takagi, H., 1978, Analgesia induced by microinjection of morphine into and electrical stimulation of the nucleus reticularis paragigantocellularis of rat medulla oblongata, Neuropharmacology, 17, 775–778.

---

*References to Byull.eksp.biol.i med. are in Russian. The journal is translated into English under the title of Bulletin of Experimental Biology and Medicine by Consultants Bureau and is published several months after the Russian issue comes out.

Akert, K., 1965, The anatomical substrate of sleep, Progress in Brain Research, 18, 9-19.

Akil, H., and Libeskind, Y. C., 1975, Monoaminoergic mechanisms of stimulation-produced analgesia, Brain Res., 94, 279-296.

Albe-Fessard, D., 1968, Notions anatomo-physiologicines sur les naies et les centres d'integration des massages doloureux, J.de Physiat.norm.et pathol. Paris, 1, No. 2-4.

Albe-Fessard, D., 1979, "Physiology of Pain - Some Recent Concepts," vol. 1, Int. Rehab. Med., Enlar Publisher, Basel. pp. 100-105.

Albe-Fessard, D., and Fessard, A., 1975, Recent advances on the neurophysiological bases of pain sensation, Acta Neurobiol.Exp., 35, No. 516, 715-740.

Albe-Fessard, D. G., and Lombard, M. C., 1980, Animal models for chronic pain, in: "Pain and Society," H. W. Kosterlitz and L. Y. Terenins, eds., Weinheim: Verlag Chemie GmbH, Dahlem, pp. 299-310.

Albe-Fessard, D., and Lombard, M. C., 1981, Animal models for pain due to central deafferentation, Pain, 10, Suppl.1, 80.

Albe-Fessard, D., Arfel, G. J., Guiot, G., Derome, P., Hertzog, E., Voourcih, G., Brown, H., Aleonard, P., De La Herran, J., and Frigo, J. C., 1966, Electrophysiological studies of some deep cerebral structures in man, J.Neurol.Sci., 3, No. 1, 37-51.

Albe-Fessard, D., Nashold, B. S., Lombard, M. C., Yamaguchi, Y., and Boureau, F., 1979, Rat after dorsal rhizotomy, a possible animal model for chronic pain, in: "Advances in Pain Research and Therapy," vol. 3, J. J. Bonica et al., eds., Raven Press, New York, pp. 761-766.

Aliev, M. N., and Kryzhanovsky, G. N., 1979, Experimental stereotypy induced by disturbance of GABA-ergic mechanisms in caudate nuclei, Byull.eksp.biol.i med. 87, No. 4, 314-317.

Aliev, A. M., Igonkina, S. I., and Kryzhanovsky, G. I., 1981, Formation of the generator of pathologically enhanced excitation in the caudate nucleus in the experimental Parkinsonian syndrome, Byull.eksp.biol.i med., No. 12, 657-659.

Allikmets, L. H., Stanley, M., and Garshon, S., 1979, The effect of lithium on chronic haloperidol enhanced apomorphine aggression in rats, Life Sci., 25, No. 2, 165-170.

Alnaes, E., Kaada, B. R., and Wester, K., 1973, EEG synchronization and sleep induced by stimulation of the medial and orbital frontal cortex in cat, Acta Physiol.Scand., 84, No. 1, 96-102.

Alving, B. O., 1968, Spontaneous activity in isolated somata of Aplysia pacemaker neurons, J.Gen.Physiol., 51, 29-45.

Ambache, N., Morgan, R. S., and Wright, G. P., 1948, The action of tetanus toxin on the rabbit's iris, J.Physiol., (Lond.), 107, 45.

Anand, B. K., 1971, Regulation of nervous drives in feeding behavior, Fotas homo, 3, No. 1, 56-66.

Anand, B. K., and Brobeck, J. R., 1951, Hypothalamic control of food intake in rats and cats, Vale J. Biol.and Med., 24, 123-140.

Anden, N. E., and Stock, G., 1973, Effect of clozapine on the turnover of dopamine in the corpus striatum and in the limbic system, J.Pharmacol., 25, 346-347.

Anden, N. E., Butcher, S. G., Corrodi, H., Fuxe, K., and Ungerstedt, U., 1970, Receptor activity and turnover of dopamine and norepinephrine after neuroleptics, Europ.J.Pharmacol., 11, 303-314.

Anderson, P., and Curtis, D. R., 1964, The pharmacology of synaptic and acetylcholine-induced excitation of ventrobasal thalamic neurons, Acta Physiol.Scand., 61, 100-120.

Andersen, P., and Lime, T., 1966, Mode of activation of hippocampal pyramidal cells by excitatory synapses on dendrites, Exp. Brain Res., 2, 247-260.

Anderson, D. J., and Mattheus, B., 1977, Pain in the Trigeminal Region, North-Holland, Biomedical Press, Amsterdam, Elsevier, New York.

Anderson, L. S., Black, G. R., Abraham, J., and Ward, A. A., 1971, Neuronal hyperactivity in experimental trigeminal deafferentation, J.Neurosurg., 35, 444-452.

Angaut, P., and Brodal, A., 1967, The projection of the vestibulocerebellum onto the vestibular nuclei of the cat, Arch. Ital.Biol., 105, 441-479.

Angrist, B. M., and Gershon, S., 1970, The phenomenology of experimentally induced amphetamine psychosis. Preliminary observations, Biol.Psychiat., 2, 95-107.

Anokhin, P. K., 1958, "Internal Inhibition as a Problem of Physiology," (in Russian), Medgiz, Moscow, p. 471.

Anokhin, P. K., 1971, "Principled Questions of the General Theory of Functional Systems," (in Russian), USSR Academy of Sciences, Moscow.

Anokhin, P. D., 1975, "Outlines of the Physiology of Functional Systems," (in Russian), Meditsina, Moscow.

Anokhina, I. P., and Kogan, B. M., 1981, Some peculiarities of dopamine exchange in schizophrenia, Zh.nevropatol.i psikhiatr., 81, No. 9, 1343-1346.

Antoniadis, A., Muller, W. E., and Wollert, U., 1980a, Central nervous system stimulating and depressing drugs as possible ligands of the benzodiazepine receptor, Neuropharmacology, 19, (1), 121-124.

Antoniadis, A., Muller, W. E., and Wollert, U., 1980b, Benzodiazepine receptor interactions may be involved in the neurotoxicity of various penicillin derivatives, Ann. Neurol., 8, (1), 71-73.

Antoniadis, A., Muller, W. E., and Wollert, U., 1980c, Inhibition of GABA and benzodiazepine receptor binding by penicillins, Neurosci.Lett., 18, 309-312.

April, R. S., and Spencer, W. A., 1969, Enhanced synaptic effective-
    ness following prolonged changes in synaptic use,
    Experientia, 25, 1272-1273.

Arduini, A., and Pompeiano, O., 1957, Microelectrode analysis of
    units of the rostral portion of the nucleus fastigii, Arch.
    Ital.Biol., 28, 56-70.

Arushanyan, E. B., and Otellin, V. A., Caudate Nucleus. Morpho-
    logical, physiological and pharmacological outlines (in
    Russian), Izd.Nauka, Leningrad.

Asanuma, H. B., and Brooks, V. B., 1963, Antidromic inhibition in
    cat's cerebral cortex, Fed.Proc., 22, No. 3, 456.

Ashmarin, I. P., 1978, Brain oligopeptides-analgesics, stimulators
    of memory and sleep, Mol.Biol., 12, No. 5, 965-979.

Aubree, J. C., and Lader, M. H., 1980, High and very high dosage
    antipsychotics: a critical review, J.Clin.Psychiatry, 41,
    No. 10, 341-350.

Averill, D. R., 1979, Idiopathic epilepsy, in: "Spontaneous Animal
    Models of Human Disease," E. J. Andrews, B. C. Ward, N. H.
    Altman, eds., Academic Press, New York, Vol. II, pp. 162-165.

Avrutsky, G. Ya., Allikmets, L.Kh., Neduva, A. A., Zharkovsky, A.
    M., Belyakov, A. V., and Matvienko, O. A., 1983, Some
    methods of overcoming secondary therapeutic resistance
    formed as a result of adaptation to psychotropic agents
    during prolonged therapy (clinicoexperimental studies),
    Zh.nevropatol.i psikhiatr., 83.

Ayala, G. F., and Vasconetto, C., 1972, Role of recurrent excitatory
    pathways in epileptogenesis, Electroenceph.Clin.Neuro-
    physiol., 33, 96-98.

Ayala, G. E., Dichter, M., Gumnit, R. J., Matsumoto, H., and
    Spencer, W. A., 1973, Genesis of epileptic interictal
    spikes: new knowledge of cortical feedback systems suggests
    a neurophysiological explanation of brief paroxysms, Brain
    Res., 52, 1-17.

Babb, T. L., Mitchell, A. G., and Crandall, P. M., 1974, Fastigio-
    bulbar and dentatothalamic influences on hippocampal cobalt
    epilepsy in the cat, Electroenceph.Clin.Neurophysiol., 36,
    141-154.

Babb, T. L., Soper, H. V., Lieb, J. P., Brown, W. J., Ottine, C. A.,
    and Grandall, P. H., 1977, Elecrophysiological studies of
    long-term electrical stimulation of the cerebellum in
    monkeys, J.Neurosurg., 47, 353-365.

Balagura, S., and Ralph, T., 1973, The analgesic effect of elect-
    rical stimulation of the diencephalon and mesencephalon,
    Brain Res., 60, 369-379.

Bancaud, J., Taiairach, J., Morel, P., Bresson, M., Bonis, A.,
    Geier, S., Hemon, S., and Buser, P., 1974, Generalized
    epileptic seizures elicited by electrical stimulation of the
    frontal lobe in man, EEG and Clin.Neurophysiol., 37, No. 3,
    275-282.

Barchas, Y. D., Akil, H., Ellio, G. R., Holman, R. B., and Watson,
    S. Y., 1978, Behavioral neurochemistry: neuroregulators and
    behavioral states, Science, 20, 964-973.
Bartholini, G., and Stadler, H., 1975, Cholinergic and GABAergic
    influences on the dopamine release in extrapyramidal
    centers, in: "Chemical Tools in Catecholamine Research. II.
    Regulation of Catecholamine Turnover," Elsevier, Amsterdam,
    pp. 235-241.
Bartholini, G., Stadler, H., and Gadea-Ciria, M., 1976, The use of
    push-pull cannula to estimate the dynamics of acetylcholine
    and catecholamine within various brain areas, Neuropharm-
    acology, 15, No. 3, 515-519.
Basbaum, A. I., 1974, Effects of central lesions on disorders pro-
    duced by multiple dorsal rhizotomy in rats, Exp.Neurop., 42,
    490.
Basbaum, A., and Wall, P. D., 1976, Chronic changes in the response
    of cells in adult rat dorsal horn following partial deaffer-
    entation, Brain Res., 116, 181-204.
Basbaum, A. I., and Fields, H. L., 1978, Endogenous pain control
    mechanisms: review and hypothesis, Ann.Neurol., 4, 451-462.
Batton, R. R., Jayaramen, A., Ruggiero, D., and Carpenter, M. B.,
    1977, Fastigial efferent projections in the monkeys: an
    autoradiographic study, J.Comp.Neurol., 174, No. 2, 281-305.
Batuev, A. S., 1978, Cortical mechanisms of the brain's integrative
    activity (in Russian), Izd.Nauka, Leningrad.
Bateuv, A. S., and Bogoslovski, M. M., 1963, Relations between
    occipital and motor cortical areas in the cat (electrophys-
    iological investigation), Sechenov Physiol.Journ.of the
    USSR, 49, No. 9, 1017-1025.
Beaton, I. M., 1976, The sedative effects of nicotinamide on gebril
    wheel-running activity, Experientia, 36, 1036-1043.
Beaton, J. M., Pegram, G. V., Smythies, J. R., and Bradley, R. J.,
    1974 The effect of nicotinamide on mouse sleep," Experi-
    entia, 30, 926-927.
Beaty, O., Collis, M. G., and Shepherd, J. T., 1981, Action of
    lithium on the adrenergic nerve ending, J.Pharm.Exp.Ther.,
    218, 309-317.
Behbehani, M. M., and Fields, H. R., 1979, Evidence that an excit-
    atory connection between the periaqueductal gray and nucleus
    raphe magnus mediates stimulation produced analgesia, Brain
    Res., 170, 85-93.
Bekhtereva, N. P., 1974, "Neurophysiological Aspects of Human Psy-
    chic Activity," (in Russian), Izd. Meditsina, Leningrad.
Bekhtereva, N. P., 1980, "Healthy and Diseased Brain in Man," Izd.
    Nauka, Leningrad.
Bekhtereva, N. P., Bondarchuk, A. N., Smirnov, V. M., and
    Melyucheva, L. A., 1972, Curative electric stimulation of
    deep-lying brain structures, Voprosy neirokhirurgii, No. 1,
    7-12.

Bekhtereva, N. P., Kambarova, D. K., and Pzodeev, V. K., 1978,
     Stable Pathologic Stage in Cerebral Diseases (in Russian),
     Izd. Meditsina, Leningrad.
Bell, D. S., 1965, Comparison of amphetamine psychosis and schizo-
     phrenia, Brit.J.of Psychiatry, 111, 701-707.
Belmaker, R. H., 1981, Receptors adenylate cyclase depression and
     lithium, Biol.Psychiatry, 16, No. 4, 330-350.
Belyakov, A. V., 1981, Experience gained in using therapy of large
     doses of neuroleptics in the case of therapeutic resistance
     in patients suffering from schizophrenia, in: "New Methods
     of Therapy of Psychic Diseases," (in Russian), Izd.
     Meditsina, Moscow.
Bennetti, G. J., and Mayer, D. J., 1979, Inhibition of spinal cord
     interneurons by narcotic microinjection and focal electrical
     stimulation in the periaqueductal central gray matter, Brain
     Res., 172, 243-257.
Berger, P., 1978, Medical treatment of mental illness, Science, 200,
     974-981.
Berger, N. A., and Sikorski, G. M., 1980, Nicotinamide stimulated
     repair of DNA damage in human lymphocytes, BBRC, 95, No. 1,
     62-72.
Bergmann, F., Costin, A., Chaimovitz, M., and Zerachia, A., 1970,
     Seizure activity evoked by implantation of ouabain and
     related drugs into cortical and subcortical regions of the
     rabbit brain, Neuropharmacology, 9, No. 5, 441-449.
Beritov, I. S., and Gedevanishvili, D. G., 1945, On the electrical
     activity of the cortex of the monkey's cerebrum, Tr.Inst.
     fiziologii im. Beritashvili, 6, 279-308.
Bernardi, G., Marciani, M. G., Morocutti, C., and Giacomini, P.,
     1976, The action of picrotoxin and bicuculine on rat caudate
     neurons inhibited by GABA, Brain Res., 102, No. 2, 379-384.
Bernheimer, H., Birkmayer, W., Hornykiewicz, O., Jellinger, K., and
     Sitelberger, F., 1973, Brain dopamine and the syndrome of
     Parkinson and Huntington. Clinical, morphological and
     neurochemical correlating, J.Neurol.Sci., 20, No. 4,
     415-455.
Bernshtein, N. A., 1966, "Outlines of the Physiology of Movements
     and the Physiology of Activity," (in Russian), Izd.
     Meditsina, Moscow.
Bertalanffy, L. von, 1962, General system theory-a critical review,
     in: "General Systems, Vol.VII," pp. 1-20.
Bess, L. N., 1970, Network model as a biological pacemaker, J.The-
     oret.Biol., 28, 58-90.
Besson, M. J., Cheramy, A., Feltz, P., and Glowinsky, J., 1971,
     Dopamine: spontaneous and drug induced release from the
     caudate nucleus in the rat, Brain Res., 1971, 32, No. 2,
     407-424.
Beurle, R. L., 1956, Properties of a mass of cells capable of re-
     generating pulses, Proc.Roy.Soc.B., Phil.Trans., 240, 55-94.

Bidzinsky, Ye., Bacha, T., Ostrovsky, K., Zyzanska, R., and Stempel,
    L., 1980, Estimation of the influence of the electric stimu-
    lation of cerebellar cortex on epileptic seizures in
    patients with the most severe forms of epilepsy, in:
    "Surgical Treatment of Epilepsy," (in Russian), Izd.
    Metsniereba, Tbilisi, pp. 81-82.
Bigakle, H., Heller, Y., Bizzini, B., and Habermann, E., 1981,
    Tetanus toxin and botulinum A toxin inhibit release and
    uptake of various transmitters as studied with particulate
    from rat brain and spinal cord, Naunyn-Schmiedeberg's
    Archives of Pharmacology, 316, 244-251.
Bird, E. D., and Iversen, L. L., 1974, Huntington's chorea: post-
    mortem measurement of glutamic acid decarboxylase, choline
    acetyltransferase and dopamine in basal ganglia, Brain, 97,
    No. 3, 457-472.
Bird, E. D., Barnes, J., Spokes, E. G., and Mackay, A. V. P., 1977,
    Increased brain dopamine and reduced glutamic acid decarb-
    oxylase and choline acetyl transferase activity in schizo-
    phrenia and related psychosis, Lancet, 2, No. 8049,
    1157-1159.
Bird, E. D., Spokes, E. G. S., and Iversen, L. L., 1979, Increased
    dopamine concentration in limbic areas of brain from
    patients dying with schizophrenia, Brain, 102, 347-360.
Bishop, G. H., 1959, The relation between nerve fibre size and
    sensory modality: phylogenetic implications of the afferent
    innervation of cortex, J.Nerv.Ment.Dis., 128, 89-114.
Bishop, B., 1980, Pain: its physiology and rationale for management,
    Psychical Therapy, 60, No. 1, 13-37.
Björklund, A, and Stenevi, U., 1979, Reconstruction of the nigro-
    striatal dopamine pathway by intracerebral nigral trans-
    plants, Brain Res., 117, No. 3, 555-560.
Black, R. G., 1974, A laboratory model for trigeminal neuralgia, in:
    "Advances in Neurology," Vol. 4, Raven Press, New York, pp.
    651-658.
Blinder, E. H., Wallach, M. B., and Gerson, S., 1971, Effect of
    lithium ions on release of ^{14}C-norepinephrine by stimulation
    from perfused cat spleen, Arch.Int.Pharmacodyn.Ther., 190,
    150-154.
Blom, S., 1963, Tic douloureux treated with new anticonvulsant.
    Experience with G-32883;, Arch.Neurol., 9, 285-290.
Bloom, F. E., Costa, E., and Salmoirghi, G. C., 1965, Anesthesia and
    the responsiveness of individual neuron of the caudate
    nucleus of the cat to acetylcholine, norepinephrine and
    dopamine administered by microelectrophoresis, J.Pharmacol.
    Exp.Ther., 150, No. 2, 224-252.
Bogatsky, A. V., 1982, (ed.), Phenazepam (in Russian), Izd. Naukova
    Dumka, Kiev.
Bogatsky, A. V., Andronati, S. A., Vikhlyaev, Yu.J., Voronina, T.
    A., Jakubovskaya, L. N., and Benko, A. V., 1977, 1,4-benz-

diazepines, their cyclic homologues and analogues. XXXII. Structure and pharmacological properties of 7-halogeno-5-(substituted phenyl)-1,2-dihydro-3H-1,4-benzdiazepines-2, Khimikofarmakologicheskii zhurnal, No. 11, 85-91.

Bogolepov, N. K., and Yerokhina, L. G., 1966, Pain hyperkineses in trigeminal neuralgia, Zh.nevropat.i psikhiatr., No. 1, 9-17.

Bogolepov, N. K., and Yerokhina, L. G., 1969, Seizure of typical trigeminal neuralgia as a multineuronal reflex, Zh. nevropat.i psikhiatr., No. 4, 487-493.

Bogoslovsky, M. M., 1968, Characteristics of evoked responses to the direct electric stimulation of the cortex isolated from subcortical neural influences, Zh.vyssh.nerv.deyat., 18, No. 3, 456-462.

Bonica, J. J., 1974, Organization and function of a pain clinic, in: "Advances in Neurology," J. J. Bonica, ed., Raven Press, New York, Vol. 4, pp. 433-443.

Bonica, J. J., 1977, Neurophysiologic and pathologic aspects of acute and chronic pain, Surgery, 112, 750-761.

Bonica, J. J., 1979, Causalgia and other reflex sympathetic dystrophies, in: "Advances in Pain Research and Therapy," Vol. 3, J. J. Bonica et al., eds., Raven Press, New York, pp. 141-166.

Bors, E, 1951, Phantom limbs of patients with spinal cord injury, Arch.Neurol.Psych (Chic), 66, 610-631.

Botterell, E. H., Callaghan, J. C., and Jousse, A. T., 1954, Pain in paraplegia: clinical management and surgical treatment, Proc.Roy.Soc.Med., 47, 281-288.

Bowen, E. P., Kosarova, J., Casella, Nicklas, W. J., and Berl, S., 1976, Focal epileptogenic activity induced by topical application of antisera to brain actomyosin-like protein, Brain Res., 102, No. 2, 363-365.

Bowers, M. B., 1974, Central dopamine in schizophrenic syndromes, Arch.Gen.Psychiatry, 31, 50-54.

Bowsher, D., 1957, Termination of the central pain pathway in man, Brain, 80, 606-622.

Bragin, E. O., Vasilenko, G. F., and Durinjan, R. A., 1982, Role of central gray in regulation of pain sensitivity during auricular electrostimulation, Bull.exp.med., No. 7, 11-13.

Braslavsky, V. E., Shchavelev, V. A., Kryzhanovsky, G. N., Nikushkin, E. V., and Germanov, S. B., 1982, Effect of nicotinamide on focal and generalized epileptic activity in the brain cortex, Byull.eksp.biol.i med., 94, No. 8, 39-42.

Brazier, M. A. B., (ed.), 1979, "Brain Mechanisms in Memory and Learning: From Single Neuron to Man," Raven Press, New York.

Bremer, F., 1935, Cerveau isolé et physiologie du sommeil, Comp. Rend.Soc.Biol., 118, 1235-1242.

Bremer, F., 1942, Le tétanos strychnique et le mécanisme de la synchronisation neuronique, Arch.Internat.Physiol., 51, No. 2, 211-260.

Bremer, F., 1953, Strychnine tetanus of the spinal cord, in: "The
    Spinal Cord," Ciba Found. Symp., Boston, pp. 78-83.
Bremer, F., 1954, The neurophysiological problem of sleep, in:
    "Brain Mechanisms and Consciousness," E. D. Adrian, F.
    Bremer, H. Jasper, eds., Blackwell, Oxford, pp. 137-162.
Bremer, F., 1973, Preoptic hypnogenic area and reticular activating
    systems, Arch.Ital.Biol., 111, No. 2, 86-111.
Briley, M., and Langer, S., 1978, Influence of GABA receptor
    agonists on the binding of ^{3}H-diazepam to the benzodiazepine
    receptor, Eur.J.Pharmacol., 52, 129-132.
Brodal, P., 1971, The corticopontine projection in the cat. 1. The
    projection from the proreate gyrus. 2. The projection from
    the orbital gyrus, J.Comp.Neurol., 142, No. 2, 127-140,
    141-152.
Brodal, A., Pompeiano, O., and Walberg, F., 1962, "The Vestibular
    Nuclei and Their Connections, Anatomy and Functional Corre-
    lations," Charles C. Thomas (publisher), Springfield,
    Illinois.
Brooks, V. B., and Asanuma, H., 1963, Pharmacological studies of
    recurrent inhibition and facilitation, Am.J.Physiol., 208,
    674-681.
Brooks, V. B., Curtis, D. R., and Eccles, J. C., 1957, The action of
    tetanus toxin on the inhibition of motoneurons, J.Physiol.
    (Lond), 135, 655-672.
Bruck, J., Gerstenbrand, F., Grundig, E., and Prosez, P., 1965,
    "Neue Therapiewege auf Ground von Stoffwechselveranderungen
    der extrapyramidalen Erkrankungen, in: "Conf.Hung.pro
    Therapia et Investigation in Pharmacologia," V. 3, Kultura:
    Budapest, pp. 149-158.
Bruggencate, G. T., Teichmann, R., and Weller, E., 1972, Neuronal
    activity in the lateral vestibular nucleus of the cat. II.
    EPSPs in Deiters' neurons mediated by fast conducting fibres
    of the spinal cord, Pflüger Arch., 337, No. 2, 135-146.
Bruyn, G. W., 1968, Huntington's chorea. Historical, clinical and
    laboratory synopsis, in: "Handbook of Clinical Neurology,
    Vol. 6, Disease of Basal Ganglia," Elsevier, Amsterdam, pp.
    298-377.
Buchtal, F., Rosenflach, P., and Erminio, F., 1960, Motor unit
    territory and fiber density in myopathies, Neurol., 10, No.
    4, 398-408.
Buchwald, N. A., Horvath, F. E., Soltysik, S., and Romero-Sierra,
    C., 1965, Caudate influences on behavior and the EEG, Elect-
    roencephalogr. Clin.Neurophysiol., 18, No. 3, 529.
Buchwald, N. A., Price, D. D., Vernon, L., and Hull, C. D., 1973,
    Caudate intracellular response to thalamic and cortical
    inputs, Exp.Neurol., 38, No. 2, 311-323.
Bunney, B. S., and Aghajanian, G. K., 1975, Evidence for drug on
    both pre-and postsynaptic catecholamine receptors in CNS,
    in: "Pre-and Postsynaptic Receptors," Dekker, New York, pp.
    311-356.

Bunney, B. S., Aghajanian, G. K., 1976, D-Amphetamine induced in-
    hibition of central dopaminergic neurons: mediation by a
    striatal-nigral feed-back pathway, Science, 192, 391-393.
Bunney, W. E., Pert, A., Rosenblatt, J., and Port, K. B., 1979, Some
    biological considerations of the mode of action of lithium
    in the treatment of affective illnesses, in: "Lithium," T.
    B. Cooper, et al., eds., Excerpta Medica, Amsterdam-
    Oxford-Princeton, pp. 675-684.
Bunyatyan, G. Kh., 1976, Latest achievements in biochemistry and
    biochemical pharmacology of gamma-aminobutyric acid, Zh.
    Vsesoyuzn.Khim.obshch.im.D.I. Mendeleeva, 21, No. 2, 130-137.
Burke, W., and Sefton, A. S., 1966, Discharge patterns of principal
    cells and interneurons in lateral geniculate nucleus of rat,
    J.Physiol.(Lond.), 187, 201-212.
Burke, R. E., Rudomin, P., Vyklicky, L., and Zajac, F. E., 1971,
    Primary afferent depolarization and flexion reflexes pro-
    duced by radiant heat stimulation of the skin, J. Physiol.,
    213, 185-214.
Burmistrov, Yu.M., 1965, Circulation of the nervous impulse in the
    system of lateral giant axones of the abdominal chain of the
    crayfish, Biofizika, 10, No. 1, 90-97.
Burns, B. D., 1968, The Uncertain Nervous System, Ed., Arnold
    (Publ.), London.
Burt, D. R., Enna, S. J., Creese, J., and Snyder, S. H., 1975,
    Dopamine receptor binding in the corpus striatum of mamalian
    brain, Proc.Natl.Acad.Sci.USA. (Wash)., 72, No. 11,
    4655-4659.
Burt, D. R., Creese, I., and Snyder, S. H., 1977, Antischizophrenic
    drugs: chronic treatment elevates dopamine receptor binding
    in brain, Science, 196, 326.
Bustamante, L., Lueders, H., Pippenger, C., and Goldensohn, F. S.,
    1981, Quantitative evaluation of anticonvulsant effects on
    penicillin-induced spike foci in cats, Neurology, 31, No. 9,
    1163-1166.
Bustos, G., Kuhar, M. J., and Roth, R. H., 1972, Effect of gamma-
    hydroxybutyrate and gamma-butyrolactone on dopamine syn-
    thesis and uptake by rat striatum, Biochem.Pharmacol., 21,
    No. 19, 2649-2652.
Butcher, S. G., and Butcher, L. L., 1974, Origin and modulation of
    acetylcholine activity in the neostriatum, Brain Res., 71,
    167-171.
Buus, L. J., 1973, The effects of amantadine and (+)-amphetamine on
    motility in rats after inhibition of monoamine synthesis and
    storage, Psychopharmacologia, 29, No. 1, 55-64.
Caccia, M., 1975, Clonazepam in facial neuralgia and cluster head-
    ache. Clinical and electrophysiological study, Europ.
    Neurol., 13, No. 6, pp. 560-563.
Calne, D. B., 1970, in: "Parkinsonism: Physiology, Pharmacology and
    Treatment," Arnold, ed., London, p. 134.

Calvin, W. H., 1972, Synaptic potential summation and repetitive
    firing mechanisms: input-output theory for recruitment of
    neurons into epileptic bursting firing patterns, Brain Res.,
    39, No. 1, 71-94.
Calvin, W. H., Ojemann, G. A., and Ward, A. A., 1973, Human cortical
    neurons in epileptogenic foci: comparison of inter-ictal
    firing patterns to those of 'epileptic' neurons in animals,
    Electroceph.Clin.Neurophysiol., 34, 337-351.
Calvin, W. H., 1975, Generation of spike trains in CNS neurons,
    Brain Res., 84, No. 1, 1-22.
Calvin, W. H., 1979, Some design features of axons and how neur-
    algias may defeat them, in: "Advances in Pain Research and
    Therapy, vol. 3," J. J. Bonica et al., eds., Raven Press,
    New York, pp. 297-309.
Calvin, W. H., 1980, Normal repetitive firing and its pathophysio-
    logy, in: "Epilepsy: a Window to Brain Mechanisms," J. S.
    Lockard and A. A. Ward, eds, Raven Press, New York. pp.
    97-121.
Cambier, Y., and Denen, H., 1971, Les douleurs fulgurantes. Etio-
    logie et phsyiopathologie, Presse Med., 79, 1419-1422.
Compochiaro, P., Schwarcz, R., and Coyle, J. T., 1977, GABA receptor
    binding in rat striatum: localization and effect of
    denervation, Brain Res., 136, No. 3, 501-511.
Cananzi, A. R., Costa, E., and Guidotti, A., 1980, Potentiation by
    intraventricular muscimol of the anticonvulsant effect of
    benzodiazepines, Brain Res., 196, No. 2, 447-453.
Cannon, W. B., and Rosenbluth, A., 1949, The Supersensitivity of
    Denervated Structures, MacMillan, New York.
Carenzi, A., Cheney, D. L., and Costa, E., 1975, Action of opiates,
    antipsychotics, amphetamine, apomorphine on dopamine recept-
    ors in rat striatum: in vivo changes of 3,5'-cyclic AMP
    content and acetylcholine turnover rate," Neuropharmacology,
    14, No. 12, 927-939.
Carlsson, A., 1974, Antipsychotic drugs and catecholamine synapses,
    J.Psychiat.Res., 11, 57-64.
Carlsson, A., 1978, Does dopamine have a role in schizophrenia,
    Biol.Psychiat., 13, 3-21.
Carlsson, A., 1979, Antipsychotic drugs and catecholamine synapses,
    J.Psychiat.Res., 11, 57-64.
Carlsson, A., Kehr, W., and Lindqvist, H., 1977, Agonist-antagonist
    interaction on dopamine receptors in brain, as reflected in
    the rates of tyrosine and tryptophan hydroxylation,
    J.Neurol.Trans. 40, No. 2, 99-113.
Carpenter, D. (ed.), 1982, "Cellular Pacemakers, Vol. 1, Mechanisms
    of Pacemaker Generation. Vol. 2, Function in Normal and
    Disease States," John Wiley & Sons New York.
Carrea, R., and Lanari, A., 1962, Chronic effect of tetanus toxin
    applied locally to the cerebral cortex of the dog, Science,
    137, 342-343.

Carstens, E., Yokota, T., and Zimmermann, M., 1979, Inhibition of spinal neuronal responses to noxious skin heating by stimulation of mesencephalic periaqueductal gray in the cat, J.Neurophysiol., 42, 558–568.

Casey, D. E., and Denney, D., 1974, Dimethylaminomethanol in tardive dyskinesia, New Engl.J.Med., No. 291, 781–791.

Cervero, F., Iggo, A., and Ogawa, H., 1976, Nociceptor driven dorsal horn neurons in the lumbar spinal cord of the cat, Pain, 2, 293–307.

Chang, H. T., 1950, The repetitive dishcarges of corticothalamic reverberating circuit, J. Neurophysiol., 13, No. 3, 234–257.

Cheramy, A., Nieoullon, A., and Glowinsky, J., 1977a, Blockade of the picrotoxin-induced in vivo release of dopamine in the cat caudate nucleus by diazepam, Life Sci., 20, No. 5, 811–816.

Cheramy, A., Nieoullon, A., and Glowinski, J., 1977b, Effects of peripheral and local administration of picrotoxin o the release of newly synthesized ^{3}H-dopamine in the caudate nucleus of the rat, Arch.Pharmacol., 297, No. 1, 31–37.

Cheramy, A., Nieoullon, A., and Glowinsky, J., 1978, Gabaergic processes involved in the control of dopamine release from nigrostriatal dopaminergic neurons in the cat, Europ.J. Phamacol., 48, No. 3, 281–295.

Cherkes, V. A., 1978, "The Forebrain and Behavioral Elements," (in Russian), Izd.Naukova Dumka, Kiev.

Chernigovskaya, V. N., 1978, "Adaptive Bioregulation in Neuralgia," (in Russian), Izd. Nauka, Leningrad.

Chernigovskaya, N. V., 1982, Movsisyants, S. A., and Timofeeva, A. N., 1982, "Clinical Significance of Adaptive Bioregulation," (in Russian), Izd. Meditsina, Leningrad.

Chiflikian, M. D., Yessian, N. H., and Manukhin, B. N., 1978, The effect of GABA on GABA on H^3-norepinephrine synthesis, uptake and release from brain slices, Sechenov Physiological Journal of the USSR, 64, No. 8, 1947–1104.

Christensen, B. N., and Perl, E. R., 1970, Spinal neurons specifically excited by noxious or thermal stimuli: marginal zone of dorsal horn, J.Neurophysiol., 33, 293–307.

Clarke, G., and Hills, R. G., 1972, Effects of a focal penicillin lesion on responses of rabbit cortical neurons to putative neurotransmitters, Br.J.Pharmacol., 44, 435–441.

Clark, B. J., Fluckiger, E., Loew, D. M., and Vigouret, J. M., 1978, How does bromcriptine work?, Triangle, 17, No. 1, 21–31.

Clarke, Th., Saunders, B., and Feldman, B., 1979, Pyridoxine-dependent seizures requiring high doses of pyridoxine for control, Amer.J.Diseases Child., 133, 963–965.

Clement-Cormier, Y. C., Kebabian, J. W., Petzold, G. L., and Greengard, P., 1974, Dopamine-sensitive adenylate cyclase in mammalian brain: a possible site of action of antipsychotic drugs, Proc.Natl.Acad.Sci.USA, 71, 1113–1117.

Cohen, N. H., 1962, The treatment of Huntington chorea with trif-
luoperazine, J.Nerv.Ment.Dis., 134, No. 1, 62–71.

Colburn, R. W., Goodwin, F. K., Bunney, W. M., and Davies, J. M.,
1967, Effect of lithium on the uptake of norepinephrine by
synaptosomes, Nature, 215, No. 5108, 1395–1397.

Collingridge, G. L., and Davies, J., 1982, Reversible effects of
tetanus toxin on striatal-evoked responses and ³H-γ-amino-
butyric acid release in the rat substantia nigra, Br.J.
Pharmac. 76, 403–411.

Collingridge, G. L., and Davies, J., 1982, The in vitro inhibition
of GABA release by tetanus toxin, Neuropharmacology, 21, No.
9, 851–857.

Collingridge, G. L., Davies, J., James, T. A., Neal, M. J., and
Tongroach, P., 1979, Effect of tetanus toxin on uptake and
potasium-evoked release of radiolabelled transmitters from
the substantia nigra and striatum of the rat, J.Physiol.,
287, 32–33.

Collingridge, G., Collins, G. G. S., Davies, J., James, T. A., Neal,
M. J., and Tongroach, P, 1980, Effect of tetanus toxin on
transmitter release from the substantia nigra and striatum
in vitro, J.Neurochem., 34, 540–547.

Connell, P. H., 1958, Amphetamine Psychosis, Chapman and Hall,
London.

Connor, J. D., 1970, Caudate nucleus neurons: correlation of the
effects of substantia nigra stimulation with iontophoretic
dopamine, J.Physiol., 208, 691–703.

Cools, A. R., 1972, Athetoid and choreiform hyperkinesias produced
by caudate application of dopamine in cat, Psychopharmacologia, 25, No. 2, 229–237.

Cools, A. R., 1974, The transsynaptic relationship between dopamine
and serotonin in the caudate nucleus of cats, Psychopharmacologia (Berl.), 36, 17–28.

Cools, A. R., and van Rossum, G. M., 1970, Caudal dopamine and
stereotype behavior of cats, Arch.Int.Pharmacodyn., 187,
163–173.

Cools, A. R., and van Rossum, J. M., 1976, Excitation-mediating and
inhibition-mediating dopamine receptors: A new concept
towards a better understanding of electrophysiological,
biochemical, pharmacological, functional and clinical data,
Psychopharmacologia, 45, No. 2, 243–254.

Cools, A. R., Lohman, A. H. M., and Van der Bercken, J. H. L.,
(eds.), 1977, Psychobiology of the Striatum, Elsevier, North
Holland Biomedical Press, Amsterdam, New York, Oxford.

Coombs, J. S., Curtis, D. R., and Landgren, S., 1956, Spinal cord
potentials generated by impulses in muscle and cutaneous
afferent fibres," J.Neurophys., 19, No. 5, 452–467.

Cooper, I. S., 1970, Clinical physiology of abnormal movements, in:
"L-dopa and Parkinsonism," A. Barbeau, ed., F. H. McDowell,
Philadelphia, pp. 170–179.

Cooper, I. S., 1978, (ed.), "Cerebellar Stimulation in Man," Raven
    Press, New York.
Cooper, I., Crighel, E., and Amin, J., 1973, Clinical and
    physiological effects of stimulation of the paleocerebellum
    in humans, J.Amer.Geriat.Soc., 21, 40-43.
Cooper, I. S., Amin, J., Gilman, S., and Waltz, J. W., 1974, The
    effect of chronic stimulation of cerebellum on epilepsy in
    man, in: "The Cerebellum, Epilepsy and Behavior," Plenum
    Press, New York, pp. 119-171.
Cooper, I. S., Upton, A. R. M., and Amin, I., 1980, Some effects of
    chronic cerebellar stimulation on EEG and epilepsy in man,
    in: "Advances in Epileptology. The Xth Epilepsy Inter-
    national Symposium," J. A. Wada, and J. K. Penry, eds.,
    Raven Press, New York, pp. 215-226.
Costa, E., 1980, Benzodiazepines and neurotransmitters, Arzneimittel-
    Forsch., 30, No. 5a, 858-861.
Costa, E., and Greengard, P., 1975, Mechanisms of action of benzo-
    diazepines, Adv. Biochem.Psychopharmacol., 14, No. 1,
    176-182.
Costa, E., and Guidotti, A., 1979, Molecular mechanisms in the
    receptor action of benzodiazepines, Ann.Rev.Pharmacol.
    Toxicol., 19, 531-545.
Costa, E., Guidotti, A., and Mao, C. C., 1975a, Evidence for in-
    volvement of GABA in the action of benzodiazepines: studies
    on rat cerebellum, in: "Mechanism of Action of Benzodiaze-
    pines," E. Costa and P. Greengard, eds., Raven Press, New
    York, pp. 113-130.
Costa, E., Guidotti, A., Mao, C. C., and Suria, A., 1975b, New
    Concepts on the mechanism of action of benzodiazepines, Life
    Sci., 17, 167-186.
Costa, E., Guidotti, A., and Toffano, G., 1978, Molecular mechanisms
    mediating the action of diazepam on GABA receptors, Br.J.
    Psychiatry, 133, 239-248.
Costall, B., and Naylor, R. J., 1974a, The importance of the ascend-
    ing dopaminergic system to the extrapyramidal and mesolimbic
    areas for the cataleptic action of the neuroleptic and
    cholinergic agents, Neuropharmacol., 13, 353-364.
Costall, B., and Naylor, R. J., 1974b, Stereotyped and circling
    behavior induced by dopaminergic agonists after lesions of
    the midbrain raphe nuclei, Eur.J.Pharmacol., 29, 206-222.
Costall, B., and Naylor, R. J., 1978, Experimental studies of dopa-
    mine function in movement disorders, in: "Neurotransmitter
    Systems and their Clinical Disorders," H. I. Legg, ed.,
    Academic Press, New York, pp. 129-150.
Costall, B., Naylor, R. J., and Olley, J. E., 1972, Stereotypic and
    anticataleptic activities od amphetamine after intracerebral
    injection, Europ.J.Pharmacol., 18, No. 1, 83-94.
Costall, B., Marsden, C. D., Naylor, R. J., and Pycock, C. J., 1977,
    Stereotyped behavior patterns and hyperactivity induced by

amphetamine and apomorphine after discrete 6-hydroxydopamine lesion of extrapyramidal and mesolimbic nuclei, Brain Res., 123, No. 1, 89–112.

Cotzias, G. C., Papavasiliou, P. R., and Gellene, R., 1969, Modification of parkinsonism, New Eng.J.Med., 280, 337–345.

Crain, S. M., 1972, Tissue culture models of epileptiform activity, in: "Experimental Models of Epilepsy," D. P. Purpura, ed., Raven Press, New York, pp. 291–316.

Crain, S. M., and Bornstein, M. B., 1972, Bioelectric activity of neonatal mouse cerebral cortex during growth and differentiation in tissue culture, Science, 176, 182–184.

Crane, G. E., 1968, Tardive dyskinesia in patients treated with major neuroleptics, Am.J.Psychiat., 124, 40–48.

Creese, I., Burt, D. R., and Snyder, S. H., 1975, Dopamine receptor binding: differentiation of agonist and antagonist states with ^{2}H-dopamine and ^{2}H-haloperidol, Life Sci., 17, No. 6, 933–1001.

Creese, I., Burt, D. R., and Snyder, S. H., 1976, Dopamine receptor binding predicts clinical and pharmacological potencies of antischizophrenic drugs, Science, 192, 481–483.

Crill, W. E., 1980, Neuronal mechanisms of seizure initiation, in: "Antiepileptic Drugs: Mechanism of Action," G. H. Glaser, J. K. Penry, and D. M. Woodbury, eds., Raven Press, new York, pp. 169–183.

Cross, A. J., Crow, I. J., and Owen, 1981, H^3-Flupentixol binding in post-mortem brains of schizophrenics: evidence for a selective increase in dopamine DA receptors, Psychopharmacol., 74, No. 2, 122–124.

Crow, T. J. 1972, Catecholamine-containing neurons and electrical self-stimulation: I. A review of some data, Psychol.Med., 2, 414–421.

Crow, T. J., 1979, What is wrong with dopaminergic transmission in schizophrenia?, Trends in Neuroscience, 2, No. 2, 52–55.

Crow, T. J., 1980, Molecular pathology of schizophrenia: more than one disease process?, Brit. Med.J., 280, 62–68.

Crow, T. J., 1982, Two dimensions of pathology in schizophrenia: dopaminergic and non-dopaminergic, Psychopharmacol.Bulletin, 18, No. 3, 22–29.

Crow, T. J., and Johnstone, E. C., 1978, Dopaminergic processes in schizophrenia and mechanism of antipsychotic effect, in: "Neurotransmitter Systems and Their Clinical Disorders," H. J. Legg, ed., Academic Press, New York, pp. 207–219.

Crow, T. J., Johnstone, E. C. Deakin, J. F. W., and Longden, A., 1976, Dopamine and schizophrenia, Lancet, 2, 563–566.

Crow, T. J., Deakin, J. F. W., and Longden, A., 1977, The nucleus accumbens-possible site of antipsychotic action of neuroleptic drugs, Psychol.Med., 7, No. 2, 213–221.

Crow, T. J., Johnstone, E. C., Longden, A., and Owen, F., 1978, Dopamine in schizophrenia, Adv.Biochem.Psychopharmacol., 15, 41–45.

Crow, T. J., Owen, F., Cross, A. J., Ferrier, I. N., Johnstone, E.
    C., McCredie, R. G., Owen, D. G. C., and Poulter, M., 1981,
    Neurotransmitter enzymes and receptors in postmortem brain
    in schizophrenia: evidence that an increase in D2 is associ-
    ated with the Type I syndrome, in: "Transmitter Biochemistry
    of Human Brain Tissue," P. Reider and E. Usdin, ed.,
    MacMillan, London, pp. 85-96.
Crue, B. L., and Carregal, J. A., 1974, Postsynaptic repetitive
    neuron discharge in neuralgic pain, Advances in Neurology,
    4, 643-649.
Curtis, D. R., 1959, Pharmacological investigation upon inhibition
    of spinal motoneurons, J.Physiol., (Lond.), 145, 175.
Curtis, D. R., 1962, Direct extracellular application of drugs,
    Biochem.Pharmacol., 9, 205-212.
Curtis, D. R., and De Groat, W. O., 1968, Tetanus toxin and spinal
    inhibition, Brain Res., 10, 208-212.
Curtis, D. R., Hosli, L., Johnston, G. A. R., and Johnston, J. H.,
    1968, The hyperpolarization of spinal motoneurons by glycine
    and related amino acids, Exp.Brain Res., 5, 235-258.
Curtis, D. R. Duggan, A. W., and Johnston, G. A. R., 1969, Glycine,
    strychnine, picrotoxin and spinal inhibition, Brain Res.,
    14, 759-762.
Curtis, D. R., Duggan, A. W., and Johnston, G. A. R., 1971, The
    specificity of strychnine as a glycine antagonist in the
    mammalian spinal cord, Exp.Brain Res., 12, No. 5, 547-565.
Curtis, D. R., Game, C. G., Johnston, G. A. R., McCulloch, R. M.,
    and McLachlan, R. H., 1972, Convulsive action of penicillin,
    Brain Res., 43, 242-245.
Curtis, D. R., Felix, D., Game, C. J., and McCulloch, R. M., 1973,
    Tetanus toxin and synaptic release of GABA, Brain Res., 51,
    358-362.
Curtis, D. R., and Johnston, G. A. R., 1974, Amino acid-transmitters
    in the mammalian central nervous system, in: "Ergebnisse der
    Physiologie, biol. Chemie u. exper. Pharmakologie," Reviews
    of Physiology Bioch. and Exper. Pharmacol., vol. 6,
    Springer-Verlag Berlin-Heidelberg, New York pp. 97-188.
Curtis, D. R., Lodge, D., and Johnston, G. A., 1976, Central actions
    of benzodiazepines, Brain Res., 118, 344-347.
Denilov, I. V., 1968, Mechanisms of the conditioned-reflex mainten-
    ance of experimental hyperkinesias in dogs, Zh.vyssh.
    nervnoi deyatelnosti, No. 6, 979-983.
Danilova, E. I., Grafova, V. N., and Kryzhanovsky, G. N., 1979,
    Analgesic effects of inhibitory mediators in pain syndrome
    of spinal origin, Byull.eksp.biol.i med., 86, No. 6,
    525-528.
Davenport, J., Schwindt, P. C., and Crill, W. E., 1977, Penicillin-
    induced spinal seizures: selective effects on synaptic
    transmission, Exper.Neurol., 56, 132-150.
Davidoff, R. A., 1972, Penicillin and inhibition in the cat spinal
    cord, Brain Res., 45, 638-642.

Davies, L., and Martin, J., 1947, Studies upon spinal cord injuries. II. The nature and treatment of pain, J. Neurosurg., 4, 483–491.

Davies, J., and Tongroach, P., 1979, Tetanus toxin and synaptic inhibition in the substantia nigra and striatum of the rat, J.Physiol., (London), 290, 23–36.

Davis, R., Engle, H., Kudzina, J., Gray, E., Ryan, T., and Dusnak, A., 1982, Update of chronic cerebellar stimulation for spasticity and epilepsy, Appl.Neurophysiol., 45, 44–51.

de Andres, J., Guitierrez-Ravis, E., and Reinsoso-Suarez, F., 1975, Modification of the sleep-wakefulness cycle after lesion at the level oral pontine gegmentum and superior cerebellar peduncle, in: "Sleep. 2nd Europ. Congress Sleep Res.," Karger, Basel, pp. 235–238.

Deguchi, T., Ichiyama, A., Nishizuka, Y., and Hayaishi, O., 1968, Studies on the biosynthesis of nicotinamide-adenine dinucleotide in the brain, Biochim.Biophys.Acta, 158, No. 2, 382–393.

Deizh, H. A., Aickin, C. C., and Lux, H. D., 1979, Decrease of inhibitory driving force in crayfish stretch reception: a mechanism of the convulsant action of penicillin, Neurosci. Lett., 11, No. 3, 347–352.

Delgado, J. M. R., 1969, Physical control of the mind, in: "Toward a Psychocivilized Society," Harper and Row Publishers, New York, Evanston, London.

Delgado, J. M. R., 1979, Inhibitory functions in the neostriatum, in: "The Neostriatum," I. Divac and G. E. Oberg, ed., Pergamon Press, Oxford, New York, pp. 241–261.

Delgado, J. M. R., and Sevillano, M., 1961, Evolution of repeated hippocampal seizures in the cat, Electroenceph.Clin.Neurophysiol., 13, 722–733.

Delgado-Escueta, A. V., and Horan, M. P., 1980, Brain synaptosomes in epilepsy: organization of ionic channels and the Na^+-K^+ pump, in: "Antiepileptic Drugs: Mechanisms of Action," G. H. Glaser, J. K. Penry, and D. M. Woodbury, eds., Raven Press, New York, pp. 85–126.

Dement, W. S., and Kleitman, N., 1957, Cyclic variations in the EEG during sleep and their relations to the eye movements, body motility and dreaming, Electroencephal.Clin.Neurophysiol., 9, 673–690.

Dencer, S. J., Enoksson, P., Johansson, R., Lundin, L., and Moden, U., 1981, Late (4–8 years) outcome of treatment with megadoses of fluphenazine enanthate in drug-refractory schizophrenias, Acta Psychiatr.Scand., 63, No. 1, 1–12.

Dennis, S. G., and Melzack, R., 1977, Pain-signalling systems in the dorsal and ventral spinal cord, Pain, 4, 97–132.

Denny-Brown, D., and Yanagisawa, N., 1973, The function of the descending root of the fifth nerve, Brain, 96, 783–814.

De Vito, R. V., Brusa, A., and Arduini, A., 1956, Cerebellar and vestibular influences on Deitersian units, J.Neurophysiol., 19, No. 3, 241-253.

Devor, M., and Wall, P. D., 1976, Type of sensory fibre sprouting to form a neuroma, Nature (Lond.), 262, 705-708.

De Wied, D., and Jolles, J., 1982, Neuropeptides derived from pro-opiocortin: behavioral, physiological and neurochemical effects, Physiological Review, 62, No. 3, 976-1059.

De Wied, D., Witter, A., and Greven, H. M., 1975, Commentary: behav-iorally active ACTH analogues, Biochem.Pharmacol., 24 (16), 1463-1468.

Dichter, M. A., and Spencer, W. A., 1969a, Penicillin-induced inter-ictal discharge from the cat hippocampus. I. Characteristic and topographical features, J. Neurophysiol., 32, 649-662.

Dichter, M., and Spencer, W. A., 1969b, Penicillin-induced inter-ictal discharges from the cat hippocampus. II. Mechanisms underlying origin and restriction, J. Neurophysiol., 32, 663-687.

Dichter, M. A., Herman, C. I., and Selzer, M., 1972, Silent cells during interictal discharges and seizures in hippocampal penicillin foci. Evidence for the role of extracellular K^+ in the transition from the interictal state to seizures, Brain Res., 48, 173-183.

Dimpfel, W., 1979, Hyperexcitability of cultured central nervous system neurones caused by tetanus toxin, Expl.Neurol., 65, No. 1, 53-65.

Dingledine, R., and Gjerstad, L., 1979, Penicillin blocks hippo-campal IPSPs, unmasking prolonged EPSPs, Brain Res., 168, No. 1, 205-209.

Divac, I., 1972, Neostriatum and function of prefrontal cortex, Acta Neurobiol.Exp., (Warsz), 32, 461-477.

Divac, I., and Oberg, G. E., (eds.), The Neostriatum, Pergamon Press, Oxford, New York.

Divry, P., Boban, J., Collard, J., Pinchard, A., and Nols, E., 1959, Etude et experimentation cliniques du R 1625 ou haloperidol, nouveau neuroleptique et neurodisleptique, Acta Neurobiol., Belg., 59, 337-366.

Dolina, S. A., and Arshavsky, V. V., 1975, Present concepts of the mechanisms of the development and cessation of a convulsive seizure, Usp.fiziol.nauk, 6, No. 2, 56-92.

Dong, W. K., and Wagman, I. H., 1976, Modulation of nociceptive responses in the thalamic posterior group of nuclei, Advances in Pain Research and Therapy, 1, 455-461.

Donlon, P. T., 1976, High dosage neuroleptic therapy, Int. Pharmaco-psychiatry, 11, No. 14, 235-246.

Dousa, T., and Hechter, O., 1970, Lithium and brain adenylcyclase, Lancet, 1, No. 7651, 834-835.

Dow, R. S., 1961, Some aspects of cerebellar physiology, J. Neuro-surg., 18, 512-530.

Dow, R. S., 1965, Extrinsic regulatory mechanisms of seizure activity, Epilepsia, 6, 122-140.

Dow, R. S., 1974, Experimental cobalt epilepsy and the cerebellum, in: "The Cerebellum, Epilepsy and Behavior," I. S. Cooper, M. Riklan, and R. S. Snider, eds., Plenum Press, New York, pp. 57-95.

Dow, R. S., 1962, Fernandez-Guardiola, A., and Manni, E., 1962, The influence of the cerebellum on experimental epilepsy, Electroencephalogr.Clin.Neurophysiol., 14, 383-398.

Dowzenko, A., Buksowicz, C., and Kuran, W., 1976, Wpływ coretalupolfa (oksprenolol) na drżenie parkinsonowskie i na drżenie samoistne, Neurol.Neurochir.pol., 10, No. 1, 49-534.

Dray, A., and Staughan, D. W., 1976, Benzodiazepines: GABA and glycine receptors on single neurons in the rat medulla, J. Pharm. Pharmacol., 28, No. 2, 314-315.

Dubner, R., Gobel, S., and Price, D. D., 1976, Peripheral and central trigeminal 'pain' pathways, Advances in Pain Research and Therapy, 1, 137-148.

Duchen, L. W., and Tonge, D. A., 1973, The effect of tetanus toxin on neuromuscular transmission and on the morphology of motor endplates in slow and fast muscle of the mouse, J. Physiol. (Lond.), 228, 157.

Duckrow, R. B., and Taub, A., 1977, The effect of diphenylhydantoin on self-mutilation in rats produced by unilateral multiple dorsal rhizotomy, Exp.Neurol., 54, 33.

Dunn, P. F., and Somjen, G. G., 1977, Membrane resistance, monosynaptic EPSPs, and the epileptogenic action of penicillin in spinal motonuerones, Brain Res, 128, 569-574.

Dyck, P. J., and O'Brien, P. C., 1976, Pain in peripheral neuropathy related to rate and kind of fibre degeneration, Neurology (Minneapolis), 22, 466-471.

Dyck, P. J., Lampert, E. H., and O'Brien, P., 1976, Pain in peripheral neuropathy related to rate and kind of fiber degeneration, Neurology, 28, 466-471.

Ebersole, J. S., and Levine, R. A., 1975, Abnormal neuronal responses during evolution of a penicillin epileptic focus in at visual cortex, J.Neurophysiol., 38, 250-266.

Eccles, J. C., 1953, The neurophysiological basis of mind, in: "The Principles of Neurophysiology," Clarendon Press, Oxford.

Eccles, J. C., 1964, The Physiology of Synapses, Springer-Verlag, Berlin, Göttingen, Heidelberg.

Eccles, J. C., Ito, M., and Szentagothai, J., 1967, The Cerebellum as a Neuronal Machine, Springer, New York.

Eccles, J. C., Kostyuk, P. G., and Schmidt, R. F., 1962, Presynaptic inhibition of the central actions of flexor reflex afferents, J.Physiol., 161, No. 2, 258-281.

Eccles, J. C., Nicoll, R. A., Schwarz, D. W., and Táboriková H., 1974, Cerebello-spinal pathway via the fastigial nucleus and the medial reticular nucleus, Brain Res., 66, No. 3, 525-530.

Eggers, Chr., 1981, Die Bedeutung limbischer Funktionstörung für die
    Atiologie kindlicher Schizophrenien, Forscher.Neurol.,
    Psychiatry., 49, 101-108.
Ehringer, H., and Hornykiewicz, O., 1960, Verteilung von Nora-
    drenalin und Dopamine im Gehirn der Menschen und ihr
    Verhalten bei Erkrankungen des extrapyramidalen Systems,
    Klin.Wschr., 38, 1236-1239.
Eichler, A. J., Antelman, S. M., and Black, C. A., 1980, Amphetamine
    stereotypy is not a homogenous phenomenon: sniffing and
    licking show distinct profiles of sensitization and toler-
    ance, Psychopharmacology, 68, 287-290.
Ellinwood, E. H., 1967, Amphetamine psychosis: I. Description of
    individuals and process, J.Nerv.Ment.Dis., 144, 274-283.
Ellinwood, E. H., Sudilovsky, A., and Nelson, L. M., 1973, Evolving
    behavior in the clinical and experimental amphetamine
    (model) psychosis, Am.J.Psychiat., 130, 1088-1093.
Endo, K., and Araki, T., 1981, Identification of excitatory inter-
    neurons in the motor cortex by the cross-correlation method,
    Electroenceph.Clin.Neurophysiol., 52, 141.
Engel, J., and Berggren, U., 1980, Effects of lithium on behavior
    and central monamines, Acta Psychiatr.Scand., 61, Suppl.280,
    133-143.
Ernst, A. M., 1967, Mode of action of apomorphine and dexamphetamine
    on gnawing compulsion in rats, Psychopharmacologia, 10, No.
    3, 316-323.
Ernst, A. M., and Smelik, P. G., 1966, Site of action of dopamine
    and apomorphine on compulsive gnawing behavior in rats,
    Experentia, 22, 837-838.
Eroglu, L., and Atamer-Simsek, S., 1980, Effect of lithium on
    stress-induced changes in the brain levels of monoamines in
    rats, Arzneimittel-Forsch., 30, No. 12, p. 2115-2117.
Fanaryan, V. V., 1975, "On the neuronal organization of the efferent
    systems of the cerebellum," (in Russian), Izd.Nauka,
    Leningrad.
Feldberg, W., 1963, "A Pharmacological Approach to the Brain from its
    Inner and Outer Surface," Edward Arnold (Publishers),
    London, p. 103.
Felpel, L. P., 1972, Effects of strychnine, bicuculine and picro-
    toxin on labyrinthine-evoked inhibition in neck motoneurones
    of the cat, Exp.Brain Res., 14, 494-502.
Feltz, P., 1969, Dopamine, aminoacids and caudate unitary responses
    to nigral stimulation, J.Physiol., (Lond.), 205, 8-9.
Felts, P., and Albe-Fessard, D., 1972, A study of an ascending
    nigro-caudate pathway, Electroenceph.Clin.Neurophysiol., 33,
    179-193.
Fernandez-Guardiola, A., Manni, E., Wilson, J. H., and Dow, R. S.,
    1962, Microelectrode recording of cerebellar and cerebral
    unit activity during convulsive afterdischarge, Exp.Neurol.,
    6, 48-69.

Fields, H. L., and Anderson, S. D., 1978, Evidence that raphe-spinal
        neurons mediate opiate and midbrain stimulation-produced
        analgesias, Pain, 5, 333-349.
Fields, H. L., Anderson, S. D., Clanton, C. H., and Basbaum, A. I.,
        1976, Nucleus raphe magnus: a common mediator of piate and
        stimulus produced analgesia, Trans.Amer.Neurol.Ass., 101,
        208-210.
Fields, H. R., Basbaum, A. J., Clanton, C. H., and Anderson, S. D.,
        1977, Nucleus raphe magnus inhibition of spinal cord dorsal
        horn neurons, Brain Res., 126, 441-453.
Flemenbaum, A., 1977, Lithium inhibition of norepinephrine and
        dopamine receptors, Biol.Psychiat., 12, No. 4, 563-572.
Fog, R. L., 1967, Role of the corpus striatum in typical behavioral
        effects in rats produced by both amphetamine and neuroleptic
        drugs, Acta Pharmacol.Toxicol. (Kbh), 25, Suppl. 4, 1-59.
Fog, R., 1972, On stereotypy and catalepsy: studies on the effect of
        amphetamines and neuroleptics in rats, Acta Neurol.Scand.,
        48, Suppl. 50, 1-73.
Fog, R., and Pakkenberg, H., 1971, Behavioral effects of dopamine
        and p-hydroxyamphetamine injected into corpus striatum of
        rats, Exp.Neurol., 31, No. 1, 75-86.
Fog, R. L., Randrup, A., and Pakkenberg, H., 1967, Aminergic mechan-
        isms in corpus striatum and amphetamine-induced stereotyped
        behavior, Psychopharmacologia, 11, No. 2, 179-183.
Fog, R. L., Randrup, A., and Pakkenberg, H., 1968, Neuroleptic
        action of quaternary chlorpromazine and related drugs in-
        jected into various brain areas in rats, Psychopharm-
        acologia, 12, No. 3, 428-432.
Fog, R. L., Randrup, A., and Pakkenberg, H., 1971, Intrastriatal
        injections of quaternary butyrophenons and oxypertine:
        neuroleptic effect in rats, Psychopharmacologia, 19, No. 3,
        224-230.
Fonnum, F., Grofova, I., Rinvik, E., Storm-Mathisen, J., and
        Walberg, F., 1974, Origin and distribution of glutamate
        decarboxylase in substantia nigra in the cat, Brain Res.,
        71, No. 1, 77-82.
Foreman, R. D., Beal, J. E., Applebaum, A. E., Coulter, J. D., and
        Willis, W. D., 1976, Effects of dorsal column stimulation on
        primate spinothalamic tract neurons, J. Neurophysiol., 39,
        534-546.
Förster, O., 1927, Die Leitungsbahnen des Schmerzgefühls und die
        chirurgische Behandlung der Schmerzzustände, Urban &
        Schwarzenberg, Wien.
Fox, S., and O'Brien, J. H., 1962, Inhibition and facilitation of
        afferent information by the caudate nucleus, Science, 137,
        No. 3812, 423-435.
Frankstein, S. J., Bijasheva, Z. G., and Smolin, L. N., 1965, In-
        hibitory synapses and inflammation, Nature, 205, 294-295.

Franz, D. N., and Iggo, A., 1968, Dorsal root potentials and ventral root reflexes evoked by nonmyelinated fibers, Science, 162, 1140-1142.

Frazier, C. H., and Russell, E. G., 1924, Neuralgia of face: analysis of 754 cases with relation to pain and other sensory phenomena before and after operation, Arch.Neurol.Psychiat., 11, 557-563.

Fraeman, L. W., and Heimburger, R. F., 1947, Surgical relief of pain in paraplegic patients, Arch.Surg., 55, 433-440.

Frigyesi, T. L., and Purpura, D. P., 1967, Electrophysiological analysis of reciprocal caudate-nigral relation, Brain Res., 6, 440-456.

Frommer, G. P., Galambes, R., and Norton, T., 1968, Visual evoked responses in cats with optic tract lesions, Exp.Neruol., 21, 346-363.

Fukuda, I., and Iwama, K., 1970, Inhibition des interneurones du corps genomillé latéral par l'activation de la formation réticulée, Brain Res., 18, 548.

Fukuda, J., Highstein, S. M., and Ito, M., 1972, Cerebellar inhibitory control of the vestibulo-coular reflex investigated in rabbit. IIIrd nucleus, Exp.Brain Res., 14, 511-526.

Fullerton, P. M., 1963, The effect of ischemia on nerve conduction in the carpal tunnel syndrome, J.Neurol.Neurosurg.Psychiatry, 26, 385-397.

Futamachi, K. J., and Prince, D. A., 1975, Effect of penicillin on an excitatory synapse, Brain Res., 100, No. 3, 589-597.

Fuxe, K., 1978, The position of dopamine among biogenic amines with neurotransmitter function, Triangle, 17, No. 1, 1-11.

Fuxe, K., and Ungerstedt, U., 1970, Histochemical, biochemical and functional studies on central monamine neurons after acute and chronic amphetamine administration, in: "Amphetamines and Related Compounds," Raven Press, New York, pp. 257-288.

Fuxe, K., Fredholm, B. B., Agnati, L. F., Orgen, S. O., Everitt, B. J. Jonsson, E., and Gustafsson, J. A., 1978, Interaction of ergot drugs with central monoamine systems. Evidence for a high potential in the treatment of mental and neurological disorders, Pharmacology, 16, Suppl. 1, 99-134.

Gabor, A. J., Scobey, R. P., and Wehril, C. J., 1979, Relationship of epileptogenicity to cortical organization, J.of Neurophysiology, 42, 1609-1625.

Gadea-Ciria, M., 1976, Plasticity of ponto-geniculooccipital waves during paradoxical sleep after frontal lobe lesions in the cat, Exp.Neurol., 53, No. 2, 328-338.

Galindo, A., 1969, GABA-picrotoxin interation in the mammalian central nervous system, Brain Res., 14, 763-767.

Gallager, D. W., 1978, Benzodiazepines: potentiation of GABA inhibitory response in the dorsal raphe nucleus, Europ.J. Pharmacol., 43, No. 1, 133-143.

Gallager, D. M., Mallorga, P., Thomas, J. W., and Tallman, J. F., 1980 GABA-benzodiazepine interactions: physiological, pharmacological and developmental aspects, Federation Proc., 39, 3043-3049.

Gasser, H. S., and Erlanger, J., 1929, The role of fiber size in the establishment of a nerve block by pressure or cocaine, Am.J.Physiol., 88, 581.

Gavish, M., and Snyder, S. H., 1981, γ-Aminobutyric acid and benzodiazepine receptors: copurification and characterization, Proc.Natl.Acad.Sci.USA., 78, 1939-1941.

Gaze, R. M., 1970, "The Formation of Nerve Connections," Academic Press, London-New York.

Gebhart, G. F., 1982, Opiate and opioid peptide effects on brainstem neurons: relevance to nociception and antinociceptive mechanisms, Pain, 12, 93-140.

Geller, H. M., Taylor, D. A., and Hoffer, B. J., 1978, Benzodiazepines and central inhibitory mechanisms, Arch. Pharmacol., 304, 81-88.

Geller, H. M., Hoffer, B. J., and Taylor, D. A., 1980, Electrophysiological actions of benzodiazepines, Federation Proc., 39, 3016-3023.

Gellhorn, E., 1943, "Autonomic regulations: Their Significance for Physiology, Psychology and Neuropsychiatry," Interscience Publ. Inc., New York.

Gellhorn, E., and Loofbourrow, G. N., 1963, "Emotions and Emotional Disorders (A Neurophysiological Study)," Harper & Row Publishers, Hoeber Medical Division, New York, Evanston, London.

Gellhorn, E., Ballin, H. M., and Kawakami, M., 1960, Studies on experimental convulsions with emphasis on the role of the hypothalamus and the reticular formation, Epilepsia, 1, 233-254.

Gerard, R. W., 1951, Physiology of pain. Abnormal neuron states in causalgia and related phenomena, Anaesthesiology, 12, 1-13.

Gernandt, B. E., and Megirian, D., 1961, Ascending propriospinal mechanims, J.Neurophysiol., 24, No. 4, 364-376.

Gernandt, B. E., and Thulin, C. A., 1952, Vestibular connections of the brain stem, Amer.J.Physiol., 171, 121-127.

Gerstenbrand, F., Patelsky, K., and Prosenz, P., 1963, Erfahrungen mit L-Dopa in der Therapie des Parkinsonism, Psychiat.Neurol. (Basel), 146, 246.

Gilliat, R. W., and Wilson, T. G., 1953, A pneumatic tourniquet test in the carpal tunnel syndrome, Lancet, 2, 295-297.

Glotzner, F. L., 1974, Intracellular recording of spontaneous local responses in the chronic epileptogenic focus of a rhesus monkey, Exp.Neurol., 42, 233-237.

Glotzner, F. L., Fetz, E. E., and Ward, A. A., 1973, Neuronal activity in the chronic and acute epileptogenic focus, Exp. Neurol., 42, 502-518.

Gobel, S., 1978a, Golgi studies of the neurons in layer I of the dorsal horn of the medulla (trigeminal nucleus caudalis), J.Comp.Neurol., 180, 375-394.

Gobel, S., 1978b, Golgi studies of the neurons of layer II of the dorsal horn of the medulla (trigeminal nucleus caudalis), J.Comp.Neurol., 180, 395-414.

Gobel, S., and Binck, J. M., 1977, Degenerative changes in primary trigeminal axons and in neurons in nucleus caudalis following tooth extirpations in the cat, Brain Res., 132, 347-354.

Gobel, S., Falls, W. M., and Hockfield, S., 1977, The division of the dorsal and ventral horns of the mammalian caudal medula into light layers using anatomical criteris, in: "Pain in the Trigeminal Region," D. J. Anderson and B. Mattews, Eds., Elsevier, Amsterdam, pp. 443-453.

Goddard, G. V., 1967, Development of epileptic seizures through brain stimulation at low intensity, Nature (London), 214, 1020-1021.

Goddard, G. V., MacIntyre, D. C., and Luch, C. K., 1969, A permanent change in brain function resulting from daily electrical stimulation, Exp.Neurol., 25, 295-330.

Godfraind, J. M., 1976, Mechanismes neurophysiologique du sommeil, Rev.Quest.Sci., 147, 505-531.

Godlevsky, L. S., and Shandra, A. A., 1983, Influence of electrical stimulation of the dentate nucleus on the complexes of the epileptic activity foci in the cerebral cortex, Byull.eksp. biol.i med., (in preparation).

Goff, V. G., 1973, A Study of Epileptic Activity and Vibrating Systems of the Cortex of the Major Hemispheres. Dissertation for a candidate's degree, Tbilisi.

Gol., A., 1967, Relief of pain by electrical stimulation of the septal area, J.Neurol.Sci., 5, 115-120.

Goldensohn, E. S., Zablow, L., and Stein, B., 1970, Interrelationship of form and latency of spike discharge from small areas of human cortex, Electroenceph.Clin.Neurophysiol., 29, No. 3, 321-322.

Goldscheider, A., 1894, Ueber den Schmerz in physiologischer und klinischer Hinsicht. Nach einem Vortrage, Hirschwald, Berlin.

Gonsales-Vegas, I. A., 1974, Antagonism of the dopamine-mediated inhibition in the nigro-striatal pathway: a mode of action of some catatonia inducing drugs, Brain Rés., 80, 219-228.

Gordon, M., Rubia, F. J., and Strata, P., 1973, The effect of pentothal on the activity evoked in the cerebellar cortex, Exp.Brain Res., 17, 50-62.

Gorgiladze, G. I., 1966, Paired work of the vestibular apparatus (on the principle of the equilibrated Khegies-Bekhterev centers), Fiziolog.Zh.SSSR, 52, No. 6, 669-676.

Gottesfeld, Z., 1975, Effect of lithium and other alkali metals on brain chemistry and behavior. I. Glutamic acid and GABA in brain regions, Psychopharmacologia (Berl.), 45, No. 3, 239-242.

Grafova, V. N., and Danilova, 1980, Action of antiepileptic drugs in myoclonia of spinal origina, Byull.eksp.biol.i med., 90, No. 11, 538-541.

Grafova, V. N., Danilova, E. I., and Kryzhanovsky, G. N., 1979, Analgesic drugs in pain syndrome of spinal origin, Byull. eksp.biol.i med., 87, No. 8, 147-151.

Granit, R., Phillips, C. G., Skoglund, S., and Steg, G., 1957, Differentiation of tonic from phasic alpha ventral horn cells by strech, pinna and crossed extensor reflexes, J.Neurophysiol., 20, No. 5, 470-481.

Grant, G., and Ardvisson, J., 1975, Transganglionic degeneration in trigeminal primary sensory neurons, Brain Res., 95, 265-279.

Griffith, J. D., Cavanauch, J., held, J., and Oates, J. A., 1972, Dextroamphetamine: evaluation of psychotomimetic properties in man, Arch.Gen.Psychiatry, 26, 97-100.

Grilliner, S., Honge, T., and Land, S., 1970, The vestibulospinal tract effects on alphamotoneurones in the lumbosacral spinal cord in the cat, Exp.Brain Res., 10, 94-120.

Grofova, I., and Rinvik, E., 1970, An experimental electronmicroscopic study on the striatonigral projection in the cat, Exp.Brain Res., 11, No. 3, 249-262.

Guidotti, A., Baraldi, M., and Schwartz, J. P., 1979, Molecular mechanisms regulating the interactions between the benzodiazepines and GABA receptors in the central nervous system, Pharmacol.Biochem.Behav., 10, 803-807.

Guidotti, A., Baraldi, M., Leon, A., and Costa, E., 1980, Benzodiazepines: a tool to explore the biochemical and neurophysiological basis of anxiety, Federation Proc., 39, 3039-3042.

Gumnit, R. J., and Takahashi, T., 1965, Changes in direct current activity during experimental focal seizures, Electroencephalogr.Clin.Neurophysiol., 19, 63-74.

Gun, A. A., 1976, Strychnine-neuronographic investigation of the projections of the orbital cortex to the adult brain structures, Fiziol.zh.SSR., 62, No. 12, 1760-1766.

Guselnikov, V. I., and Supin, A. Ya., 1968, "Rhythmic Activity of the Brain," (in Russian), Moscow State University Publishers, Moscow.

Guselnikova, K. G., and Guselnikov, V. I., 1975, "Electrophysiology of the Olfactory Analyzer of Vertebrates," (in Russian), Moscow State University Publishers, Moscow.

Gushchin, I. S., Kozhechkin, S. N., and Sverdlov, Yu. S., 1970, The hyperpolarizing effect of glycine on 'tetanus' motoneurones, Byull.eksp.biol.i med., 70, No. 8, 29.

Guyenet, P. G., Agid, V., Javoy, F., Beaujouan, J. C., Rossier, J., and Glowinskii, J., 1975, Effects of dopaminergic receptor agonists and antagonists on the activity of the neostriatal cholinergic system, Brain Res., 84, No. 2, 227-244.

Gyvels, J., Hees, J. V., and Peluso, F., 1976, Modulation of experimentally produced pain in man by electrical stimulation of

some cortical, thalamic and basal ganglic structures, <u>Adv.
in Pain Res.and Therapy</u>, 1, 475-478.

Habermann, E., 1981, Tetanus toxin and botulinum A neurotoxin inhibit and at higher concentrations enhance noradrenaline
outflow from particulate brain cortex in batch, <u>Naunyn-
Schmiedeberg's Arch.Pharmacol.</u>, 318, No. 2, 105-112.

Hablitz, J. J., 1976, Intramuscular penicillin epilepsy in the cat:
effects of chronic cerebellar stimulation, <u>Exp.Neurol.</u>, 50,
505-514.

Hablitz, J. J., and Rea, G., 1976, Cerebellar nuclear stimulation in
generalized penicillin epilepsy, <u>Brain Res.Bull.</u>, 8,
599-601.

Hablitz, J., McSherry, J., and Kellaway, F., 1975, Cortical seizures
following cerebellar stimulation in primates, <u>Electro-
encephalogr. Clin.Neurophysiol.</u>, 38, 423-426.

Haefely, W. E., Kulcsar, A., Mohler, H., Pieri, L., Polc, P., and
Schaffner, R., 1975, Possible involvement of GABA in the
central action of benzodiazepines, <u>Adv.Biochem.Psycho-
pharmacol.</u>, 14, 113-120.

Halazs, P., 1972, The generalized epileptic spike-wave mechanism and
the sleep-wakefulness system, <u>Acta Physiol.Acad.Sci.Hung.</u>,
42, 293-314.

Handwerker, H. O., Iggo, A., and Zimmermann, M., 1975, Segmental and
supraspinal actions on dorsal horn neurons responding to
noxious and non-noxious skin stimuli, <u>Pain</u>, 1, 147-165.

Hanley, J., Rickles, W. R., Grandall, P. H., and Walter, R. D.,
1972, Anatomic recognition of EEG correlates of behavior in
a chronic shcizophrenic patient, <u>Am.J.Psychiatry</u>, 128,
1524-1528.

Harding, A. E., and Le Fann, J., 1977, Carpal tunnel syndrome related to antebrachial Cimino-Brescia fistula, <u>J.Neurol.
Neurosurg.Psychiatry</u>, 40, 511-513.

Harris, A. B., 1980, Structural and chemical changes in experimental
epileptic foci, <u>in</u>: "Epilepsy: a Window to Brain Mechanisms," J. S. Lockward and A. A. Ward, eds., Raven Press, New
York, pp. 149-164.

Harris, W., 1940, An analysis of 1433 cases of paroxysmal trigeminal neuralgia (trigeminal tic) and the end results of
Gasserian alcohol infection, <u>Brain</u> 63, 209-224.

Hartmann, E., 1976, Schizophrenia: a theory, <u>Psychopharmacologia
Bull.</u> (Berl.), 49, No. 1, 1-15.

Hassler, R., 1970, <u>in</u>: "Trigeminal Neuralgia," R. Hassler and A. E.
Walker, ed., Georg Thieme Verlag, Stuttgart, pp. 123-138.

Hayes, R. L., Price, D. D., and Dubner, R., Behavioral and physiological studies of sensory coding and modulation of trigeminal nociceptive input, <u>in</u>: "Advances in Pain Research and
Therapy", J. J. Bonica, ed., Vol. 3, Raven Press, New York,
pp. 219-243.

Hazouri, L. A., and Mueller, A. D., 1950, Pain threshold studies on
paraplegic man, <u>Arch.Neurol.Psychiatry (Chic.)</u>, 64, 607-613.

Head, H., 1920, Studies in Neurology, London.

Head, H., and Holmes, G., 1911-12, Sensory disturbances from cerebral lesions, Brain, 34, 102-254.

Heal, D. J., Green, A. R., and Buylaert, W. A., 1980, Inhibition of apomorphine-bromcriptine, and lergotril-induced circling behavior in rats by subsequent haloperidol administration, Neuropharmacology, 19, No. 1, 133-137.

Heath, R. G., 1954, Studies in Schizophrenia, Harvard University Press, Cambridge.

Hebb, D. O., 1949, "The Organization of Behavior," Wiley, New York.

Hebb, D. O., 1972, "Textbook of Psychology," Saunders, Philadelphia-London.

Heikkila, R. E., Shapiro, B. S., and Duvoisin, R. C., 1981, Relationship between loss of dopamine nerve terminals, striatal /H^3/ spiroperidol binding and rotational behavior in unilaterally 6-hydroxydopamine-lesioned rats, Brain Res., 211, No. 2, 285-292.

Herberg, L. J., and Wishart, T. B., 1981, Selfstimulation and locomotor changes indicating latent anticholinergic activity by an atypical neuroleptic (thioridazine), Neuropharmacology, 20, No. 1, 55-60.

Hernandez-Peon, R., 1963, Sleep induced by localized electrical or chemical stimulation of the forebrain, EEG and Clin.Neurophysiol., Suppl., 24, 188-198.

Herz, A., Zieglglänsberger, W., and Freitag-Loringhohoven, H., 1970, Development of fields of focal potentials in the caudate nucleus following micro-electrophoretic application of glutamic acid and GABA, Electroenceph.Clin.Neurophysiol., 28, 247-258.

Hess, W. R., 1944, Das Schalfsyndrom als Folge diencephaler Reizung, Helv.Physiol.Pharmacol.Acta, 2, 305-344.

Hess, W. R., 1949, Le sommeil comme fonction physiologique, J.Physiol.(Paris), 41, 61-67A.

Hess, W. R., 1954, The diencephalic sleep centre, in: "Brain Mechanisms and Consciousness," J. F. Delafreshaye, Blackwell, Oxford, pp. 117-136.

Hess, W. R., 1957, "The Functional Organization of the Diencephalon," Grune and Stratton, New York.

Heyer, E. J., Nowak, L. M., and McDonald, R. L., 1982, Membrane depolarization and prolongation of calcium-dependent action potentials of mouse neurons in cell culture by two convulsants: bicuculine and penicillin, Brain Res., 232, 41-56.

Hillman, P., and Wall, P. D., 1969, Inhibitory and excitatory factors influencing the receptive fields of lamina 5 spinal cord cells, Exp.Brain Res., 9, 284-306.

Hobson, J. A., and McCarley, R., 1972, Spontaneous discharge rates of cat cerebellar Purkinje cells in sleep and waking, Electroenceph.Clin.Neurophysiol., 33, 457-469.

Hocher, B., Spira, M. E., and Werman, R., 1976, Penicillin decreases chloride conductance in crustacean muscle: a model for the epileptic neuron, Brain Res., 197, 85-103.

Honge, T., Jankowska, E., and Lundberg, A., 1965, Effects evoked from the rubrospinal tract in cats, Experientia, 21, 525-526.

Hongo, T., Jankowska, E., and Lundberg, A., 1968, Post-synaptic excitation and inhibition from primary afferents in neurons of the spinocervical tract, J.Physiol., 199, 569-592.

Hornykiewicz, O. D., 1966, Dopamine (3-hydroxytyramine) and brain function, Pharmacol.Revs., 18, No. 2, 925-964.

Hösli, L., and Haas, H. L., 1972, The hyperpolarization of neurones of the medulla oblongata by glycine, Experientia (Basel), 28, 1057-1058.

Hösli, L., and Tebecis, A. K., 1970, Action of amino acids and convulsant on bulbar reticular neurones, Exp.Brain Res., 11, 111-127.

Hosobuchi, Y., Adams, J. E., and Rutkin, B., 1973, Chronic thalamic stimulation for the control of facial anesthesia dolorosa, Arch.Neurol., 29, 158-161.

Hosobuchi, Y., Adams, J. E., and Linchitz, R., 1977, Pain relief by electrical stimulation of the central gray matter in humans and its reversal by naloxone, Science, 197, 183-186.

Howe, J. F., Loeser, J. D., and Calvin, W. H., 1977, Mechanosensitivity of dorsal root ganglia and chronically injured axons: physiological basis for the radicular pain of nerve root compression, Pain, 3, 25-41.

Hrbek, Y., 1980, "Propriocepti-Motor Circuits Governing Striated Muscles. Structure, Function and Disorders. The Higher Nervous Activity - XXXV Monograph Series," Vol. 6, Statni Pedagogicke Nakladatelstvi, Prague.

Hubel, D. H., 1960, Electrocorticograms in cats during natural sleep, Arch.Ital.Biol., 98, 171-181.

Hunt, A. D., Stokes, I., and McCrory, W. W., 1954, Pyridoxine dependency: report of case of intractable convulsions in an infant controlled by pyridoxine, Pediatrics, 13, 140-145.

Hunter, R. A., 1959, Status epilepticus: history, incidence and problem, Epilepsia, 1, 162-188.

Hutton, T. J., Frost, J. D., and Foster, J., 1972, The influence of the cerebellum in cat penicillin epilepsy, Epilepsia, 13, 401-408.

Hyttel, J., 1980, Further evidence that ^{3}H-cis(Z)-flupenthixol binds to the adenylate cyclase associated dopamine receptor (D-1) in rat corpus striatum, Psychopharmacology, 61, No. 1, 107-110.

Igonkina, S. I., and Kryzhanovsky, G. N., 1977, Analgesia under electrostimulation of midbrain nuclei in rats with pain syndrome of spinal origin, Byull.eksp.biol.i med., 84, No. 7, 16-19.

Igonkina, S. I., and Kryzhanovsky, G. N., 1979, Analysis of anal-
    gesia induced by the generator of excitation in the dorsal
    raphe nucleus, Byull.eksp.biol.i med., 88, No. 9, 278-281.
Inoue, N., Tsukada, Y., and Barbeau, A., 1975, Behavioral effects in
    rats following intrastriatal microinjection of manganese,
    Brain Res., 95, No. 1, 103-124.
Ito, M., Hongo, T., and Okada, Y., 1969a, Vestibular-evoked potent-
    ials in Deiters' neurones, Exp.Brain Res., 7, 214-230.
Ito, M., Kawai, N., Udo, M., and Mano, N., 1969b, Axon reflex act-
    ivation of Deiters' neurones from cerebellar cortex through
    collaterals of the cerebellar afferents, Exp.Brain Res., 8,
    No. 3, 249-268.
Ito, M., Udo, M., Mano, N., and Kawai, N., 1970, Synaptic action of
    the fastigiobulbar impulses upon neurons in the medullary
    reticular formation and vestibular nuclei, Exp.Brain Res.,
    11, 29-47.
Iversen, L. L., 1975, Dopamine receptors in the brain, Science, 188,
    No. 4193, 1084-1089.
Iversen, L. L., 1977, Catecholamine-sensitive adenylate cyclase in
    nervous system, J.Neurochem., 29, No. 1, 5-12.
Iversen, L. L., 1978, Biochemical psychopharmacology, of GABA, Adv.
    Biochem.Psychopharmacol., 19, 25-38.
Iversen, L. L., and Johnston, G. A. R., 1971, GABA uptake in rat
    central nervous system: comparison of uptake in slices and
    homogenates and the effects of some inhibitors, J.Neuro-
    chem., 18, 1939-1950.
Iversen, S. D., 1971, The effect of surgical lesions to frontal
    cortex and substantia nigra on amphetamine responses in rat,
    Brain Res., 31, No. 2, 295-311.
Iversen, S., 1977, Behavioral implications of dopaminergic neurons
    in mesolimbic system, Adv.Biochem.Psychopharmacol., 16,
    209-214.
Jancowska, E., Jukes, M. G., Lund, S., and Lundberg, A., 1967, The
    effect of DOPA on the spinal cord. 6. Half-centre organiz-
    ation of interneurones transmitting effects from the flexor
    reflex afferents, Acta Physiol.Scand., 70, 389-402.
Jänig, W., and Zimmerman, M., 1971, Presynaptic depolarization of
    myelinated afferent fibres evoked by stimulation of cu-
    taneous C fibres, J.Physiol., 214, 29-50.
Janowsky, D. S., El-Yousef, M. K., Davis, J. M., and Sekerke, H. J.,
    1973, Provocation of schizophrenic symptoms by intravenous
    administration of methylphenylate, Arch.Gen.Psychiat., 28,
    185-191.
Johnson, A. M., Loew, D. M., and Vigouret, J.-M., 1976, Stimulant
    properties of bromcriptine on central dopamine receptors in
    comparison to apomorphine, (+)-amphetamine, and L-DOPA,
    Brit.J.Pharmacol., 56, No. 1, 59-68.
Johnston, D., and Brown, T. H., 1981, Giant synaptic potential
    hypothesis for epileptiform activity, Science, 211, 294-297.

Johnston, G. A. R., and Willow, M., 1981, Barbiturates and GABA receptors, in: "GABA and Benzodiazepine Receptors," Costa et al., eds., Raven Press, New York, pp. 191-198.

Johnstone, E. C., Crow, T. J., and Mashiter, K., 1977, Anterior pituitary hormone secretion in chronic schizophrenia: an approach to neurohumoral mechanisms, Psychol.Med., 7, 223-228.

Jouvet, M., 1965a, Paradoxical sleep-a study of its nature and mechanisms, in: "Sleep Mechanisms," Akert, C. Bally, and J. P. Schade, Elsevier Publ. Co., Amsterdam, pp. 20-62.

Jouvet, M., 1965b, Etude de la dualité des etats de sommeil et des mécanismes de la phase paradoxale, in: "Aspects Anatomo-fonctionnels de la Physiologie du Sommeil." A Symposium, M. Jouvet, ed., Centr.Natl.Rech.Sci., Paris pp. 393-442.

Jouvet, M., 1967, Neurophysiology of the state of sleep, Physiol. Rev., 47, 117-127.

Jouvet, M., 1978, Le sommeil paradoxal est-il responable d'une programmation genetique du cerveau, C.R.Soc.Biol., 172, 9-32.

Julien, R. M., 1972, Cerebellar involvement in the antiepileptic action of diazepam, Neuropharmacology, 11, No. 5, 683-691.

Julien, R., and Halpern, L., 1972, Effects of diphenylhydantoin and other antiepileptic drugs on epileptiform activity and Purkinje cell discharge rates, Epilepsia, 13, 387-400.

Julien, R. M., and Laxer, K. D., 1974, Cerebellar responses to penicillin-induced cerebral cortical epileptiform discharge, Electroencephalogr. Clin.Neurophysiol., 37, 123-132.

Jung, R., and Hassler, R., 1960, The extrapyramidal motor system, in: "Handbook of Physiology," J. Field et al., eds., Amer. Physiol. Soc., Washington, pp. 863-927.

Jurna, I., 1968, Reserpine rigidity in the rat - a model for the analysis of anti-parkinson drugs, Arch.Pharmacol.Exp. Pathol., 259, No. 1, 181-192.

Jurna, I., 1976, The cholinergic rigidity, Pharm.and Ther.B., 2, No. 2, 413-421.

Jurna, I., 1980, Effect of stimulation in the periaqueductal gray matter on activity in ascending axons of the rat spinal cord: selective inhibtion of activity evoked by afferent A and C fibre stimulation and failure of naloxone to reduce inhibition, Brain Res., 196, 33-42.

Kaada, B. R., 1951, Somato-motor, autonomic and electrocortico-graphic responses to electrical stimulation of 'rhinence-phalic' and other structures in primates, cat and dog, Acta Physiolog.Scand., 24, Suppl. 83, 1-285.

Kaada, B., 1976, Neurophysiology and acupuncture: a review, in: "Advances in Pain Research and Therapy," J. J. Bonnica and D. Albe-Fessard, eds., vol. 1, Raven Press, New York.

Kambarova, D. K., 1977, Neurophysiological mechanisms of some epil-eptic reactions. Report II, Fiziol.cheloveka, 3, No. 2, 211-225.

Kandel, E. R., and Spencer, W. A., 1961, Electrophysiology of hippo-
    campal neurons. 11. Afterpotentials and repetitive firing,
    J.Neurophysiol., 24, 243-259.
Kano, M., and Ishikawa, K., 1972, Effect of tetanus toxin on the
    inhibitory neuromuscular junction of crayfish muscle, Exper.
    Neurol., 37, 550-560.
Kao, L. I., and Crill, W. E., 1972a, Penicillin induced segmental
    myoclonus. I. Motor responses and intracellular recording
    from motoneurons, Arch.Neurol., 26, 156-161.
Kao, L. I., and Crill, W. E., 1972b, Penicillin induced segmental
    myoclonus. II. Membrane properties of cat spinal motoneur-
    ons, Arch.Neurol., 26, 162-168.
Karamyan, A.Ch., 1970, "Functional Evolution of the Vertebrate
    Brain," (in Russian), Izd. Nauka, Leningrad.
Karlov, V. A., 1974, "Epileptic Status," (in Russian), Izd.
    Meditsina, Moscow.
Karlov, V. A., Marchenko, D. A., and Savitskaya, O. N., 1973,
    Changes in the bioelectric activity of the brain of patients
    with trigeminal neuralgia, Zh.nervopatol.i psikhiatr., No.
    3, 331-335.
Karmanova, I. G., Maksimchuk, V. F., Panov, A. N., et al., 1978,
    Role of the paradoxical phase in the organization of the
    wakefulness-sleep cycle in rats, Fiziol.Zh.SSSR, 64, No. 8,
    1974-1081.
Karmanova, I. G., Khomutetskaya, O. Ye., and Shilling, N. V., 1981,
    Comparative physiological analysis of the stages of evo-
    lution of sleep and the mechanisms of its regulation,
    Uspekhi fiziol.nauk., 12, No. 2, 3-19.
Karobath, M., and Sperk, G., 1979, Stimulation of benzodiazepine
    receptor binding by γ-aminobutyric acid, Proc.Natl.Acad.
    Sci.USA, 76, 1004-1006.
Karobath, M., Placheta, P., Lippstisch, M., and Krogsgaard-Larsen,
    P., 1980, Characterization of GABA-stimulated benzodiazepine
    receptor binding, Adv.Biochem.Psychopharmacol., 21, 313-320.
Karpiak, S. E., Graf, L., and Rapport, M. M., 1976, Antiserum to
    brain gangliosides produces recurrent epileptiform activity,
    Science, 194, 736-737.
Kassil, G. N., Kryzhanovsky, G. N., Matlina, E.Sh., Pukhova, G. S.,
    and Grafova, N. N., 1972, Catecholamine exchange in painful
    tetanus, 204, No. 1, 246-249.
Katz, R. I., and Kopin, I. J., 1969, Release norepinephrine-^{3}H and
    serotonin-^{3}H evoked from brain slices by electrical-field
    stimulation. Calcium dependency and the effect of lithium
    ouabain and tetrodotoxin, Biochem.Pharmacol., 18, 1935-1939.
Kazennikov, O. V., Shik, M. L., and Yakovleva, T. V., 1980, On the
    two paths of locomotor influence of the brainstem on the
    spinal cord, Fiziol.Zh.SSR, 66, 1260-1263.
Kebabian, J. W., and Calne, D. B., 1979, Multiple receptors for
    dopamine, Nature, 277, No. 5989, 93-96.

Kebabian, J. W., Petzold, G. L., and Greengart, P., 1972, Dopamine-sensitive adenylate cyclase in the caudate nucleus of rat brain and its similarity to the 'dopamine' receptors, Proc. Nat.Acad.Sci.(USA), 69, 2145-2149.

Kelly, P. H., and Iversen, S. D., 1975, Amphetamine and apomorphine responses in the rat after lesions of mesolimbic or striatal dopamine neurons, in: "Abst. of 6th Int.Congr.Pharmacol.," SI, Helsinki.

Kelly, P. H., and Miller, R. J., 1975, The interaction of neuro-leptic and musculinic agents with central dopaminergic system, Br.J.Pharmacol., 54, 5-121.

Kemp, J. M., 1968, Observation of the caudate nucleus of the cat impregnated with the Golgi method, Brain Res., 11, 467-470.

Kempinsky, W. H., Boniface, W. R., Morgan, P. P., and Bush, A. K., 1960, Reserpine in Huntington's chorea, Neurology, 10, No. 1, 38-42.

Kennedi, B., and Leonard, B. E., 1980, Similarity between action of nicotinamide and diazepam on neurotransmitter metabolism in the rat, Biochem.Soc.Trans., 8, 59-60.

Kerr, F. L., 1963, The etiology of trigeminal neuralgia, Arch. Neurol., 8, 15-25.

Kerr, F. W. L., 1976, Segmental circuitry and spinal cord nocicept-ive mechanisms, in: "Advances in Pain Research and Therapy, J. J. Bonica and D. Albe-Fessard, eds., Vol. 1, Raven Press, New York, pp. 75-89.

Kerr, F. W., 1980, The structural basis of pain: circuitry and pathways, in: "Pain Discomfort and Humanitarian Care. Development in Neurology, Vol. 4," L. R. Y., Ng. and J. Bonica, eds., Elsevier, North Holland, pp. 49-60.

Kerr, F. W., and Miller, R. H., 1966, The ultrastructural pathology of trigeminal neuralgia, Arch.Neurol., 15, 308.

Kety, S. S., and Matthysse, S., 1972, Prospects for research of schizophrenia, I. Introduction, Neurosci.Res.Progr.Bull., 10, No. 1, 375-376.

Khanababyan, M. V., 1981, "Noradrenergic Mechanisms of the Brain," (in Russian) Izd. Nauka, Leningrad.

Khananashvili, M. M., 1978, "Experimental Pathology of Higher Nervous Activity," (in Russian), Izd.Meditsina, Moscow.

Khananashvili, M. M., 1983, "Pathology of Higher Nervous Activity Behavior," (in Russian), Izd. Meditsina, Moscow.

Khodorov, B. I., 1976, "General Physiology of Excitable Membranes," (in Russian), Izd.Nauka, Moscow.

Khomulo, P. S., and Timofeeva, O. A., 1980, Intensification of food motivation in rabbits of the lateral hypothalamic structures, Fiziol.Zh.SSR, No. 6, 810-815.

Kim, J. S., Bak, I. J., Hasler, R., and Okada, Y., 1971, Role of gamma-aminobutyric acid (GABA) in the extrapyramidal motor system. 2. Some evidence of a type of GABA rich strio-nigral neurons, Exp.Brain Res., 14, No. 1, 95-104.

Kim, J. S., Jornhuber, H. H., Holzmiller, B., Schmid-Burgk, W.,
    Mergner, T., and Krzepinski, G., 1980, Reduction of
    cerebrospinal fulid glutamic acid in Huntington's chorea and
    in schizophrenic patients, Arch.Psychiat.Nervenkr., 229, No.
    1, 1-16.
King, R. J., Raese, J. D., Huberman, B. A., 1982, and Barchas, J.
    D., Dopamine neuronal instability: a model for the
    schizophreniform psychoses, Psychopharmacol.Bulletin, 18,
    70-72.
Kirk, E. J., 1975, Impulses in dorsal spinal rootlets in cats and
    rabbits arising from dorsal root ganglia isolated from the
    periphery, J.Comp.Neurol., 115, 165-176.
Kitai, S. T., Koksis, J. D., Preston, R. J., and Sugimori, M., 1976,
    Nonsynaptic inputs to caudate neurons identified by
    intracellular injection of horseradish peroxydase, Brain
    Res., 109, No. 3, 601-606.
Kjerulf, T. D., and Loeser, J. D., 1973, Neuronal hyperactivity
    following deafferentation of the alteral cuneate nucleus,
    Exp.Neurol., 39, 70-85.
Kjerulf, T. D., O'Neal, J. T., Calvin, W. H., Loeser, J. D., and
    Westrum, L. E., 1973, Deafferentation effects in lateral
    cuneate nucleus of the cat: correlation of structural alter-
    ations with firing pattern changes, Exp.Neurol., 39, 86-102.
Klaus, R., 1937, Die bioelektrische Tätigkeit der Grosshirnrinde im
    normalen Schlaf und in der Narkose durch Schlafmittel,
    J.Psychol.Neurol., (Leipzig), 4, 510-531.
Klawans, H. L., 1973, The Pharmacology of Extrapyramidal Movement
    Disorders, Karger, Basel.
Klawans, H. L., and McKendall, R. R., 1971, Observation on the
    effect of levodopa on tardive lingual-facial-buccal dys-
    kinesia, J.Neurol.Sci., 14, 189-192.
Klawans, H. L., and Rubowitz, R., 1974, The effect of cholinergic
    and anticholinergic agents on tardive dyskinesias, J.Neurol.
    Neurosurg.Psychiatr., 37, 941-947.
Klawans, H. L., and Weiner, W. J., 1974, The effects of d-ampheta-
    mine on choreiform movement disorders, Neurology (Minneap.),
    24, 312-318.
Klawans, H. L., and Weiner, W. J., 1976, The pharmacology of chorea-
    tic movement disorders, in: "Progress in Neurobiology, Vol.
    6," Pergamon Press, Oxford, pp. 49-80.
Klawans, H. L., Goetz, C., and Westheimer, R., 1972, Pathophysiology
    of the schizophrenia and the striatum, Dis.Nerv.Syst., 33,
    No. 4, 711-719.
Klawans, H. L., Weiner, W. J., and Nausieda, P. A., 1977, The effect
    of lithium on an animal model of tardive dyskinesia, Prog.
    Neuro-Psychopharmacol., 1, 53.
Kleitman, N., 1964, The evolutionary theory of sleep and wakeful-
    ness, Perspect.Biol.Med., 7, 169-178.

Klepner, C. A., Lippa, A. S., Benson, D. I., Sano, M. C., and Bear, B., 1979, Resolution of two biochemically and pharmacologically distinct benzodiazepine receptors, Pharmacol., Biochem., Behav., 11, No. 4, 457-462.

Knyihar, E., and Csillik, B., 1976, Effect of peripheral axotomy on the fine structure of the Rolando substance, Exp.Brain Res., 26, 73-87.

Koenigstein, H., 1948, Experimental study of itch stimuli in animals, Arch.Derm.Syph., N.Y., 57, 828-849.

Koneiskey, H. L., Bossart, J. F., Miller, D. D., and Patil, P. N., 1978, Conformation of dopamine at the dopamine receptor, Proc.Nat.Acad.Sci., (Wash.), 75, No. 6, 2641-2643.

Kopa, J., Szabö, S., and Grastyan, E., 1962, A dual behavioral effect from stimulating the same thalamic point with identical stimulus parameters in different conditional reflex situations, Acta Physiol.Sci., Hung., 21, 207-214.

Kornetsky, C., and Markowitz, R., 1978, Animal model of schizophrenia, Adv.Biochem.Psychopharm., 19, 583-593.

Kostyuk, P. G., 1959, "Bineuronal Reflex Arch," (in Russian), Izd. Medgiz, Moscow.

Kostyuk, P. G., 1960, Peculiarities of polysynaptic excitation and inhibition of separate motor neurons, Fiziol.Zh.SSR, 46, No. 4, 398-407.

Kostyuk, P. G., 1974, Some general questions of neuronal integration, in: "Mechanisms of Neuronal Unification in the Nerve Center," (in Russian), Izd.Nauka, Leningrad, pp. 6-11.

Kostyuk, P. G., and Preobrazhensky, N. N., 1975, "Mechanisms of the Integration of Visceral and Somatic Afferent Signals," (in Russian), Izd. Nauka, Leningrad.

Kostyuk, P. G., Pyatigorsky, B. A., and Lang, 1970, Synaptic processes in the neurons of Clarke's column caused by an antidromic wave from the dorsolateral cord, Neirofiziologiya, 2, No. 3, 269-278.

Kotov, Yu.B., and Tseitlin, M. L., 1966, Simple model of the generation of impulses of the nerve cell, Biofizika, 11, No. 3, 547-549.

Kovler, M. A., Avakumov, V. M., 1980, Kruglikova-Lvova, R. P., 1980, Pantogam a new psychopharmacological agent, Khim.-farm.zh., 14, No. 9, 118-122.

Kozhechkin, S. H., and Ostrovskaya, R. U., 1977, Are benzodiazepines GABA antagonists, Nature, 269, No. 5623, 72-73.

Kozlovskaya, M. M., 1964, On the morphofunctional organization of the defensive type of reactions which originate during the stimulation of the hypothalamus, Fiziol.Zh.SSR, 50, No. 10, 1218-1226.

Krauthamer, G. M., and Albe-Fessard, D., 1964, Electrophysiologic studies of the basal ganglia and striapallidal inhibition of nonspecific afferent activity, J.Neurophysiol., 2, No. 1, 73-83.

Kreindler, A., 1962, Active arrest mechanisms of epileptic seizures, Epilepsia, 3, 329-337.

Krishtal, O. A., and Pidoplichko, V. I., 1981, A 'receptor' for protons in small neurons of trigeminal ganglia: possible role in nociception, Neuroscience Letters, 24, 243-246.

Kristiansen, K., and Courtois, G., 1949, Rhythmic electrical activity from isolated cerebral cortex, Electroenceph.Clin. Neurophysiol., 1, 265-272.

Krnjevic, K., 1975, Electrophysiology of dopamine receptors, Adv. Neurol., 9, 13-24.

Krnjevic, K., Radnic, M., and Straughan, D. W., 1966, Nature of a cortical inhibitory process, J.Physiol., 184, 49-77.

Krol, M. B., and Fedorova, Ye. A., 1966, "Principal Neuropathologic Syndromes," (in Russian), Izd. Meditsina, Moscow, 59-94.

Kruglikov, R. I., 1981, "Neurochemical Mechanisms of Instruction and Memory," (in Russian), Izd.Nauka, Moscow.

Kruglikov, R. I., Myshobodsky, M. S., and Ezrokhi, V. I., 1970, "Convulsive Activity," (in Russian), Izd. Nauka, Moscow.

Kryzhanovsky, G. N., 1957, Some peculiar changes in the functional condition of CNS in experimental tetanus and the mechanisms of tetanus toxin action. Report I. On certain specific features of the change in the functional condition of the central nervous system in experiments with tetanus toxin action, Byull.eskp.biol.i med., No. 12., 43-51.

Kryzhanovsky, G. N., 1966, "Tetanus," (in Russian), Izd. Meditsina, Moscow.

Kryzhanovsky, G. N., 1971, The natural pathway of toxin: its transport to the central nervous system and the state of the spinal reflex apparatus in tetanus intoxication, in: "Principles on Tetanus. Proceedings, 2nd International Conference on Tetanus," L. Eckmann, ed., Hans Huber Publ., Berne-Stuttgart, pp. 155-168.

Kryzhanovsky, G. N., 1968, Some peculiarities of the integrative activity of the spinal reflex apparatus under the conditions of disturbed inhibitory mechanisms, in: "Integrative Activity of the Nervous System under Normal Conditions and in a Pathologic State," (in Russian), Izd. Meditsina, Moscow, pp. 21-35.

Kryzhanovsky, G. N., 1973, The mechanism of action of tetanus toxin: effect on synaptic processes and some particular features of toxin binding by nerve tissue, Naunyn-Schmiedeberg's Arch. Pharmacol., 276, 247-270.

Kryzhanovsky, G. N., 1974, Principle of 'dispatch station' in activity of the nervous system, in: "Proc. Intern.Union of Physiol. Science, Vol.. XXVI Intern. Congress," New Delhi, Abstss. p. 475.

Kryzhanovsky, G. N., 1975, Principle of determinative 'dispatch station' in the pathology of the nervous system, in: "Second International Congress on Pathological Physiology," Prague, p. 215.

Kryzhanovsky, G. N., 1976, The experimental central pain and itch syndromes: modeling and general theory, in: "Advances in

Pain Research and Therapy," vol. 1, J. J. Bonica and D. G. Albe-Fessard, eds., Raven Press, New York, pp. 225-231.

Kryzhanovsky, G. N., 1978a, Disinhibition and disintegration of systems, in pathology, Arkhiv patologii, 40, 1, 4-14.

Kryzhanovsky, G. N., 1978b, Disinhibition and disintegration in biological systems, Uspekhi sovremennoi biologii, 85, No. 3, 447-462.

Kryzhanovsky, G N., 1979, Analgesia induced by generator of excitation in midbrain gray matter, in: "Advances in Pain Research and Therapy", Vol. 3, J. J. Bonica et al., Raven Press, New York, pp. 473-478.

Kryzhanovsky, G. N., 1980a, Generator mechanisms of central pain syndromes and analgesia, Vestnik Akad.Med.Nauk SSSR, No. 9, 33-38.

Kryzhanovsky, G. N., 1980b, Some principles of diagnosis and treatment of neuropathologic syndromes, in: "Theoretical Principles of Pathologic States," (in Russian) N. P. Bekhtereva, Izd.Nauka, Leningrad, pp. 85-100.

Kryzhanovsky, G. N., 1981a, Pathophysiology of Tetanus, in: "Tetanus. Important new Concepts," R. Veronesi, ed., Excerpta Medica, Amsterdam-Oxford-Princeton, pp. 109-182.

Kryzhanovsky, G. N., 1981b, Behavioral changes in animals on creation of hyperactive determinant structures in some portions of the central nervous system, in: "Psychophysiology Today and Tomorrow," N. P. Bechtereva, Pergamon Press, London, pp. 237-247.

Kryzhanovsky, G. N., 1981c, General pathology of the nervous system (current status and prospects), Vestnik Akad.Med.Nauk SSR, No. 8, 4-14.

Kryzhanovsky, G. N., 1981d, System relations in neuropathology and psychiatry (some theoretical aspects), Zh.nevropatol.i psikhiatr., 81, No. 7, 961-1969.

Kryzhanovsky, G. N., 1981e, System relationships in neuropathology and psychiatry (some problems of therapy), Zh.nevropatol.i psikhiatr., 81, No. 8, 1121-1129.

Kryzhanovsky, G. N., and Aliev, M. N., 1976a, Experimental neuropathologic syndromes in the creation of hyperactivity determinant dispatch stations in the caudate nuclei, Byull.eksp. biol. i med., 81, No. 4, 397-399.

Kryzhanovsky, G. N., and Aliev, M. N., 1976b, Experimental Parkinsonism, IRCS Medical Science Nervous System, 4, 272.

Kryzhanovsky, G. N., and Aliev, M. N., 1978, Experimental Parkinson's syndrome, in: "Pathogenesis, Clinical Practice and Treatment of Parkinsonism," (in Russian), G. V. Morozov, ed., Izd.Meditsina, Moscow, pp. 26-29.

Kryzhanovsky, G. N., and Aliev, M. N., 1979, Pathogenesis of stereotyped behavior, Zh.nevropat.i psikhiatr., 79, No. 9, 1347-1355.

Kryzhanovsky, G. N., and Aliev, M. N., 1981, The stereotyped behavior syndrome: a new model and proposed therapy, Pharmacol. Biochem.Behav., 14, No. 3, p. 273-281.

Kryzhanovsky, G. N., and Grafova, V. N., 1972, Spinal pain syndrome, Pathol.Physiol., No. 4, 46-51.

Kryzhanovsky, G. N., and Igonkina, S. I., 1976, Experimental pain and itch syndromes of thalamic origin, Byull.eksp.biol.i med., 81, No. 6, 651-653.

Kryzhanovsky, G. N., and Dyakonova, M. V., 1964, Changes in the throughput capacity of the spinal cord efferent output in tetanus intoxication, Byull.eksp.biol. i med., 58, No. 9, 12-16.

Kryzhanovsky, G. N., and Igonkina, S. I., 1978, Analgesia caused by the creation of an excitation generator in the midbrain, Byull.eksp.biol.i med, No. 2, 145-148.

Kryzhanovsky, G. N., and Lutsenko, V. K., 1975, The intercentral relations in rat spinal cord with local depression of the inhibitory process, Neurofiziologia (Kiev), 7, No. 3, 234-242.

Kryzhanovsky, G. N., and Pivovarov, Yu. I., 1982, Alterations in heart rhythm during hyperactivation of the anterior amygdaline nucleus, Byull.eksp.biol. i med., 93, No. 5, 26-29.

Kryzhanovsky, G. N., and Russev, V. V., 1976, The effect of injuries of the medial forebrain bundle and the preoptic region on the strychnine-induced epileptiform focus activity (concerning the phenomenon of the hyperactive determinant dispatch station), Byull.eksp.biol.i med., 81, No. 10, 1155-1158.

Kryzhanovsky, G. N., and Shandra, A. A., 1981, The use of nicotinamide and pyridoxal-5-phosphate for eliminating experimental epilepsy, Zh.nevropathol.i psikhiatr., 81, 6, 801-809.

Kryzhanovsky, G. N., and Sheikhon, F. D., 1968, Inhibitory and facilitatory influences from medulla oblongata in tetanus intoxication, Byull.eksp.biol.i med., 66, No. 11, 9-14.

Kryzhanovsky, G. N., and Sheikhon, F. D., 1973, Descending supraspinal effects in tetanus toxin intoxication of the spinal cord, Exp.Neurol., 38, No. 1, 110-121.

Kryzhanovsky, G. N., and Sheikhon, F. D., 1976, Descending supraspinal effects under conditions of disturbance of the inhibitory process in the medulla. The formation of the generators of excitation, Exp.Neurol., 50, No. 2, 387-401.

Kryzhanovsky, G. N., Pevnitsky, L. A., Grafova, V. N., and Polgar, A. A., 1961, Routes of tetanus toxin entrance into the central nervous system and some problems of experimental tetanus pathogenesis, Byull.exp.biol.i med., No. 3, 42-49; No. 8, 31-37; No. 12, 30-35; 35-43.

Kryzhanovsky, G. N., Grafova, V. N., Danilova, Ye. N., Igonkina, S. I., and Sakharova, O. P., 1973a, Pain syndrome of spinal origin, Byull.eksp.biol.i med., 76, No. 9, 31-35.

Kryzhanovsky, G. N., Kurchavy, G. G., and Sheikhon, F. D., 1973b, Supraspinal effect on motor neurons in local tetanus, Byull. eksp.biol.i med., 75, No. 4, 36-39.

Kryzhanovsky, G. N., Sheikhon, F. D., and Igonkina, S. I., 1973c, Temporary course of the descending inhibitory monosynaptic and polysynaptic reflexes in local tetanus, Byull.eksp. biol.i med., 76, No. 11, 35-38.

Kryzhanovsky, G. N., Grafova, V. N., Danilova, E. I., and Igonkina, S. I., 1974a, Investigation of the pain syndrome of spinal origin (contributions to the concept of the generator mechanism of the pain syndrome), Byull.eksp.biol.i med., 78, No. 7, 15-20.

Kryzhanovsky, G. N., Igonkina, S. I., Grafova, V. N., and Danilova, E. I., 1974b, Experimental trigeminal neuralgia (on the conception of the generator mechanism of pain syndrome, Byull.eksp.biol.i med., 78, No. 11, 16-20.

Kryzhanovsky, G. N., Pozdnyakov, O. M. and Polgar, A. A., 1974c, "Pathology of the Synaptic Apparatus of the Muscle," (in Russian), Izd. Meditsina, Moscow.

Kryzhanovsky, G. N., Makulkin, R. F., and Shandra, A. A., 1976a, Mechanisms transmitting excitation generated in the orbito-frontal cortex, Byull.eksp.biol.i med., 81, No. 5, 522-525.

Kryzhanovsky, G. N., Rekhtman, M. B., Konnikov, B. A., and Petluk, V. Ch., 1976b, Photogenic epilepsy under conditions of localization of the excitation generator in the lateral geniculate body (on the phenomenon of determinative dispatch station), Byull.eksp.biol.i med., 81, No. 1, 22-25.

Kryzhanovsky, G. N., Rekhtman, M. B., Konnikov, B. A., and Sheikhon, F. D., 1976c, Experimental vestibulopathy as a result of localization of the generator of pathologically enhanced excitation in the vestibular nucleus (the phenomenon of "determinative dispatch station"), Byull.eksp.biol.i med., 81, No. 2, 147-150.

Kryzhanovsky, G. N., Grafova, V. N., and Danilova, Ye. I., 1977a, Generators of pathologically enhanced excitation as determinant structures in the spinal cord, Byull.eksp.biol.i med., 83, No. 5, 515-519.

Kryzhanovsky, G. N., Kotov, A. V., Kuligina, O. A., Tolpigo, S. M., and Sudakov, K. V., 1977b, The modelling of the neuropathological syndrome by formation of long-term generators of pathologically enhanced excitation in the hypothalamus of rabbits, Byull.eksp.biol.i med., 84, No. 10, 405-408.

Kryzhanovsky, G. N., Makulkin, R. F., and Shandra, A. A., 1977c, The role of the hyperactive determinant structures in the generation of functional complexes of epileptic activity in the cerebral cortex, Byull.eksp.biol. i med., 83, No. 1, 5-10.

Kryzhanovsky, G. N., Makulkin, R. F., and Goon, A. A., 1978a, Effect of the generator of pathologically enhanced excitation in the orbital cortex on the sleep-wakefulness cycle, Zh.vyssh. nerv.deyat., 28, 4, 782-791.

Kryzhanovsky, G. N., Makulkin, R. F., Shandra, A. A., and Boiko, D. V., 1978b, Mechanism of formation of epileptic activity complexes in the cerebral cortex under influence of the determinant focus, Byull.eksp.biol.i med., 86, No. 7, 14-19.

Kryzhanovsky, G. N., Makulkin, P. F., and Shandra, A. A., 1978c, The principle of determinant and the formation of epileptic activity complexes, Zh.nevropatol. i psikhiatr., 78, 4, 547-556.

Kryzhanovsky, G. N., Okujava, V. M., Rekhtman, M. B., and Mzhavia, I. A., 1978d, Neuronal activity in the epileptic focus created by the injection of tetanus toxin into the cat motor cortex, Neurofiziologia (Kiev), 10, No. 6, 582-590.

Kryzhanovsky, G. N., Rekhtman, M. B., and Konnikov, B. A., 1978e, Changes in the functional organization of the lateral geniculate body after tetanus toxin injection, Neurofiziologia (Kiev), 10, No. 1, 38-43.

Kryzhanovsky, G. N., Konnikov, B. A., and Rekhtman, M. B., 1979a, Concerning the problem of focal pathology in 'genuine' epilepsy, Zh.nevropatol.i psikhiatr., 79, No. 6, 720-725.

Kryzhanovsky, G. N., Makulkin, R. F., and Shandra, A. A., 1979b, Formation of epileptic activity complexes under the influence of the determinant focus in cortex isole, Byull.eksp. biol. i med., 87, No. 10, 408-411.

Kryzhanovsky, G. N., Makulkin, R. F., Shandra, A. A., and Okhtishkin, N. E., 1979c, Formation of epileptic activity in the brain cortex under the effect of the determinant focus caused by acetylcholine, Byull.eksp.biol.i med., 87, No. 2, 117-122.

Kryzhanovsky, G. N., Khomulo, P. S., Kotov, A. B., and Timofeeva, O. A., 1980a, Food hypermotivation syndromes in hyperactivity of lateral hypothalamic structures in rabbits, Byull.eksp. bio.i med., 89, No. 3, 281-284.

Kryzhanovsky, G. N., Makulkin, R. F., and Shandra, A. A., 1980b, Formation of determinant structures and functional complexes in the neocortex under the conditions of brain section at different levels, Fiziol.Zh.SSR, 66, No. 6, 791-800.

Kryzhanovsky, G. N., Makulkin, R. F., Shandra, A. A., and Lobasyuk, B. A., 1980c, The influence of caudal reticular pontis nucleus electrical stimulation on foci of epileptic activity in the brain cortex, Byull.eksp.biol.i med., 90, No. 11, 533-536.

Kryzhanovsky, G. N., Makulkin, R. F., Shandra, A. A., Lobasyuk, B. A., and Godlevsky, L. S., 1980d, The nicotinamide influence on epileptic activity in the brain cortex, Byull.exp.biol.i med., 90, No. 7, 37-41.

Kryzhanovsky, G. N., Makulkin, R. F., Shandra, A. A., Lobasyuk, B. A., and Lebedyuk, M. N., 1980e, Effect of diazepam on the epileptic complex in the brain cortex at different levels of its neuronal isolation, Byull.eksp.biol.i med., 90, No. 9, 281-286.

Kryzhanovsky, G. N., Nikushkin, E. V., Braslavsky, V. E., and Glebov, R. N., 1980f, Lipoperoxidation in the hyperactive focus of rat cerebral cortex, Bull.Exp.Biol., 89, No. 6, 14-16.

Kryzhanovsky, G. N., Rodina, V. I., Glebov, R. N., and Bazyan, A.
    S., 1980g, Effect of tetanus toxin on noradrenaline release
    from rat brain synaptosomes, Byull.eksp.biol.i med., 89, No.
    2, 148-150.
Kryzhanovsky, G. N., Torshin, V. I., and Rekhtman, M. B., 1980h
    Effect of stimulation of the mesencephalic reticular form-
    ation on the epileptic activity of rats with experimental
    photogenic epilepsy, Byull.eksp.biol.i med., 90, No. 10,
    409-414.
Kryzhanovsky, G. N., Grafova, V. N., and Danilova, E. I., 1981a,
    Detection of latent postural asymmetry by a synthetic hexa-
    peptide, Byull.eksp. biol.i med., 92, No. 7, 38-42.
Kryzhanovsky, G. N., Grafova, V. N., and Danilova, E. I., 1981b,
    Effect of $ACTH_{4-7}$ and lysine-vasopressin on the activity of
    the generator of pathologically enhanced excitation, Byull.
    eksp.biol. i med., 92, No. 8, 14-17.
Kryzhanovsky, G. N., Makulkin, R. F., and Lobasyuk, B. A., 1981c,
    Effect of paleocortex stimulation on multifocal epileptic
    complex of brain cortex, Byull.eksp.biol.i med., 92, No. 10,
    398-399.
Kryzhanovsky, G. N., Polgar, A. A., Zinkevich, V. A., and Smirnova,
    V. S., 1981d, Effect of tetanus toxin on the mechanisms by
    which calcium ions regulate transmitter secretion in the
    neuromuscular junction, Byull.eksp.biol.i med., 92, No. 12,
    648-650.
Kryzhanovsky, G. N., Shandra, A. A., and Godlevsky, L. S., 1981e,
    Influence of pyridoxal-5-phosphate on epileptic activity in
    the brain cortex, Farmacol.i toksikol., 1, 136-140.
Kryzhanovsky, G. N., Shandra, A. A., Godlevsky, L. S., and Belyaeva,
    A. I., 1981f, Further investigation of the nicotinamide
    antiepileptic action, Byull.eksp.biol.i med., 91, No. 1,
    42-45.
Kryzhanovsky, G. N., Grafova, V. N., and Danilova, E. I., 1982a,
    effects of phenazepam in some experimental neuropathologic
    syndromes, in: "Phenazepam," (in Russian), Izd. Naukova
    Dumka, Kiev, pp. 181-188.
Kryzhanovsky, G. N., Lutsenko, V. K., Sakharova, O. P., and
    Lutsenko, N. G., 1982b, Tetanus toxin disturbance of ^{3}H-GABA
    transport through synaptosomes, Byull.eksp.biol.i med., 94,
    No. 7, 49-51.
Kryzhanovsky, G. N., Makulkin, R. F., and Shandra, A. A., 1982c,
    Interaction of excitatory foci and their complexes in the
    cerebral cortex. Zh.vish.nerv.deyat., 3, 487-495.
Kryzhanovsky, G. N., Shandra, A. A., and Nikushkin, E. V., 1982d,
    Antiepileptic effects of combined therapy with vitamins and
    anticonvulsants (experimental data), Zh.nevropatol.i
    psikhiatr., 82, No. 6, 84-90.
Kryzhanovsky, G. N., Makulkin, R. F., Shandra, A. A., and Godlevsky,
    L. S., 1983a, The influence of the electrical stimulation of
    the cerebellar nucleus dentatus on epileptic foci in the
    brain cortex, Byull.eksp.biol.i med., No. 3, 26-29.

Kryzhanovsky, G. N., Ponomarchuk, V. S., and Russev, V. V., 1983b,
    Ophthalmotonus regulation disturbance in forming the
    generator of pathologically enhanced excitation in the
    hippocampus, Byull.eksp.biol.i med., 95, No. 1, 14-17.
Kugelberg, E., and Lingblom, U., 1959, The mechanism of the pain in
    trigeminal neuralgia, J.Neurol.Neurosurg.Psychiatry, 22,
    36-43.
Kumazawa, T., Perl, E. R., Burgess, P. R., and Whitehorn, D., 1975,
    Ascending projections from marginal zone (lamina 1) neurons
    of the spinal dorsal horn, J.Comp.Neurol., 162, 1-12.
Kuznetsova, G. D., and Koroleva, V. I., 1978, "Stationary Excitation
    Foci in the Cerebral Cortex," (in Russian), Izd. Nauka, Moscow.
Ladinsky, H., Consolo, S., Bianchi, S., Samahin, R., and Ghezzi, D.,
    1975, Cholinergic-dopaminergic interaction in the striatum:
    the affects of 6-hydroxydopamine on pimozide treatment on
    the increased striatal acetylcholine levels by apomorphine,
    piribedil and d-amphetamine, Brain Res., 84, No. 1, p.
    221-226.
La Grutta, V., Amato, G., and Zagami, M. T., 1971, The importance of
    the caudate nucleus in the control of convulsive activity in
    the amygdaloid cortex and the temporal cortex of the cat,
    Electroenceph.Clin.Neurophysiol., 31, 57-69.
Lal, S., de la Vega, C. E., and Garelis, E. A., 1973, Apomorphine,
    pimozide, L-dopa and the probenecid test in Huntington's
    chorea, Psych.Neurol.Neurochir., 76, No. 1, p. 113-117.
Latash, L. P., 1968, "Hypothalamus, Adaptive Activity and Electro-
    encephalogram," (in Russian), Izd. Nauka, Moscow.
Latash, L. P., 1978, Sleep as a state and as a process, in:
    "Functional State," (in Russian), Ye. N. Sokolov, and N. N.
    Danilova, eds., Moscow State University Publishers, Moscow,
    pp. 38-40.
Lavy, S., Marks, E. S., and Abramsky, O., 1974, Parkinsonism and
    hyperthyroidism, Europ.Neurol., 12, No. 1, 20-27.
Lee, K. S., and Klaus, W., 1971, The subcellular basis for the
    mechanism of inotropic action of cardial glycosides, Pharm-
    acol.Res., 23, 193-261.
Lee, T., Seeman, P., Tourtellotte, W. W., Farley, I. J., and
    Hornykiewicz, O., 1978, Binding of ^{3}H-neuroleptics and
    ^{3}H-apomorphine in schizophrenic brains, Nature, 274, 897-900.
Lee, H. K., Dunwiddle, T. V., and Hoffer, B. J., 1979, Interaction
    of diazepam with synaptic transmission in the in vitro rat
    hippocampus, Naunyn-Schmiedeberg's Arch.Pharmacol., 309,
    131-136.
Lemaine, G., Clemencon, M., Gomis, A., Pollin, B., and Salvo, B.,
    1977, Strategies et choix dans la recherche a propos des
    travaux sur le sommeil, Mouton, Hage-Paris.
Leontovich, T. A., 1978, "Neuronal Structure of the Subcortical
    Formations of the Forebrain," (in Russian), Izd. Meditsina,
    Moscow.

Leontovich. T. A., 1980, On the morphofunctional mechanisms of the regulation of impulse conduction and the formation of the functional neuronal systems in CNS, Uspekhi fiziol. nauk, 11, No. 3, 64–84.

Leriche, R., 1939, La Chirurgie de la Douleur, Masson & Cie, Paris.

Levitt, M., and Heybach, J. P., 1981, The deafferentation syndrome in genetically blind rats: a model of the painful phantom limb, Pain, 10, 67–73.

Levitt, M., and Levitt, J. H., 1981, The deafferentation syndrome in monkeys: dysanesthesias of spinal origin, Pain, 10, No. 2, 129–149.

Lewis, T., Pickering, G. W., and Rothshild, P., 1931, Centripetal paralysis arising out of arrested bloodflow to the limb, including notes on a form of tingling, Heart, 16, 1.

Li, C. L., and Elvidge, A. R., 1951, Observations on phantom limb in a paraplegic patient, J.Neurosurg., 8, 524–527.

Liebeskind, J. C., 1976, Pain modulation by central nervous system stimulation, Adv.Pain Res.Ther., 1, 445–453.

Liles, S. O., 1974, Single unit responses of caudate neurons to stimulation of frontal cortex, substantia nigra and entopeduncular nucleus in the cats, J.Neurophysiol., 37, No. 1, 254–265.

Lindeman, R. R., Zlobina, G. P., and Mukhin, A. G., 1980, Problem of neuronal reception and dopamine hypothesis of schizophrenia (review), Zh.nevropatol.i psikhiatr., 80, 762–769.

Lindsley, D. B., 1950, Brainstem influences on spinal motor activity, A.Res.Nerv.Ment.Dis.Proc., 30, No. 1952, 174–195.

Lineberry, C. G., and Vierck, C. J., 1975, Attenuation of pain reactivity by caudate nucleus stimulation in monkeys, Brain Res., 98, 110–134.

Livanov, M. N., 1965, Neurocybernetics, in: "Problems of Modern Neurophysiology," (in Russian), Izd.Nauka, Moscow-Leningrad, pp. 37–72.

Livanov, M. N., 1975, Neuronal mechanisms of memory, Uspekhi fiziol. nauk, 6, 66–89.

Livingston, W. K., 1943, Pain Mechanisms. A physiologic interpretation of causalgia and its related states, Macmillan, New York.

Livingston, W. K., 1948, The vicious circle in causalgia, Ann.N.Y. Acad.Sci., 50, 247.

Llinas, R., and Hess, R., 1976, Tetrodotoxin-resistant dendritic spikes in avian Purkinje cells, Proc.Nat.Acad.Sci., 73, 2520–2523.

Lloyd, D. P. C., 1943, Conduction and synaptic transmission of the reflex responses to stretch in cat, J.Neurophysiol., 6, 317–326.

Lloyd, K. G., and Hornykiewicz, O., 1973, L-glutamic acid decarboxylase in Parkinson's disease: effect of L-dopa therapy, Nature, 243, No. 5409, 521–523.

Lockard, J. S., 1980, A primate model of clinical epilepsy: mechanisms of action through quantification of therapeutic effects, in: "Epilepsy: A Window to Brain Mechanisms," J. S. Lockard and A. A. Ward, eds., Raven Press, New York, pp. 11-49.

Lockard, J. S. Ojemann, G. A., Congdon, W. C., and DuCharme, L. L., 1979, Cerebellar stimulation in aluminia-gel monkey model: increase relationship between clinical seizures and EEG interictal bursts, Epilepsia, 20, 223-234.

Lodge, D., and Curbis, D. R., 1978, Time course of GABA and glycine actions on cat spinal neurones: effect of pentobartbitone, Neuroscience Letters, 8, 125-129.

Loeser, J. D., and Ward, A. A. Jr., 1967, Some effects of deafferentation on neurons of the cat spinal cord, Arch.Neurol., 17, 629.

Loeser, J. D., and Howe, J. F., 1980, Deafferentation and neuronal injury, in: "Epilepsy: A Window to Brain Mechanisms," J. S. Lockard and A. A. Ward, eds., Raven Press, new York, pp. 123-135.

Loeser, J. D., Ward, A. A., and White, L. E., 1968, Chronic deafferentation of human spinal cord neurons, J.Neurosurg., 29, 48-50.

Loh, L., and Nathan, P. W., 1978, Painful peripheral states and sympathetic blocks, J.Neurol.Neurosurg.Psychiatry, 41, 664-671.

Lo Mathew, M. S., Strittmatter, S. M., and Snyder, S., 1982, Physical separation and characterization of two types of benzodiazepine receptors, Proc.Nat.Acad.Sci.USA, Biol.Sci., 79, No. 2, 680-684.

Lombard, M. C., Washold, B. S. Jr., and Albe-Fessard, D., 1979a, Deafferentation hypersensitivity in the rat after rhizotomy. A possible animal model for chronic pain, Pain, 6, 163.

Lombard, M. C., Washold, B. S., and Pelissier, T., 1979b, Thalamic recordings in rats with hyperalgesia, in: "Advances in Pain Research and Therapy," J. J. Bonica, ed., vol. 3, Raven Press, New York, pp. 767-772.

Lombard, M. C., Larabi, Y., and Albe-Fessard, D., 1981, Electrophysiological study of cervical dorsal horn cells in partially deafferented rats, Pain, 10, Suppl. 1, 129.

Lorente de Nó, R., 1933, Vestibulo-ocular reflex, Arch.Neurol.and Psychiatry, 30. No. 2, 245-291.

Lorente de Nó, R., 1938, Analysis of the activity of the chains of internuncial neurons, J.Neurophysiol., 1, No. 3, 207-244.

Lothman, E. W., and Somjen, G. G., 1976, Functions of primary afferents and responses of extracellular K^+ during spinal epileptiform seizures, Electroencephalogr.Clin.Neurophysiol., 41, 253-267.

Lott, I. T., Conlombe, Th., Di Paolo, R. V., Richardson, E., and Levy, N. L., 1978, Vitamin B_6-dependent seizures: pathology and chemical findings in brain, Neurology, 28, No. 1, 47-54.

Ludwig, B., Ajmeme-Marsan, C. A., and Van Buren, J., 1975, Cerebral
     seizures of probable orbitofrontal origin, Epilepsia, 16,
     No. 1, 141-158.
Leuders, H., Bustamante, L., Krinsky, A., and Goldensohn, E. S.,
     1980, Quantitative studies of spike foci induced by minimal
     concentrations of penicillin, Electroenceph.Clin.Neurophys-
     iol., 48, 80-89.
Lutsenko, V. K., and Kryzhanovsky, G. N., 1973, Mechanisms of the
     pathologically enhanced scratch reflex in animals with
     experimental tetanus, Byull.eksp.biol.i med., 76, No. 8,
     20-24.
Lutsenko, V. K., and Kryzhanovsky, G. N., 1975, Electrical activity
     of the spinal cord with local depression of the inhibitory
     process, Neurophysiology, 7, No. 5, 509-518.
Lutsenko, V. K., Kryzhanovsky, G. N., 1982, Sakharova, O. P.,
     Rebrov, G. I., and Barshevskaya, T. N., Effect of tetanus
     toxin on the content of potasium and sodium in synaptosomes,
     Byull.eksp.biol.i med., 94, No. 9, 21-24.
Magni, F., and Willis, S. W., 1964, Subcortical and peripheral
     control of brain stem reticular neurons, Arch.Ital.Biol.,
     102, 434-448.
Magoun, H. W., 1950, Caudal and cephalic influences on the brainstem
     reticular formation, Physiol.Rev., 30, No. 4, 459-474.
Magoun, H. W., 1965, Wakeful Brain (Russian translation), Izd.Mir,
     Moscow, p. 210.
Magnus, R., 1924, Körperstellung, Springer, Berlin.
Maisov, N. I., Tolmacheva, N. S., and Raevsky, K. S., 1975, Influ-
     ence of psychotropic agents on the uptake of H^1-gamma-amino-
     butyric acid by synaptosomes of the rat brain, Farmakol.
     toksikol., 38, No. 5, 537-540.
Maisov, N. I., Sandalov, Yu. G., Glebob, R. N., and Raevsky, K. S.,
     1976, The effect of psychotropic drugs upon the ^{3}H-GABA
     uptake by synaptosomes and the Na, K, ATPase activity,
     byull.eksp.biol.i med., 81, No. 1, 45-47.
Maiti, A., and Snider, R. S., 1975, Cerebellar control of basal
     forebrain seizures: Amygdala and hippocampus, Epilepsia, No.
     6, 521-533.
Majkowski, J., 1980, Cerebellum and epileptic activity of cerebral
     hemisphere, in: "Neurophysiological Mechanisms of Epilepsy,"
     Metsniereba, Tbilisi, pp. 156-165.
Majkowski, J., Karlinski, A., and Klimovicz, I., 1980, Effect of
     cerebellar stimulation in hippoocampal epileptic discharges
     in kindling preparation, in: "Epilepsy. A Clinical and
     Experimental Research. Proceedings of the Second European
     Regional Conference on Epilepsy," Warsaw, October 5-7, 1978,
     J. Majkowski, pp. 40-45.
Makulkin, R. F., Shandra, A. A., and Boiko, D. N., 1978, Functional
     interrelations between the determinant and other epileptoid
     activity foci created in the cortex of both hemispheres,
     Byull.eksp.biol.i med., 86, No. 8, 142-146.

Mancia, M., Mariotti, M., and Spreafico, R., 1975, Interaction between structures which induce sleep and wakefulness in the brain stem, in: "Sleep 1974," 2nd Europ. Congr. Sleep Res., Rome 1974, Karger, Basel, pp. 267-269.

Markham, C. H., Precht, W., and Shimazu, H., 1966, Effect of stimulation of interstitial nucleus of Cajal on vestibular unit activity in the cat, J.Neurophysiol., 29, 493-507.

Martin, I., and Candy, J., 1978, Facilitation of benzodiazepine binding by sodium chloride and GABA, Neuropharmacology, 17, No. 11, 993-998.

Martres, M. P., Costentin, J., Baudry, M., Marcais, H., Protais, P., and Schwartz, J. C., 1977, Long-term changes in the sensitivity of pre-and postsynaptic dopamine receptors in mouse striatum evidenced by behavioral and biochemical study, Brain Res., 136, No. 2, 319-337.

Maruyama, S., and Kawasaki, T., 1975, Synergism between γ-aminobutyric acid and butyrophenones administered microelectrophoretically in the Purkinje cells of the cat cerebellum, Jap.J.Pharmacol., 25, No. 2, 209-213.

Matlina, E. Sh., Pukhova, G. S., Grafova, V. N., Kassil, G. N., and Kryzhanovsky, G. N., 1973, Effect of DOPA on metabolism of catecholamines in spinal pain syndrome, Vopr.med.khimii, 19, No. 2, 173-176.

Matsumoto, H., and Ajmone-Marsan, C., 1964, Cortical cellular phenomena in experimental epilepsy: interictal manifestations, Exp.Neurol., 9, 286-304.

Matsura, T., and Bures, J., 1971, The minimum volume of depolarized neural tissue required for triggering cortical spreading depression in rat, Exp.Brain Res., No. 12, 238-243.

Matsumoto, H., Ayala, G. F., and Gumnit, R. G., 1969, Neuronal behaviro and triggering mechanisms in cortical epileptic focus, J.Neurophysiol., 32, 688-703.

Matthies, H., Pohle, W., Popov, N., Lössner, B., Ruthrich, H.-L., Jork, R., and Ott, T., 1978, Biochemical mechanisms correlated with learning and memory formation - facts and hypotheses, in: "Neural and Neurohumoral Organization of Motivated Behavior," K. Lissak, ed., Akad. Kiado, Budapest, pp. 85-105.

Matthysse, S., 1974, Schizophrenia: relationships to dopamine transmission. Motor control and feature extraction, in: "The Neurosciences. 3-Study Programme," F. O. Smitt and F. O. Worden, eds., MIT Press, Cambridge, Mass.-London, pp. 733-737.

Mayer, D. J., 1979, Endogenous analgesia systems: neural and behavioral mechanisms, in: "Advances in Pain Research and Therapy," J. J. Bonica, eds., J. C. Liebeskind, and D. G. Albe-Fessard, vol. 3, Raven Press, New York, pp. 385-410.

Mayer, D. J., and Liebeskind, J. C., 1974, Pain reduction by focal electrical stimulation of the brain: an anatomical and behavioral analysis, Brain Res., 68, 73-93.

Mayer, D. J., and Price, D. D., 1976, Central nervous system mechanisms of analgesia, Pain, 2, 379–404.

Mayer, D. J., 1971, Wolfle, T. L., Akil, H., Carder, B., and Liebeskind, J. C., 1971, Analgesia from electrical stimulation in the brainstem of the rat, Science, 174, 1351–1354.

Mayer, D. J., Price, D. D., and Becker, D. P., 1975, Neurophysiological characterization of the anterolateral spinal cord neurons contributing to pain perception in man, Pain, 1, 59–72.

Mayer, D. J., Price, D. D., Barber, J., and Raffii, A., 1976, Acupuncture analgesia: evidence for activation of a pain inhibitory system as a mechanism of action, Advances in Pain Res. and Therapy, 1, 751–755.

Mayer, D. J., Price, D. D., and Rafii, A., 1977, Antagonism of acupuncture analgesia in man by the narcotic antagonist naloxone, Brain Res., 360–373.

Mazars, G. J., Merienne, L., and Cioloca, C., 1976, Contribution of thalamic stimulation to the physiopathology of pain, Adv.in Pain Res.and Therapy, 1, 483–485.

Mazzari, S., Massotti, M., Guidotti, A., and Costa, E., 1981, GABA receptors as supramolecular units, in: "GABA and Benzodiazepine Receptor," E. Costa et al., ed., Raven Press, New York, pp. 1–8.

Mazzuchelli-O'Flacherty, A., O'Flacherty, Z., and Hernandez-Peon, R., 1967, Sleep and other behavioral responses, induced by acetylcholinic stimulation of frontal and medial cortex, Brain Res., 4, 268–283.

McCarthy, P. S., Walker, K. J., and Woodruff, G. V., 1977, On the depressant action of dopamine in rat caudate nucleus and nucleus accumbens, Br.J.Pharmacol., 59, 469–470.

McGeer, E. G., and McGeer, P. L., 1976, Duplication of biochemical changes of Huntington's chorea by intrastriatal injections of glutamic and cainic acid, Nature, 263, No. 5554, 517–519.

McGeer, P. L., McGeer, E. G., and Wada, J. A., 1971, Glutamic acid decarboxylase in Parkinson's disease and epilepsy, Neurology, 21, No. 10, 1000–1007.

McGeer, P. L., McGeer, E. G., and Fibiger, H. C., 1973, Choline acetylase and glutamic acid decarboxylase in Huntington's chorea, Neurology, 23, No.5, 912–917.

McKhann, G. M., Alberts, R. W., Sokoloff, T., Mickelsen, O., and Tower, D. B., 1960, The quantitative significance of the gamma-aminobutyric acid pathway in cerebral oxidative metabolism, in: "Inhibition in the Nervous System and gamma-Aminobutyric Acid," Roberts, ed., Pergamon, New York, p. 169.

McLean, P. D., 1966, The limbic and visual cortex in phylogeny: further investigation from anatomic microelectrode studies, in: "Evolution of the Forebrain," Thieme, Stuttgart, p. 443.

McLennan, H., and York, D. H., 1967, The action of dopamine on neurons of the caudate nucleus, J.Physiol., 189, 393–402.

Meech, R. W., and Standen, N. B., 1975, Potassium activation in
    Helix aspersa neurones under voltage clamp: a component
    mediated by calcium influx, J.Physiol., (London), 249,
    211-259.
Mehes, J., 1938, Experimentelle Untersuchungen über den Juckreflex
    am Tier; Auslösen heftiger Juckanfälle bei der Katze durch
    intracisternale Injektionen von Morphium und einiger seiner
    Derivate, Arch.Exp.Path.Pharmak., 188, 650-656.
Mellanby, J., and Green, J, 1981, How does tetanus toxin act?,
    Neuroscience, 6, No. 3, 281-300.
Mellanby, J., and Thompson, P. A., 1972, The effect of tetanus toxin
    at the neuromuscular junction in the goldfish, J.Physiol.
    (Lond.), 224, 407.
Meller, E., and Friedman, E., 1981, Lithium dissociates
    haloperidol-induced behavioral supersensitivity from reduced
    DOPAC increase in rat striatum, Europ.J.Pharmacol., 76,
    25-29.
Meltzer, H. Y., and Stahl, M., 1976, The dopamine hypothesis of
    schizophrenia: a review, Schizophrenia Bull. 2, 19-76.
Meltzer, H. Y., Sachar, E. J., and Frantz, A. G., 1974, Serum
    prolactin levels in unmedicated schizophrenic patients,
    Arch.Gen.Psychiatr., 31, 564-569.
Melzack, R., 1971, Phantom limb pain: implications for treatment of
    pathological pain, Anaesthesiology, 35, 409.
Melzack, R., 1973, "The Puzzle of Pain," Basic Books, Inc., New
    York.
Melzack, R., and Loeser, J. D., 1978, Phantom body pain in parap-
    legics: evidence for a central 'pattern generating mechan-
    ism' for pain, Pain, 4, 195-210.
Melzack, R., and Melinkoff, D. F., 1974, Analgesia produced by brain
    stimulation. Evidence for prolonged onset period, Exptl.
    Neurol., 43, 369-374.
Melzack, R., and Torgerson, W. S., 1971, On the language of pain,
    Anesthesiology, 34, 50.
Melzack, R., and Wall, P. D., 1965, Pain mechanisms: a new theory,
    Science, 150, 971-979.
Melzack, R., and Wall, P. D., 1970, Psychophysiology of pain,
    Internat.Anesthesiol.Clinics, 8, 3.
Melzack, R., Stillwell, D. M., and Fox, E. J., 1977, Trigger points
    and acupuncture points for pain: correlations and impli-
    cations, Pain, 3, No. 1, 3-25.
Mendell, L. M., 1966, Physiological properties of unmyelinated fiber
    projections to the spinal cord, Exp.Neurol., 16, 316-332.
Menon, M. K., Fleming, R. M., and Clark, W. G., 1974, Studies of the
    biochemical mechanisms of central effects of gamma-hydroxy-
    butyric acid, Biochemical Pharmacol., 23, 875-879.
Merril, E. G., and Wall, P. D., 1978, Plasticity of connection in
    the adult nervous system, in: "Neuronal Plasticity," Carl W.
    Cotman, ed., Raven Press, new York, pp. 97-111.
Meyer, H., and Ransom, F., 1903, Untersuchungen über den Tetanus,
    Arch.Exp.Pathol.und Pharmacol., 49, No. 6, 369-416.

Mikheev, V. V., and Rubin, L. R., 1966, "Somatoneurological Syndromes," (in Russian), Izd. Meditsina, Moscow, pp. 5-83.

Miller, N. E., 1969, Learning of visceral and glandular responses, Science, 163, 434-445.

Miller, N. E., 1971, Selected Papers, Aldine-Atherton, Chicago-New York.

Miller, W. E., 1972, Interactions between learned and physical factors in mental illness, Semin.Psychiatry, No. 4, 239-254.

Miller, E., 1974, Deanol: a solution for tardive dyskinesia, New Engl.J.Med., 291, 796-797.

Miller, R. J., and Hiley, C. R., 1974, Antimuscarinic properties of neuroleptics and drug-induced Parkinsonism, Nature, 248, 596-597.

Miller, R. J., Horn, A. S., and Iversen, L. L., 1974, The action of neuroleptic drugs on dopamine-stimulated adenosine cyclic 3',5'-monophosphate production in neostriatum and limbic forebrain, Mol.Pharmacol., 10, 759-766.

Möhler, H., and Ovada, T., 1977, Benzodiazepine receptor: demonstration in the central nervous system, Science, 198, No. 43(9), 849-851.

Möhler, H., Polc, P., Cumin, R., and Pieri, L., 1979, Nicotinamide is a brain constituent with benzodiazepine-like actions, Nature, 278, 563-565.

Mohrland, J. S., McManus, D. O., and Gebhart, G. F., 1982, Lesions of nucleus reticularis gigantocellularis: effect on the autinociception produced by microinjection of morphine and focal electrical stimulation in the periaqueductal gray matter, Brain Res., 231, 143-152.

Monnier, M., 1963, Moderating brainstem system inducing synchronization of the neocortex and sleep, Electroencephalog.Clin.Neurophysiol., Suppl., 97-112.

Monnier, M., 1980, Biology of sleep. An interdisciplinary survey. 6. Comparative electrophysiology of sleep in some vertebrates, Experientia, 36, No. 1, 16-19.

Monroe, D., 1950, Two-year end results in the total rehabilitation of veterans with spinal-cord and caudaequina injuries, New Eng.J.Med., 242, 1-16.

Moradian, G. P., and Rustioni, A., 1977, Transganglionic degeneration in the dorsal horn and dorsal column nuclei of adult rats, Anat.Res., 187, 660.

Morrow, T. J., and Casey, K. R., 1976, Analgesia produced by mesencephalic stimulation: effect on bulboreticular neurons, in: "Advances in Pain Research and Therapy," J. J. Bonica and D. H. Albe-Fessard, eds., vol. 1, Raven Press, New York, pp. 503-510.

Moruzzi, G., 1962, The midpontine pretrigeminal cat, Arch.Int.Pharmacol., 140, 227-230.

Moruzzi, G., 1972, The sleep-waking cycle, Ergebn.Physiol., 64, 1-165.

Moruzzi, G., and Magoun, H. W., 1949, Brain stem reticular formation and activation of the EEG, Electroencephalog.Clin.Neuro.Physiol., 1, 455-473.

Mosketi, K. V., Mokhovikov, A. N., Makulkin, R. F., Aksentev, S. B.,
    Bartsevich, L. B., and Goikhberg, I. G., 1982, Clinicopatho-
    physiological analysis of the use of fenazepam together with
    neuroleptics and biologically active substances in the
    treatment of schizophrenia, in: "Fenazepam," (in Russian),
    Izd. Naukova Dumka, Kiev, pp. 245-260.
Mosso, I. A., and Kruger, L., 1973, Receptor categories represented
    in spinal trigeminal nucleus caudalis, J.Neurophysiol., 36,
    472-488.
Muller, P., and Seeman, P., 1977, Brain neurotransmitter receptors
    after long-term haloperidol: dopamine, acetylcholine, sero-
    tonin, α-noradrenergic, and naloxone receptors, Life Sci.,
    21, 1751-1758.
Muller, P., and Seeman, P., 1978, Dopaminergic supersensitivity
    after neuroleptic: time-course and specificity, Psychopharm-
    acol., 60, No. 1.
Murphy, J. T., and Sabah, N. H., 1970, Spontaneous firing of cere-
    bellar Purkinje cells in decerebrate and barbiturate anesthe-
    tized cats, Brain Res., 17, 515-519.
Mutani, R., 1969, Experimental evidence for the existence of an
    extrarhinecephalic control of the activity of the cobalt
    rhinencephalic epileptogenic focus. Part I., The role
    played by the caudate nucleus, Epilepsia (Amst.), 10,
    337-350.
Mutani, R., and Fariello, R., 1969, Effect of low frequency caudate
    stimulation on the EEG of epileptic neocortex, Brain Res.,
    14, 179-753.
Mutani, R., Doriguzzi, T., and Fariello, R., 1969, Caudate spindle
    induced by introduction of cobalt in the head of the caudate
    nucleus, Brain Res., 11, No. 2, 273-275.
Mutani, R., Bergamini, L., and Doriguzzi, T., 1969, Experimental
    evidence for the existence of an extrarhinencephalic control
    of the activity of the cobalt rhinencephalic epileptogenic
    focus. Part II: Effect of palecerebellar stimulation,
    Epilepsia, 10, 351-362.
Nanobashvili, Z. I., Khizanishvili, N. A., and Ioseliani, T. K.,
    1975, Influence of adrenaline and the desynaptization of the
    adrenals on the convulsive activity of the spinal cord,
    Neurofiziolog., 7, No. 2, 149-155.
Nashold, B. S., 1981, Neurosurgical therapy of deafferentation pain,
    Pain, 10, Suppl. 1, 83.
Nashold, B. S., Urban, B., and Zorub, D. S., 1976, Phantom pain
    relief by focal substantia gelatinosa rolando, in: "Advances
    in Pain Research and Therapy," J. J. Bonica and D. Albe-
    Fessard, eds., vol. 1, Raven Press, New York, pp. 959-964.
Nathan, P. W., 1976, The gate-control theory of pain. A critical
    review, Brain, 99, 123-158.
Nathan, P. W., and Rudge, P., 1974, Testing the gate-control theory
    of pain in man, J.Neurol.Neurosurg.Psychiatry, 37, 1366.

Nicoll, R. A., 1978, Pentobarbital: differential postsynaptic
     actions on sympathetic ganglion cells, Science, 199,
     451-452.
Niedermeyer, E., Singer, H. S., Folstein, S. E., Allen, R. P.,
     Mizanda, F., Fineyre, F., and Bird, B. L., 1979, Hypersomnia
     with simultaneous waking and sleep patterns in the electro-
     encephalogram. A case report with neurotransmitter studies,
     J.Neurol., 221, No. 1, 1-13.
Nikushkin, E. V., Braslavsky, V. E., and Kryzhanovsky, G. N., 1980,
     Anticonvulsant effect of the antioxidant ionol, Byull.eksp.
     biol.i med., 90, No. 12, 696-698.
Nikushkin, E. V., Braslavsky, V. E., and Kryzhanovsky, G. N., 1981,
     Activation of lipid peroxidation in neuronal membranes as a
     pathogenetic mechanism of epileptic activity, Zh.nevropatol.
     i psikhiat., 81, 810-815.
Noordenbos, W., 1959, Pain, Elsevier, Amsterdam.
Noordenbos, W., 1960, Some theoretical remarks on central pain, Acta
     Neurochirurgia (Wien), 8, 113-120.
Norcross, K., and Spehllmann, R., 1978, A quantitative analysis of
     the exitatory and depressant effects of dopamine on the
     firing of caudatal neurons: electrophysiological support for
     the existence of two distinct dopamine-sensitive receptors,
     Brain Res., 156, No. 1, 168-169.
Nose, T., and Takemoto, H., 1974, Effect of oxotremorine on homovan-
     illic acid concentration in the striatum of the rat, Europ.
     J.Pharmacol., 25, No. 1, 51-55.
Oakley, J., and Ojemann, G., 1980, Effect of stimulation of caudate
     and ventral anterior thalamus on seizure frequency in the
     chronic primate model of epilepsy, in: "Epilepsy: Xth Inter-
     national Symposium," J. Wada and J. K. Penry, eds., Raven
     Press, New York, pp. 242-257.
Obata, K., and Highstein, S. M., 1970, Blocking by picrotoxin of
     both vestibular inhibition and GABA action on rabbit oculo-
     motor neurons, Brain Res., 18, 538-541.
Obata, K., and Yoshida, M., 1973, Caudate evoked inhibition and
     action of GABA and other substances on cat pallidal neurons,
     Brain Res., 64, 455-459.
Ochoa, J., and Noordenbos, W., 1979, Pathology and disordered sen-
     sation in local nerve lesions: an attempt at correlation,
     in: "Advances in Pain Research and Therapy, vol. 3," J. J.
     Bonica et al., eds., Raven Press, New York, pp. 67-90.
Ogata, N., 1975, Ionic mechanisms of the depolarization shift in the
     hippocampal slices, Exp.Neurol., 46, 147-155.
Ojemann, G. A., 1980, Basic mechanisms implicated in surgical treat-
     ments of epilepsy, in: "Epilepsy: a Window to Brain Mech-
     anisms," J. S. Lockard and A. A. Ward, eds., Raven Press,
     New York, pp. 261-277.
Okujava, V. M., 1969, Principal Neurophysiological Mechanisms of
     Epileptic Activity, (in Russian), Ganatleba, Tbilisi.
Okujava, V. M., Mestvirishvili, N., and Bagashvili, T. I., 1979, The
     influence of brain stem reticular formation on seizure

activity in the cerebral cortex, Physiol.J.USSR, No. 10, 1465-1472.

Oleson, T. D., and Liebeskind, J. C., 1976, Modification of midbrain and thalamic evoked responses by analgesic brain stimulation in the rat, in: "Advances in Pain Research and Therapy," J. J. Bonica and D. H. Albe-Fessard, eds., vol. 1, Raven Press, New York, pp. 487-494.

Oleson, T. D., Twombly, D. A., and Liebeskind, J. C., 1978, Effects of pain-attenuating brain stimulation and morphine on electrical activity in the raphe nuclei of the awake rat, Pain, 4, 211-230.

Oliveras, J. R., Besson, J. M., Guilbaud, G., and Liebeskind, J. C., 1974, Behavioral and electrophysiological evidence of pain inhibition from midbrain stimulation in the cat, Exp.Brain Res., 20, 32-44.

Oliveras, J. L., Redjemi, F., Guilbaud, G., and Besson, J. M., 1975, Analgesia induced by electrical stimulation of the inferior centralis nucleus of the raphe in the cat, Pain, 1, 139-145.

Olney, J. W., Sharpe, L. G., and De Gubareff, T., 1975, Excitoxic amino acids, Neurosci.Abstr., 5, No. 3, 371.

Olsen, R. W., 1981, GABA-benzodiazepine-barbiturate receptor interactions, J.Neurochem., 37, No. 1, 1-13.

Olsen R. W., and Leeb-Lundberg, F., 1981, Convulsant and anticonvulsant drug binding sites related to GABA-regulated chloride ion channels, in: "GABA and Benzodiazepine Receptors," E. Costa et al., eds., Raven Press, New York, pp. 93-102.

Olszewski, J., 1950, On the anatomical and functional organization of the spinal trigeminal nucleus, J.Comp.Neurol., 92, 401-413.

Oniani, T. N., 1978, On the functional state of the brain in different phases of sleep, in: "Functional states" (in Russian), Ye. N. Sokolov and N. N. Danilova, eds., Moscow State University Publishers, Moscow, pp. 46-47.

Oniani, T. N., Koridze, M. G., and Abzenidze, Ye. V., 1972, Significance of the reverberation of excitation in the neuronal networks of the limbic system in the regulation of training and short-term memory of animals, in: "Current Problems of the Activity and Structure of CNS," (in Russian), third edition, Metsniereba, Tbilisi, pp. 37-55.

Oniani, T. N., Koridze, M. G., and Kavkasidze, M. G., 1974, Dynamics of the electric activity of the neo- and archipaleocrotex during different phases of the wakefulness-sleep cycle, in: "Problems of the Neurophysiology of Emotions and the Wakefulness-Sleep Cycle," (in Russian), first edition, Metsniereba, Tbilisi, pp. 85-119.

Orlov, A. A., and Pirogov, A. A., 1975, Heterosensory interaction on the neurons of the frontal associative cortex of the cat, Fiziol.zh.SSR, 61, No. 7, 991-999.

Osborne, S. R., 1977, The free food (contrafree loading) pheomenon: a review and analysis, Anim.Learn.and Behav., 5, No. 3, 221-235.

Ostrovskaya, R. U., and Molodavkin, G. M., 1980, On the partici-
    pation of GABAergic structures in the realization of halo-
    peridol effects, Byull.eksp.biol.i med., 89, No. 3, 313–315.
Otsuka, M., 1972, Gamma-aminobutyric acid in the nervous system, in:
    "Structure and Function of Nervous Tissue," N. Y. Bourne,
    ed., pp. 249–289.
Owen F., Cross, A. J., Crow, I. J., Longden, A., Poulter, M., and
    Riley, G. J., 1978, Increased dopamine receptor sensitivity
    in schizophrenia, Lancet, 2, 223–225.
Owen, F., Cross, A. J., Crow, T. J., Lofthouse, R., and Poulter, M.,
    1980, Neurotransmitter receptors in brain in schizophrenia,
    Acta Psychiatr.Scand.(Suppl.), 291, 20–26.
Pear, G. H., and Wellhöner, H. H., 1973, The action of tetanus toxin
    on preganglionic sympathetic reflex discharges, Naunyn-
    Schmiedeberg's Arch.Pharmacol., 276, 437.
Pavlov, I. P., 1923, 1951, Twenty years' experience in the object-
    ive study of higher nervous activity (behavior) of animals,
    Complete Works (in Russian), Vol. 3, Books I and II.
Pavlov, I. P., 1927, 1951, Lectures on the work of the major hemis-
    pheres of the brain, Complete Works (in Russian), Vol. 4.
Papavasiliou, P. S., Cotzias, G. C., Düby, S., Steck, A. J.,
    Fehling, C., and Bell, M. A., 1972, Levodopa in parkinson-
    ism: potentiation of central effects with a peripheral
    inhibitor, New Engl.J.Med., 286, No. 1, 8–14.
Papuashvili, N. S., Okujava, V. M., and Mestvirishvili, L. P., 1980,
    Influence of the chronically produced cobalt epileptic focus
    on the structure and duration of separate phases of the
    wakefulness-sleep cycles, Izv.Akad.Nauk Gruz.SSR.Seria
    Biologia, 6, No. 5, 402–411.
Parry, C. B. W., 1981, Therapies of pain due to spinal root avul-
    sion, Pain, 10, Suppl. 1, 84.
Pellegrino, L. Y., and Cushman, A. Y., 1967, "A Stereotoxic Atlas of
    the Rat Brain," Appleton Century Crofts, New York.
Penfield, W., and Jasper, H., 1954, Epilepsy and the Functional
    Anatomy of the Human Brain, Little Brown and Company,
    Boston.
Perez de la Mora, M., Feria-Velasco, A., and Tapia, R., 1973, Pyri-
    doxal phosphate and glutamate decarboxylase in subcellular
    particles of mouse brain and their relationship to convul-
    sions, J.Neurochem., 20, 1575–1587.
Perry, Th.L., Buhanan, J., and Kish, S. J., 1979, γ-Aminobutyric
    acid deficiency in brain of schizophrenic patients, Lancet,
    1, No. 8110, 237–239.
Perry, T., Hansen, S., and Kloster, M., 1973, Huntington's chorea:
    deficiency of gamma-aminobutyric acid in brain, New Engl.
    J.Med., 288, 337–342.
Persianov, L. S., Kryzhanovsky, G. N., Pobedinsky, N. M., Orlova, V.
    G., and Volobuev, A. I., Technique of determing the disturb-
    ances of the contractile activity of the uterine tubes,
    Inventor's Certificate No. 774543 (June 30, 1978), regis-
    tered on July 7, 1980.

Pert, A., Rosenblat, J. E., Sivit, C., Pert, C. B., and Bunney, W.
    E., 1978, Long-term treatment with lithium prevents the
    development of dopamine receptor supersensitivity, Science,
    201, No. 4351, 171-173.

Pert, C. B., Pert, A., Rosenblatt, J. E. Tallman, J. F., and Bunney,
    W. E., 1979, Catecholamine receptor stabilization: a poss-
    ible mode of lithium's antimanic action, in: "Catecholamines:
    Basic and Clinical Frontiers, Vol. 1," F. Usdin,
    ed., Pergamon Press, New York, pp. 583-585.

Petsche, H., Rappelberger, P., Lapins, R., and Vollmer, R., 1979,
    Rhythmicity in seizure patterns-topographical aspects, in:
    "Origin of Cerebral Field Potentials," E. J. Speckmann and
    H. Caspers, eds., Thieme, Stuttgart, pp. 60-79.

Pijnenburg, A. J. J., Honig, W. M. M., Van der Heyden, J. A. M., and
    Van Possum, J. M., 1976, Effects of chemical stimulation of
    the mesolimbic dopamine system upon locomotor activity,
    Europ.J.Pharmacol., 35, 45-58.

Pivovarov, Yu.I., and Kryzhanovsky, G. N., 1982, Effect of hyper-
    activation of the anterior amygdaline nucleus on the heart
    work under altered reactivity, Byull eksp.biol.i med., 93,
    No. 6, 27-36.

Polc, P., and Haefely, W., 1976, Effects of two benzodiazepines,
    phenobarbiton and baclofen on synaptic transmission in the
    cat cuneate nucleus, Arch.Pharmacol., 294, 121-131.

Polc, P., and Haefely, W., 1977, Effect of intravenous kainic acid,
    N-methyl-D-aspartate and (-)-luciterin on the cat spinal cord,
    Arch.Pharmacol., 300. 199-203.

Polc, P., Möhler, H., and Haefely, W., 1974, The effect of diazepam
    on spinal cord activities: possible sites and mechanisms of
    action, Arch.Pharmacol., 284, 319-337.

Polc, P., Schneeberger, J., and Haefely, W., 1979, Effect of several
    central acting drugs on the sleep-wakefulness cycle of cats,
    Neuropharmacology, 18, No. 3, 259-267.

Pomeranz, B., Wall, P. D., and Weber, W. V., 1968, Cord cells re-
    sponding to fine myelinated afferents from viscera, muscle,
    and skin, J.Physiol.(Lond.), 199, 511-532.

Pomeroy, A., and Rand, M. J., 1971, Facilitation of noradrenaline
    uptake by lithium, Aust. N.C.J.Psychiatry, 5, No. 2,
    280-285.

Post, R. M., Fink, E., Carpenter, W. T., and Goodwin, F. K., 1975,
    Cerebro-spinal fluid amine metabolites in acute schizo-
    phrenia, Arch.Gen.Psychiatry, 32, 1063-1069.

Pozdnyakov, O. M., Polgar, A. A., Zinkevich, V. A., and Smirnova, V.
    S., 1981, Variation in the relationships between quantum
    vesicular parameters caused by the disturbed synaptic trans-
    mitter release in the mammalian neuromuscular junctions,
    Byull eksp.biol.i med., 92, No. 12, 738-740.

Precht, W., and Shimazu, H., 1965, Functional connection of tonic
    and kinetic vestibular neurons with primary vestibular
    afferents, J.Neurophysiol., 28, No. 6, 1914-1028.

Precht, W., and Yoshida, M., 1971, Blockage of caudate-evoked in-
    hibition of neurons in the substantia nigra by picrotoxin,
    Brain Res., 32, No. 1, 229-233.
Precht, W., Schwindt, P. C., and Baker, R., 1973, Removal of
    vestibular commissural inhibition by antagonists of GABA and
    glycine, Brain Res., 62, 222-226.
Price, D. D., and Dubner, R., 1977, Neurons that subserve the
    sensory-discriminative aspects of pain, Pain, 3, 307-338.
Price, D. A., and Mayer, D. J., 1974, Physiological laminar organ-
    ization of the dorsal horn of M. mulatta, Brain Res., 79,
    321-325.
Price, D. A., and Mayer, D. J., 1975, Neurophysiological character-
    ization of anterolateral quadrant neurons subserving pain in
    M. mulatta, Pain, 1, 59-72.
Price, D. D., and Wagman, I. H., 1970, The physiological roles of A-
    and C-fiber input to the dorsal horn of M. mulatta, Exp.
    Neurol., 29, 373-390.
Price, D. D., Dubner, R., and Hu, J. W., 1976, Trigeminothalamic
    neurons in nucleus caudalis responsive to tactile, thermal,
    and nociceptive stimulation of the monkey's face, J.Neuro-
    physiol., 39, 936-953.
Price, K. S., Farely, I. J., and Hornykiewicz, O., 1978a, Neuro-
    chemistry of parkinson's disease:relation between striatal and
    limbic dopamine, Adv.Biochem.Psychopharmacol., 19, 720-741.
Price, D. D., Hayes, R. L., and Ruda, M. A., 1978b, Spatial and
    temporal summation of input to spino-thalamic tract neurons
    and their relation to somatic sensation, J.Neurophysiol.,
    41, 933-947.
Prince, D. A., 1967, Electrophysiology of 'epileptic' neurons,
    Electroenceph.Clin.Neurophysiol., 23, 83-84.
Prince, D. A., 1968a, Inhibition in 'epileptic' neurons, Exp.
    Neurol., 21, 307-321.
Prince, D. A., 1968a, The depolarization shift in 'epileptic' neur-
    ons, Exp.Neurol., 21, 467-485.
Prince, D. A., 1971, Cortical cellular activities during cyclically
    occurring interictal epileptiform discharges, Electroen-
    cephalogr.Clin.Neurophysiol., 31, 649-484.
Prince, D. A., 1978, Neurophysiology of epilepsy, Ann.Rev.neurosci.,
    1, 395-415.
Prince, D. A., and Futamachi, K. G., 1976, Intracellular recording
    from chronic epileptogenic foci in the monkey, Electroen.
    Clin.Neurophysiol., 29, No. 5, 446-510.
Prince, D. A., and Gutnick, M. J., 1972, Neuronal activities in
    epileptogenic foci of immature cortex, Brain Res., 45, 455-468.
Prince, D. A., and Wilder, B. J., 1967, Control mechanisms in
    cortical epileptogenic foci. 'Surround' inhibition, Arch.
    Neurol., 16, 194-202.
Prince, D. A., and Wong, R. V. S., 1981, Human epileptic neurons
    studied in vitro, Brain Res., 210, 323-333.

Procacci, P., Lopp, M., Maresca, M., and Komano, S., 1974, Studies
on the pain threshold in man, Advances in Neurology, 4,
107-115.

Purpura, D. P., and Housepian, E. M., 1961, Morphological and physiological properties of chronically isolated immature neocortex, Exp.Neurol., 4, No. 5, 377-401.

Racagni, G., Bruno, F., Gattabeni, F., Maggi, A., DiGuido, A. M.,
Parenti, M., and Groppetti, A., 1977, Functional interaction
between rat substantia nigra and striatum: GABA and dopamine
interrelation, Brain Res., 134, No. 2, 353-358.

Raeva, S. N., 1977, Microelectrode Investigations of Neuronal Activity of the Human Brain (in Russian), Izd. Nauka, Moscow.

Ramos, S., Grollman, E. F., Lazo, P. S., Dyer, S. A., Habig, W. H.,
Hardegree, M. C., Kaback, H. R., and Kohn, L. D., 1979,
Effect of tetanus toxin on the accumulation of the permanent
lipophilic cation tetraphenylphosphonium by guinea-pig brain
synaptosomes, Proc.Natl.Acad.Sci.USA. 76, 4783.

Randrup, A., 1970, Role of brain dopamine in the antipsychotic
affect of neuroleptics. Evidence from studies of ampheta-
mine-neuroleptic interaction, Mod.Probl.Pharmacopsychiatry,
5, 60-65.

Randrup, A., and Munkvad, I., 1966, Behavioral stereotypies induced
by pharmacological agents, Pharmacopsyhiatr.Neuro-Psycho-
pharmacol., 1, No. 1, 18-27.

Randrup, A., and Munkvad, I., 1967, Stereotyped activities produced by
amphetamine in several animal species and man, Psychopharmacologia (Berl.), 11, No. 3, 300-310.

Randrup, A., and Munkvad, I., 1972, Evidence indicating an association between schizophrenia and dopaminergic hyperactivity in
the brain, Orthomol.Psychiatr, 1, 1-7.

Ransom, B. R., and Barker, J. L., 1976, Pentobarbital selectivity
enhances GABA-mediated postsynaptic inhibition in tissue
cultured mouse spinal neurons, Brain Res., 114, 530-535.

Rasmussen, T., 1975, Cortical resection in the treatment of focal
epilepsy, Adv.Neurol., 8, 139-154.

Reches, A., Ebstein, R. P., and Belmeker, R. H., 1978, The differential effects of lithium on noradrenaline-dopamine-sensit-
ive accumulation of cyclic AMP in guinea pig brain, Psychopharmacology, 58, No. 2, 213-216.

Reichental, E., and Hocherman, S., 1977, The critical cortical area
for development of penicillin-induced epilepsy, Electro-
enceph.Clin.Neurophysiol., 42, No. 2, 248-251.

Rekhtman, M. B., Konnikov, B. A., and Kryzhanovsky, G. N., 1978,
Neuronal activity analysis in the lateral geniculate body of
cat, Neurophysiology, 10, No. 1, 30-38.

Rekhtman, M. B., Konnikov, B. A., and Kryzhanovsky, G. N., 1979,
Effect of diazepam on the epileptic activity of rats with
experimental photogenic epilepsy, Byull eksp.biol.i med.,
87, No. 2, 160-164.

Rekhtman, M. B., Samsonova, N. A., and Kryzhanovsky, G. N., 1980a,
Electrical activity and sodium-potassium ATPase level in

penicillin-induced epileptogenic focus in the rat brain cortex and diasepam effect on them, Neurophysiology (Kiev), 12, No. 4, 349-356.

Rekhtman, M. B., Samsonova, N. A., and Kryzhanovsky, G. N., 1980b, Diazepam effect on cyclic spike driving in the cortical epileptogenic focus, Neurophysiology (Kiev), 12, No. 6, 563-570.

Rexed, B., 1964, Some aspects of the cytoarchitechonics and synaptology of the spinal cord, Progress in Brain Research, 11, 58.

Reynolds, D. V., 1969, Surgery in the rat during electrical analgesia induced by focal brain stimulation, Science, 164, 44-445.

Richardson, T. L., Miller, I. I., and McLennan, H., 1977, Mechanisms of excitation and inhibition in the neostriatal system, Brain Res., 127, 219-234.

Richelson, E., 1977, Lithium ion entry through sodium channel of cultured mouse neuroblastoma cells: a biochemical study, Science, 196, No. 4293, 1001-1001.

Riddal, D. R., and Leavens, W. J., 1978, Affinities of drugs for the agonist and antagonist states of the dopamine receptor, Europ.J.Pharmacol., 51, No. 2, 187-188.

Riesen, A. H., 1975, (ed.), "The Developmental Neuropsychology of Sensory Deprivation," Academic Press, New York, San Francisco, London.

Roberts, E., and Frenkel, S., 1951, Glutamic decarboxylase in brain, J.Biol.Chem., 188, 789.

Roberts, E., Chase, T. N., and Tower, D. B., 1976, (eds.), "GABA, in Nervous System Function," Raven Press, New York.

Roitbak, A. I., 1955, "Bioelectric Phenomena in the Cortex of the Major Hemispheres," (in Russian), Part I, Izd. Akad. Nauk Gr. SSR, Tbilisi.

Roitbak, A. I., 1973, Neuroglia and the formation of new neural connections in the cerebral cortex, in: "Mechanisms of the Formation and Inhibition of Conditioned Reflexes," (in Russian), V. S. Rusinov, ed., Izd. Nauka, Moscow, pp. 82-94.

Roitbak, A. I., 1979, Phsyiology of the glia, in: "General Physiology of the Nervous System," (in Russian), Izd. Nauka, Leningrad, pp. 607-702.

Roitrub, B. A., and Oleshko, N. N., 1977, The role of DOPA and dopamine in the activation of acetylcholinesterase of rat neostriatum, Dokl.Akad.Nauk SSSR, Phsyiol., 234, No. 1, 239-241.

Roitrub, B. A., Oleshko, N. N., and Cherkes, V. A., 1973, The state of excitability and the activity of acetylcholinesterase of neostriatum after destruction of black substance in rats, Dokl.Akad.Nauk SSSR, 212, No. 6, 1482-1484.

Roitrub, B. A., Oleshko, N. N., Cherkes, B. A., Man'kovsky, M. B., Vainshtok, A. B., and Lapogonov, O. A., 1979, Interaction of DOPA with blood acetylcholinesterase in patients suffering from parkinsonism, Fiziolog.Zh.(Kiev), 25, No. 3, 239-241.

Rossi, G. F., 1963a, An experimental study of the hypnogenic mechanisms of the brain stem, Arch.Ital.Biol., 101, 470-492.

Rossi, G. F., 1963b, Sleep-inducing mechanisms in the brain stem, EEG and Clin.Neurophysiol., 24, suppl., 113-132.

Rossi, G. F., 1975, Considerations of the principles of surgical treatment of partial epilepsy, Brain Res., 95, 395-402.

Ryabokon, N. S., Bozhik, V. P., and Yarkina, T. G., 1980, Results of treating epilepsy by electrically stimulating the cerebellar cortex, in: "Surgical Treatment of Epilepsy," (in Russian), Izd.Metsniereba, Tbilisi, pp. 83-85.

Sakai, Y., Nishijima, Y., Mikuni, N., and Iwata, N., 1979, An experimental model of hyperirritability in the trigeminal skin field of the rat, Pain, 147-157.

Sakai, Y., Nishijima, Y., Mikuni, N., and Iwata, N., 1981, Inhibitory mechanisms of the hyperirritability caused by picrotoxin in the rat, Pain, 11, No. 1, 21-35.

Sakurai, T., Kuwahara, T., and Matsuda, M., 1980, Vitamin, B_6 vitamers in mouse brain and their relationship to convulsions, Jikeikai med. J., 27, No. 1, 13-21.

Samoilova, Z. T., Petelin, L. S., Tretyakova, K. A., and Chechulin, Yu. S., 1973, An experimental model of parkinsonism, Patol., fiziol., 24, No. 4, 78-80.

Samsonova, N. A., Rekhtman, M. B., Glebov, R. N., and Kryzhanovsky, G. N., 1979, Effect of diazepam on Na, K-ATPase in penicillin-induced focus of hyperactivity in the brain cortex, Byull. eksp.biol.i med., 88, No. 12, 655-659.

Sarajishvili, P. M., and Bibileishvili, Sh. I., 1975, Some peculiarities of sleep in the case of frontal epileptic foci, in: "Functional and Structural Foundations of the Systemic Activity and Mechanisms of the Plasticity of the Brain," (in Russian), Fourth Edition, USSR Academy of Medical Sciences, Brain Institute, Moscow, pp. 443-449.

Sarajishvili, P. M., and Geladze, T. S., 1977, "Epilepsy," (in Russian), Izd. Meditsina, Moscow.

Sasaki, K., and Tanaka, T., 1963, Effects of stimulation of cerebellar and thalamic nuclei upon spinal alpha motoneurons of the cat, Japan J.Physiol., 13, No. 1, 64-83.

Sato, I., amd Kawamori, N., 1975, Reticulo-reticular relationship during sleep and Waking, Physiol.and Behav., 15, No. 3, 333-337.

Saudberg, D. E., and Segal, M., 1978, Pharmacological analysis of analgesia and self-stimulation elicited by electrical stimulation of catecholamine nuclei in the rat brain, Brain Res., 152, 529-542.

Sawa, H., Maruyama, N., and Kaji, S., 1963, Intracellular potential during electrically induced seizures, Electroenceph.Clin. Neurophysiol., 15, 209-220.

Scheibel, M. E., and Scheibel, A. B., 1958, Structural substrates for integrative patterns on the brain stem reticular core, in: "Reticular Formation of the Brain," H. H. Jasper et al., ed., Little and Brown, Boston, p. 31.

Scheibel, M. E., and Scheibel, A. B., 1961, On circuit patterns of the brain stem reticular core, Ann.N.Y.Acad.Sci., 857-865.

Schildkraut, J. J., 1973, Pharmacology, the effects of lithium on biogenic amines, in: "Lithium," S. E. Gershon and B. Shopsin, eds., Plenum Press, New York, pp. 51–73.

Schildkraut, J. J., Logue, M. A., and Dodge, G. A., 1969, The effects of lithium salts on the turnover and metabolism of norepinephrine in rat brain, Psychopharmacologia, 14, No. 1, 135–141.

Schmidt, E. M., Mutsuga, N. M., and Intosh, I. S., 1976, Chronic recording of neurons in epileptogenic foci of monkey during seizures, Exp.Neurol., 52, 459–466.

Schultz, W., and Ungerstedt, U., 1977, A method to detect and record from striatal cells of low spontaneous activity by stimulating the corticostriatal pathways, Brain Res., 142, No. 2, 357–362.

Schultz, W., and Ungerstedt, U., 1978, Short-term increase and long-term inversion of striatal cell activity after degeneration of the nigrostriatal dopamine system, Exp.Brain Res., 33, No. 2, 159–171.

Schwartz, M. A., Wyatt, R. J., Yang, H.-Y.T., and Nett, N. H., 1974, Multiple forms of brain monoamine oxidase in schizophrenic and normal individuals, Arch.Gen.Psychiat., 31, 557–560.

Schwartzkroin, P. A., 1982, Epilepsy: a result of abnormal pacemaker activity in central nervous system neurons?" in: "Cellular Pacemakers. Vol. 2. Function in Normal and Disease States," D. Carpenter, ed., John Wiley & Sons, New York, pp. 323–344.

Schwartzkroin, P. A., van Duijn, H., and Prince, D. A., 1974a, Effect of projected cortical epileptiform discharges on field potentials in the cat cuneate nucleus, Exp.Neurol., 43, No. 1, 88–105.

Schwartzkroin, P. A., van Duijn, H., and Prince, D. A., 1974b, Effect of projected cortical epileptiform discharges on unit activity in the cat cuneate nucleus, Exp.Neurol., 43, No. 1, 106–123.

Schwartzkroin, P. A., 1975, Characteristic of CA_1 neurons recorded intracellularly in the hippocampal in vitro slice preparation, Brain Res., 85, 423–436.

Seeman, P., Chau-Wong, M., Tedesco, J., and Wong, K., 1975, Brain receptors for antipsychotic drugs and dopamine: direct binding assays, Proc.Nat.Acad.Sci.USA, 72, No. 11, 4376–4380.

Sefton, A. J., and Burke, W., 1965, Reverberatory inhibitory circuits in the lateral geniculate nucelus of the rat, Nature, 27, 1325–1326.

Segal, M., and Saudberg, D., 1977, Analgesia produced by electrical stimulation of catecholamine nuclei in the rat brain, Brain Res., 123, No. 2, 369–372.

Seltzer, Z., and Devor, M., 1979, Ephaptic transmission in chronically damaged peripheral nerves, Neurology, 29, No. 7, 1061–1064.

Serkov, F. N., 1977, "Electrophysiology of the Higher Segments of the Auditory System," (in Russian), Izd. Naukova Dumka, Kiev.

Sessle, B. I., and Greenwood, F., 1973, Influence of trigeminal nucleus caudalis on the response of cat trigeminal brain stem neurones with orefacial mechanoreceptive fields, Brain Res., 67, 330-333.

Sethy, V. H., and von Woert, M. H., 1974, Regulation of striatal ach concentration by dopamine receptors, Nature, 251, No. 5442, 528-530.

Shandra, A. A., and Godlevsky, L. S., 1983, Influence of the electric stimulation of the fastigial nucleus on epileptic activity in the cerebral cortex, Patol.fiziol.i eksp.terapiya, (in press).

Shandra, A. A., Godlevsky, L. S., Makulkin, R. F., and Kryzhanovsky, G. N., 1982a, Combined treatment of epilepsy by vitamins and anticonvulsants, in: "New Methods of Diagnosis, Treatment and Prophylaxis of the Main Forms of Nervous and Psychotic Diseases," (in Russian), Institute of Neurology and Psychiatry, Kharkov, pp. 295-296.

Shandra, A. A., Lobasyuk, B. A., Makulkin, R. F., and Kryzhanovsky, G. N., 1982, Influence of phenazepam on determinant and dependent epileptic complex foci in the brain cortex, in: "Phenazepam," Kiev, Izd. Naukova Dumka, pp. 188-193.

Shandra, A. A., Godlevsky, L. S., and Semenyuk, N. D., 1983, The formation of generalized seizure activity in mice during daily pentylenetetrazol administration in subthreshold doses, Byull.eksp.biol.i med., (in press).

Shaposhnikov, V. S., Oleinik, A. V., and Umurzakova, S. T., 1982, Experience in using complex specific pathogenetic therapy (CSPT) in psychic clinics, in: "New Methods of Diagnosis, Treatment and Prophylaxis of the Main Forms of Nervous and Psychotic Diseases," (in Russian), Institute of Neurology and Psychiatry, Kharkov, pp. 230-231.

Shapovalov, A. I., 1966, "Cellular Mechanisms of Synaptic Transmission," (in Russian), Izd.Meditsina, Moscow.

Shapovalov, A. I., and Arushunyan, E. B., 1965, Influence of the stimulation of the brainstem and the motor cortex on neuronal activity of the spinal cord, Fiziol.Zh.SSR, 51, No. 6, 670-680.

Shavolina, V. A., 1978, Pharmacological and experimental models of Parkinsonism, in: "Pathogenesis, Clinical Picture and Treatment of Parkinsonism," (in Russian), G. V., Morozov, ed., Society of Neurologists and Psychiatrists, Moscow, pp. 78-81.

Shchelkunov, E. L., 1964, The technique of phenamine stereotypy for evaluating the effect produced by remedial agents on the central adrenergic processes, Farmakol.i Toksikol., 27, No. 5, 628-633.

Sheard, M. H., 1980, The biological effects of lithium, Trends Neurosci., 3, No. 4, 85-86.

Sherrington C. S., 1906, The Integrative Action of the Nervous System. Cambridge University Press, Cambridge.

Sherwin, I., 1970, Burst activity of single units in penicillin
    epileptogenic focus, Electroenceph.Clin.Neurophysiol., 29,
    No. 3, 373-382.
Shevchenko, D. G., 1971, Neurophysiological mechanisms of sleep,
    Uspekhi fiziol.nauk, 2, No. 4, 73-99.
Shevelev, I. A., 1971, "Dynamics of the Visual Sensory Signal," (in
    Russian), Izd. Nauka, Moscow.
Shibata, M., and Bures, J., 1972, Reverberation of cortical spread-
    ing depression along closed-loop pathways in cat cerebral
    cortex, J.Neurophysiol., 351, 381-388.
Shik, M. L., 1981, Control of locomotion, in: Advances of
    Physiological Sciences, Vol. 1," J. Szentagothai, ed.,
    Pergamon Press, London, pp. 143-148.
Shimazu, H., and Precht, W., 1965, Tonic and clonic responses of
    cat's vestibular neurons to horizontal angular acceleration,
    J.Neurophysiol., 28, 991.
Shimazu, H., and Precht, W., 1966, Inhibition of central vestibular
    neurons from the contralateral labyrinth and its mediating
    pathway, J.Neurophysiol., 29, No. 3, 467-492.
Shuranova, Zh. P., 1977, "Study of the Elementary Operant Mechanisms
    in the Cerebral Cortex of Mammals," (in Russian), Izd.
    Nauka, Moscow.
Sidney, G., 1977, The role of prefrontal cortex in grand mal con-
    vulsion, Arch.Neurol., 26, 109-119.
Sieghart, W., and Karobath, M., 1980, Molecular heterogeneity of
    benzodiazepine receptors, Nature, 286, No. 5770, 285-287.
Siggins, G. R., Hoffer, B. J., Bloom, F. E., and Ungerstedt, U.,
    1976, Cytochemical electrophysiological studies of dopamine
    in the caudate nuceleus, in: "Basal Ganglia," Raven Press,
    New York, pp. 227-248.
Silfenius, H., Olofsson, S., and Ridderheim, P. A., 1980, Induced
    epileptiform activity evoked from dendrites of hippocampal
    neurones, Acta Physiol.Scand., 108, 109-111.
Smirnov, V. A., 1973, On the pathogenesis of trigeminal neuralgia,
    Zh.nevropatol.i psikhiatr., No. 3, 328-331.
Smirnov, V. A., 1976, "Diseases of the Facial Nervous System," (in
    Russian), Izd.Meditsina, Moscow.
Smirnov, V. M., 1976, "Stereotaxic Neurology," (in Russian),
    Izd.Meditsina, Leningrad.
Smirnov, V. M., and Borodkin, Yu.S., 1975, Artificial stable func-
    tional connections as a method of forming matrices of long-
    term memory in man (on the theory of long-term memory),
    Fiziol.cheloveka, 1, No. 3, 525-533.
Smirnov, V. M., and Borodkin, Yu.S., 1979, "Artificial Stable Func-
    tional Connections," (in Russian), Izd. Meditsina,
    Leningrad.
Smith, T. L., and Purpura, D. P., 1960, Electrophysiological studies
    on epileptogenic lesions of cat cortex, Electroenceph.Clin.
    Neurophysiol., No. 12, 59-82.
Smolin, L. N., 1981, Effect of acute noxious stimulation on trans-
    mission of signals from the low threshold cutaneous afferent

through the somatosensory system, Neurophysiologia (Kiev),
    13, No. 6, 621–627.

Smythies, J. R., Receptor Modeling for Anticonvulsant and Convulsant
    Drugs, in: "Antiepileptic Drugs: Mechanisms of Action," G.
    H. Glaser, J. K. Penry, and D. M. Woodbury, eds., Raven
    press, New York, pp. 207–222.

Snider, R. A., 1975, A cerebellar – coruleus pathway, Brain Res, 88,
    59–63.

Snider, R. S., 1979, Focal cerebellar hyperthermia: effects on
    cerebral paroxysmal afterdischarges, Epilepsia, 20, 115–125.

Snider, R. S., Thomas, W., and Snider, S. R., 1978, Focal brain
    hyperthermia. I. The cerebellar cortex, Experientia, 34,
    479–481.

Snyder, S. H., 1972, Catecholamines in the brain as mediators of
    amphetamine psychosis, Arch.Gen.Psych., 27, 169–179.

Snyder, S. H., 1973, Amphetamine psychosis: a 'model' schizophrenia
    mediated by catecholamines, Am.J.Psychiatry, 130, 61–67.

Snyder, S. H., 1974, Catecholamines as mediators of drug effects in
    szhizophrenia, in: "The Neurosciences, 3–Study Programme,"
    F. O. Schmitt and F. O. Worden, eds., MIT Press, Cambridge,
    Mass.– London, pp. 721–732.

Snyder, S. H., 1975, Neurotransmitter and drug receptors in the
    brain, Biochem.Pharmacol., 24, 1371–1374.

Snyder, S. H., 1976, The dopamine hypothesis of schizophrenia: focus
    on the dopamine receptor, Am.J.Psychiatr., 133, 197–202.

Snyder, S. H., Aghajanian, G. K., and Matthysse, S., 1972, Prospects
    for research on schizophrenia. V. Pharmacological observa-
    tions, Neurosci.Res.Program Bull. 10, 430–445.

Snyder, S. H., Banerjee, S. P., Yamamura, H. I., and Greenberg, D.,
    1974a, Drugs, neurotransmitters, and schizophrenia. Pheno-
    thiazines, amphetamines, and enzymes synthesizing psycho-
    tomimetic drugs aid schizophrenia research, Science, 184,
    No. 4143, 1243–1253.

Snyder, S., Greenberg, D., and Yamamura, H. J., 1974b, Antischizo-
    phrenic drugs and brain cholinergic receptors, Arch.Gen.Psy-
    chiatr., 31, 58–61.

Spano, P. F., Memo, M., Stefanitri, F., Fresio, P., and Trabucahi,
    M., 1980, Detection of multiple receptors for dopamine,
    Adv.Biochem.Psychopharmacol., 21, 243–251.

Speckmann, E. J., and Caspers, H., 1973, Paroxysmal depolarization
    and changes in action potentials induced by pentylenetetra-
    zol in isolated neurons of Helix pomatia, Epilepsia, 14,
    397–408.

Spehlman, R., 1975, The effects of acetylcholine and dopamine: the
    caudate nucleus depleted of biogenic amines, Brain, 98, No.
    2, 219–230.

Speth, R. S., Johnson, R. W., Regan, J., Reisine, T., Kobayashi, R.
    M., Brezolin, W., Roeske, R., and Yamamura, H. I., 1980, The
    benzodiazepine receptor of mammalian brain, Federation
    Proc., 39, 3032–3038.

Spiegel, E. A., Wycis, H. T., Szekely, E. G., Constantinovici, A.,
    Egyed, I. I., Gildenberg, P., Lehman, R., and Werthan, M., 1965,
    Role of the caudate nucleus in parkinsonian bradykinesia,
    Confin.Neurol., 26, No. 2, 336-341.
Squires, R. F., and Braestrup, C., 1977, Benzodiazepine receptors in
    rat brain, Nature, 266, 732-734.
Squires, R. F., Benson, D. J., Braestrup, C., Coupet, J., Klepner,
    C. A., Myers, V., and Beer, B., 1979, Some properties of
    brain specific benzodiazepine receptors: new evidence for
    multiple receptors, Pharmacol.Biochem.Behav., 10, 825-830.
Sramks, M., Fitz, G., Galanda, M., and Nadvornik, P., 1976, Some
    observations in treatment stimulation of epilepsy, Acta
    Neurochir.(Suppl.), 23, 257-262.
Stadler, H., Lloyd, K. G., Gadea-Ciria, M., and Bartholini, G.,
    1973, Enhanced striatal acetylcholine release by chlorpro-
    mazine and its reversal by apomorphine, Brain Res., 55, No.
    3, 476-480.
Stavraky, G. W., 1961, Supersensitivity Following Lesions of the
    Nervous System, University of Toronto Press, Toronto.
Stefans, C., and Jasper, H., 1964, Recurrent collateral inhibitions
    in pyramidal tract neurons, J.Neurophysiol., 27, No. 5,
    855-877.
Steg, G., 1969, Striatal cell activity of the putamen after system-
    atic administration of monoaminergic and cholinergic drugs,
    in: "Third Symposium on Parkinson's Disease," Edinburgh, pp.
    26-29.
Steg, G., 1972, Biochemical aspects of rigidity, in: "Parkinson's
    Disease," Huber, Bern-Stuttgart-Vienna, pp. 48-63.
Sternbach, R. A., 1978, "The psychology of Pain," Raven Press, New
    York.
Stevens, J. R., 1969, Deep temporal stimulation in man, Arch.
    Neurol., 21, 157-169.
Stevens, J. R., and Livermore, A., Jr. 1978, Kindling of the meso-
    limbic dopamine system: animal model of psychosis, Neuro-
    logy, 28, 36-46.
Stevens, J., Wilson, K., and Foote, W., 1974, GABA blockade, dopa-
    mine and schizophrenia: experimental studies in cat, Psy-
    chopharmacologia (Berl.), 39. 105-119/
Stille, G., and Christ, W., 1978, Dopaminergic transmission and
    mental disease, Triangle, 17, 13-19.
Stille, G., and Sayers, A., 1967, Concerning the effect of bul-
    bocapnine on the caudate loop, Experientia, 23, 1028-1029.
Stookey, B., and Ransohoff, 1959, Trigeminal Neuralgia. Its History
    and Treatment, Charles C. Thomas, Springfield, Illinois.
Strain, C. M., Van Meter, W. G., and Brockman, W. H., 1978, Ele-
    vation of seizure thresholds. A comparison of cerebellar
    stimulation, phenobabital and diphenylhydantoin, Epilepsia,
    19, 494-504.
Strang, R. R., 1965, Imipramine in treatment of Parkinsonism: a
    double-blinded placebo study, Brit.Med.J., 2, No. 5452,
    33-34.

Straughan, D. W., 1978, Barbiturates, benzodiazepines and the GABA system, in: "Advances in Pharmacology and Therapeutics, vol. 2. Neurotransmitters," (Proc. 7th Intern. Congress of Pharmacol.), Paris pp. 19-27.

Study, R. E., and Barker, J. L., 1981, Diazepam and (-)pentobarbital-fluctuation analysis reveals different mechanisms for potentiation of GABA responses in cultured central neurons, Proc.Natl.Acad.Sci.USA, 78, 7180-7184.

Study, R. E., and Barker, J. L., 1982, Cellular mechanisms of benzodiazepine action, JAMA, 247, No. 15, 2147-2151.

Sudakov, K. V., 1971, "Biological Motivations," (in Russian), Izd. Meditsina, Moscow.

Sunderland, S., 1968, Nerves and Nerve Injuries, Livingstone, Edinburgh.

Sunderland, S., 1976, Pain mechanisms in causalgia, J.Neurol.Neurosurg.Psychiatry, 39, 471-480.

Supavilai, P., and Karobath, M., 1979, Stimulation of benzodiazepine receptor binding by SQ-20009 is chloride-dependent and picrotoxin sensitive, Europ.J.Pharmacol., 60, No. 1, 111-113.

Supavilai, P., and Karobath, M., 1980a, Interaction of SQ-20009 and GABA-like drugs as modulator of benzodiazepine receptor binding, Eur.J.Pharmacol., 62, 229-233.

Supavilai, P., and Karobath, M., 1980b, The effect of temperature and chloride ions on the stimulation of ^{3}H -flunitrazepam binding by the muscimol analogues THYP and piperidine-4-sulfonic acid, Neurosci.Lett., 19, 337-341.

Supavilai, P., and Karobath, M., 1980c, Heterogeneity of benzodiazepine receptors in rat cerebellum and hippocampus, Europ.J. Pharmacol., 64, No. 1, 91-93.

Supavilai, P., and Karobath, M., 1981, Action of pyrazolpyridines as modulators of (^{3}H) flunitrazepam binding to the GABA/benzodiazepine receptor complex of the cerebellum, Eur.J. Pharmacol., 70, 183-193.

Suvorov, N. F., 1973, Role of the striothalamocortical system in conditioned-reflex activity, in: "Striopallidal System," (in Russian), Izd. Nauka, Leningrad, pp. 3-13.

Sverdlov, Yu.S., 1960, The spinal cord reflex activity in local tetanus (an electrophysiological study), Fiziol.Zh.SSR, 46, 941.

Sverdlov, Yu.S., 1963, Action of tetanus toxin upon the inhibitory mechanism in the spinal cord of the cat, in: "Recent Advances in the Pharmacolog of Toxins," H. W. Randonat, ed., Pergamon Press and Czechoslovak Medical Press, Oxford and Prague, pp. 113-121.

Sverdlov, Yu.S., 1969, Potentials of spinal motoneurons in cats with experimental tetanus, Neurophysiology (Kiev), 1, No. 1, 25-34.

Sypert, G. W., Oakley, J., and Ward, A. A., 1970, Single-unit analysis of propagated seizures in neocortex, Exp.Neurol., 28, No. 3 308-325.

Szabo, J., 1962, Topical distribution of the striatal efferents in the monkey, Exptl.Neurol., 5, No. 1, 21-36.

Szekely, E. G., Zivanovic, B., and Spiegel, E. A., 1969, The influence of stimulation of the hypothalamus and of the dorsomedial nucleus upon experimental bradykinesia and akinesia, J.Neurol.Sci., 9, No. 2, 255-260.

Szentágothai, J., 1964, Propriospinal pathways and their synapses, Progress in Brain Res., 11, 155-177.

Szentágothai, J., and Arbib, M. A., 1974, Conceptual Models of Neural Organization, Yvone M. Homsy, ed., WRP Writer, Boston.

Takahashi, K., Kubota, K., and Uno, M., 1967, Recurrent facilitation in cat pyramidal tract cells, J.Neurophysiol., 30, 22-34.

Takano, K., and Kano, M., 1973, Gamma-bias of the muscle poisoned by tetanus toxin, Naunyn-Schmiedeberg's Arch.Pharmacol., 276-413.

Takeshige, C., Luo, C. P., Kamada, Y., Oka, K., Murai, M., and Hisamitsu, T., 1979, Relationship between midbrain neurons (periaqueductal central grey and midbrain reticular formation) and acupuncture analgesia, animal hypnosis, in: "Advances in Pain Research and Therapy," J. J. Bonica et al., eds., vol. 3, Raven Press, New York, pp. 615-621.

Tallman, J. F., and Gallager, D. W., 1979, Modulation of benzodiazepine binding site sensitivity, Pharmacol.Biochem.Behav., 10, 809-813.

Tallman, J. F., Thomas, J. W., and Gallager, D. W., 1978, GABAergic modulation of benzodiazepine binding site sensitivity, Nature, 274, 383-385.

Tapia, R., 1974, The role of gamma-aminobutyric acid and its metabolism in the cerebral excitability, in: "Neurohumoral Coding of Brain Function," D. Myers and R. R. Drucker-Collin, eds., Plenum Press, New York, London, pp. 2-26.

Tapia, R., Sandoval, M. E., and Contreras, P., 1975, Evidence for a role of glutamate decarboxylase activity as a regulatory mechanism of cerebral excitability, J.Neurochem., 24, 1283-1285.

Tasker, R. R., Tsuda, T., and Hawylyshyn, P., 1981, Clinical neurophysiology of deafferentation pain, Pain, 10, Suppl. 1, 82.

Tebécis, A. K., 1973, Transmitters and reticulospinal neurons, Exp.Neurology, 40, No. 2, 297-308.

Temkov, I., and Kirov, K., 1971, "Clinical Psychopharmacology," (in Russian), Izd. Meditsina, Moscow.

Testa, G., and Gloor, P., 1974, Generalized penicillin epilepsy in the cat: effect of midbrain cooling, Electroenceph.Clin. Neurophysiol., 36, 517.

Teyer, K. M., and Synder, S., 1971, Differential effects od d-amphetamine on behavior and on catecholamine neurons of rat brain, Brain Res., 28, 295-309.

Thomas, R. C., 1972, Electrogenic sodium pump in nerve and muscle cells, Physiol.Rev., 52, No. 3, p. 563-584.

Thomas, P. K., 1979, Painful neuropathies, in: "Advances in Pain
    Research and Therapy," J. J. Binica et al., ed., vol. 3,
    Raven Press, New York, pp. 103-110.

Thompson, R. F., Johnson, R. H., and Hoopes, J., 1963, Organization
    of auditory, somatic sensory and visual projection to as-
    sociation fields of cerebral cortex in the cat, J.Neurophys-
    iol., 26, 343-364.

Titeler, M., Weinreich, D., Sinclair, D., and Seeman, R., 1978,
    Multiple receptors for brain dopamine, Proc.Natl.Acad.Sci.
    USA 75, No. 3, 1153-1156.

Tolbert, L. C., 1980, Mechanism of action of benzodiazepines, J.
    Med.Sci., 17, No. 2, 168-170.

Torebjörk, H. E., and Hallin, R. G., 1979, Microneurographic studies
    of peripheral pain mechanisms in man, in: "Advances in Pain
    Research and Therapy," J. J. Bonica et al., ed., vol. 3,
    Raven Press, New York, pp. 121-131.

Trabucchi, M., Cheney, D. L., and Racagni, G., 1975, In vivo in-
    hibition of striatal acetylcholine turnover by L-DOPA,
    apomorphine and (+)amphotamine, Brain Res., 85, No. 1,
    130-134.

Traub, R. D., and Wong, R. K. S., 1981, Penicillin-induced epilepti-
    form activity in the hippocampal slice: a model of synchron-
    ization f CA₃ pyramidal cell bursting," Neuroscience, 6,
    223-230.

Travell, J., 1976, Myofascial trigger points: clinical view, in:
    "Advances in Pain Research and Therapy," J. J. Bonica and
    D. Albe-Fesard, eds., Vol. 1, Raven Press, New York, pp.
    919-926.

Treiser, S., and Kellar, K. J., 1979, Lithium effects on adrenergic
    receptor supersensitivity in rat brain, Eur.J.Pharmacol.,
    58, No. 1, 85-86.

Tsubekawa, T., Nishimoto, H., Kotani, A., and Moriyasu, N., 1975,
    The modulating mechanism of the thalamic relay nucleus
    stimulation upon the nociceptive thalamic neuron, in: "Pain,
    1st World Congress," Abstracts, Florence, p. 198.

Tsukahara, N., 1972, The properties of the cerebello-pontine reverb-
    erating circuit, Brain Res., 40, 67-71.

Ugryumov, V. M., and Zotov, Yu.V., 1971, Tactics of the surgical
    treatment of multifocal epilepsy, in: "First All-Union Congress
    of Neurosurgeons," (in Russian), vol. 3, Moscow, pp. 149-152.

Ukhtomsky, A. A., 1978, "Selected Works," (in Russian), Izd.Nauka,
    Leningrad.

Umurzakova, S. T., Shaposhnikov, V. S., and Oleinik, A. V., 1982,
    Experience in using complex specific pathogenetic therapy in
    a psychiatric clinic, in: "New Methods of Diagnosis, Treat-
    ment and Prophylaxis of the Main Forms of Neural and Psychic
    Diseases," (in Russian), Institute of Neurology and Psy-
    chiatry, Kharkov, pp. 230-231.

Ungerstedt, U., 1971, Mechanism of action of L-dopa studied in an
    experimental Parkinson model, in: "Monoamines Noyaux Cris

Centraux et Syndrome de Parkinson," Symposium Bel-Air IV, Paris, pp. 165-170.

Ungerstedt, U., Butcer, L. L., 1969, Butcher, S. G., Anden, N.-E., and Fuxe, K., 1969, Direct chemical stimulation of dopaminergic mechanisms in the neostriatum of the rat, Brain Res., 14, No. 2, 461-471.

Ungerstedt, U., Avemo, A., and Avemo, E., 1973, Animals Models of Parkinsonism, Advances in Neurol., 3, No. 2, 257-271.

Ungerstedt, U., Ljundberg, T., and Schultz, W., 1978, Dopamine receptor mechanisms: behavioral and electrophysiological studies, Dopamine.Adv.Bioch.Psychopharmacol., 19, 311-321.

Urca, G., and Liebeskind, J. C., 1979, Electrophysiological indices of opiate action in awake and anesthetized rats, Brain Res., 161, No. 1, 162-166.

Urca, G., and Nahin, R. L., 1978, Morphine-induced multiple unit changes in analgesic and rewarding brain sites, Pain Abstr., 1, 261.

Urca, G., Frenk, H., Liebeskind, J. C., and Taylor, A. N., 1977, Morphine and enkephalin: analgetic and epileptic properties, Science, 197, No. 1, 83-86.

Uretsky, N. J., and Shodgrass, R. S., 1977, Studies of the mechanism of stimulation of dopamine synthesis by amphetamine in striatal slicers, J.Pharmacol.Exp.Ther., 202, No. 3, 565-580.

Ushima, J., Koizumi, K., and Brooks, C. M. C., 1960, Excitability of spinal neurons and changes resulting from reticular formation stimulation, Amer.J.Physiol., 198, No. 2, 393-398.

Vaisberg, M., and Saunders, J. S., 1963, Treatment of dyskinesias including Huntington's chorea with thiopropozate and R 1625", Dis.Nerv.Syst., 24, No. 3, 499-500.

Valdman, A. V., and Ignatov, Yu.D., 1976, "Central Mechanisms of Pain," (in Russian), Izd.Nauka, Leningrad.

Valdman, A. V., Evartau, E. E., and Kozlovskaya, M. M., 1976, "Psychopharmacology of Emotions," (in Russian), Izd. Meditsina, Moscow.

Valli, M., and Pringuelj, D., 1980, Actualités concentrant le mécanisme d'action biochimique des benzodiazépines, Thérapie, 35, No. 5, 561-569.

VanGilder, J. C., and O'Leary, J. L., 1971, Effect of Nembutal anesthesiaupon Purkinje cell activation in the cat, Electroencephalogr.Clin.Neurophysiol., 30, No. 2, 173-188.

Van Hees, J., and Gybels, J. M., 1972, Pain related to single afferent C-fibres in human skin, Brain Res., 48, No. 2, 397-400.

Van Rossum, J. M., 1967, The significance of dopamine receptor blockade for the action of neuroleptic drugs, in: "Neuropsychopharmacology," H. Brill, J. O. Cole, and P. B. Bradley, eds., Excerta Medica, Amsterdam, pp. 321-329.

Vastola, E. F., 1959, After-positivity in lateral geniculate body, J.Neurophysiol., 22, No. 2, 258-272.

Vastola, E. F., 1960, After-positivity in lateral geniculate body
    during repetitive stimulation, J.Neurophysiol., 2, No. 1,
    54-61.

Vein, A. M., Yakhno, N. N., Rotenberg, V. S., Vlasov, N. A., and
    Sumsky, L. I., 1971, Sleep. neurophysiological, vegetative,
    psychological, chemical and pathophysiological aspects, Uspekhi
    fiziol. nauk, 2, 24-72.

Vein, A. M., Golubev, V. L., and Berzinsh, Yu.E., 1981, "Parkinson-
    ism," (in Russian), Izd. Znaniye, Riga.

Velasco, M., Skinner, F. E., Asaro, K. D., and Lindsley, D. B., 1968,
    Thalamo-cortical systems regulating spindle bursts and
    recruiting responses. I. Effect of cortical ablations, EEG
    and Clin.Neurophysiol., 25, No. 4, 463-470.

Verimer, T., Goodale, D. B., Long, J. P., and Flynn, J. R., 1980,
    Lithium effects on haloperidol-induced pre- and postsynaptic
    dopamine receptor supersensitivity, J.Pharmacol., 32, No. 5,
    665-666.

Verzeano, M., and Negishi, K., 1960, Neuronal activity in cortical
    and thalamic networks. A study with multiple microelectrodes,
    J.Gen.Physiol., 43, Suppl., 77-195.

Villablanca, I. R., Marcus, R. I., and Olmstead, Ch.E., 1976,
    Effects of caudate nuclei or frontal cortex ablations in
    cats. II. Sleep-wakefulness EEG and motor activity, Exp.
    Neurol., 53, No. 1, 31-50.

Von Voightlander, P. F., and Moore, K. E., 1971a, The release of
    ^{3}H-dopamine from cat brain following electrical stimulation
    of substantia nigra and caudate nucelus, Neuropharmacology,
    10, No. 5, 733-741.

Von Voightlander, P. F., and Moore, K. E., 1971b, Nigrostriatal
    pathway: stimulation-evoked release of ^{3}H-dopamine from
    caudate nucleus, Brain Res., 35, No. 3, 580-583.

Von Voightlander, P. F., and Moore, K. E., 1973, Involvement of
    nigrostriatal neurons in the in vivo release of dopamine,
    amantadine and tyramine, J.Pharmacol., Exp.Ther., 184, No.
    3, 542-552.

Wada, J. A., 1976, "Kindling," Raven Press, New York.

Waddington, J. L., and Crow, T., 1979, Rotational responses to
    serotoninergic and dopaminergic agonists after unilateral
    dihydroxytryptamine lesions of the medial forebrain bundle:
    co-operative interaction of serotinin and dopamine in neo-
    striatum, Life Sci., 25, No. 7, 1307-1314.

Wagman, I. H., and Price, D. D., 1969, Responses of dorsal horn
    cells of M. mulata to cutaneous and sural nerve A and C
    fibre stimuli, J.Neurophysiol., 32, No. 7, 803.

Wagner, H. R., Feeney, D. M., Gullotta f. P., and Cote, I. L., 1975,
    Suppression of cortical epileptiform activity by generalized
    and localized ECoG desynchronization, Electroenceph.Clin.
    Neurophysiol., 39, No. 4, 499-506.

Walberg, F. O., Westum, L. E., and Hauglie-Hansen, E., 1962,
    Fastigioreticular fibres in the cat. An experimental study
    with silver method, J.Comp.Neurol., 119, No. 2, 187-199.

Wall, P. D., 1974, Physiological mechanisms involved in the pro-
    duction and relief of pain, in: "Recent Advances in Pain:
    Pathophysiology and Clinical Aspects," J. J. Bonica, P.
    Procacci, and C. A. Pagni, eds., Charles C Thomas,
    Springfield, pp. 36-63.

Wall, P. D., 1978, The gate control theory of pain mechanisms. A
    re-examination and re-statement, Brain, 101, No. 1, 1-18.

Wall, P. D., 1979, Changes in damaged nerve and their sensory con-
    sequence, in: "Advances in Pain Research and Therapy," J. J.
    Bonica et al., eds., vol. 3, Raven Press, New York, pp.
    39-52.

Wall, P. D., 1981, Alterations in the central nervous system after
    deafferentation, Pain, 10, Suppl. 1, 75.

Wall, P. D., and Gutnier, A., 1974, Ongoing activity in peripheral
    nerves: with physiology and pharmacology of impulse origin-
    ating from a neuron, Exp.neurol., 43, No. 6, 580-593.

Wall, P. D., Waxman, S., and Basbaum, A. I., 1974, Ongoing activity
    in peripheral nerve: III. Injury discharge, Exp.Neurol., 45,
    No. 6, 576-589.

Wall, P. D., Devor, M., Inbal, R., Scadding, J. W., Schonfeld, D.,
    Seltzer, L., and Tomkiewicz, 1979, Autotomy following peri-
    pheral nerve lesions: experimental anaesthesia dolorosa,
    Pain, 7, No. 1, 103-115.

Walsh, G. O., 1971, Penicillin iontophoresis in neocortex of cat:
    effects on the spontaneous and induced activity of single
    neurons, Epilepsia, 12, No. 1, 1-11.

Ward, A. A., Jr., 1972, Topical convulsant metals, in: "Experimental
    Model of Epilepsy," D. P. Purpura, J. K. Penry, D. Tower, D.
    M. Woodbery, and R. Walter, eds., Raven Press, New York, pp.
    13-35.

Ward, A.A., 1980, The physiological basis for the surgical therapy
    of epilepsy, in: "Limbic Epilepsy and Discontrol Syndrome,
    Proceedings of the First International Symposium," Sydney,
    pp. 239-254.

Warner, A. M., and Pieroxynsky, G., 1977, Pseudocatatonia associated
    with abuse of amphetamine and cannabis, Postgrad.Med., 61,
    No. 1, 275-277.

Wastek, G., Speth, R., Reisine, T., and Yamamura, H., 1978, The
    effect of gamma-aminobutyric acid on ^{3}H-flunitrazepam bind-
    ing in rat brain, Eur.J.Pharmacol., 50, No. 4, 445-447.

Webb, W. B., and Cartwright, R. D., 1978, Sleep and dreams, Ann.Rev.
    Psychol., 29, No. 2, 223-252.

Weiner, W. J., Goetz, C., and Westheimer, R., 1973, Serotoninergic
    and antiserotoninergic influences on amphetamine-induced
    stereotyped behavior, J.Neurol.Sci., 20, No. 4, 373-379.

Wellhöner, H. H., 1982, Tetanus Neurotoxin, Rev.Physiol.Biochem.
    Pharmacol., 93, No. 1, 2-68.

Werman, R., Davidoff, R. A., and Aprison, M. H., 1968, Inhibitory
    action of glycine on spinal neurons in the cat, J.Neurophys-
    iol., 31, No. 1, 81-95.

Westrum, L. E., Canfield, R. C., and Black, R. G., 1976, Transgang-
       lionic degeneration in the spinal trigeminal nucleus follow-
       ing removal of tooth pulps in adult cats, Brain Res., 101,
       No. 1, 137–140.
Wiesenfeld, Z., and Lindblom, U., 1982, Behavioral and electrophy-
       siological effects of various types of peripheral nerve
       lesions in the rat: a comparison of possible models for
       chronic pain, Pain, 8, No. 3, 285–298.
Willer, J. C., Boureau, F., and Albe-Fessard, D., 1978, Role of
       large diameter cutaneous afferents in transmission of no-
       ciceptive message: electrophysiological study in man, Brain
       Research, 152, No. 2, 358–364.
Willer, J. C. Boureau, F., and Albe-Fessard, D., 1980, Human no-
       ciceptive reactions: effects of spatial summation of affer-
       ent from relatively large diameter fibers, Brain Research,
       201, No. 3, 465–470.
Williamson, T. L., and Crill, W. E., 1976, The effects of pentylene-
       tetrazol on molluscan neurons. I. Intracellular recording
       and stimulation, Brain Res., 116, No. 2, 217–229.
Willis, W. D., Trevino, D. L., Coulter, J. D., and Maunz, R. A.,
       1974, Responses of primate spinothalamic tract neurons to
       natural stimulation of hindlimb, J.Neurophysiol., 37, No. 2,
       358–372.
Wilson, S. A. K., 1954, in: "Neurology," Second Edition, Butterworth,
       London.
Wilson, V. J., Kato, M., Thomas, R. C., and Peterson, B. W., 1966,
       Excitation of lateral vestibular neurons by peripheral
       afferent fibers, J.Neurophysiol., 29, No. 5, 508–529.
Winter, K., 1973, Central pains and treatment by stereotaxic
       thalamotomy, Eur.Neurol., 10, No. 2, 65–74.
Wolf, P., and Hass, H. L., 1977, Effects of diazepines and
       barbiturates on hippocampal recurrent inhibition,
       Naunyn-Schmiedeberg's Arch.Pharmacol., 199, No. 2, 211–218.
Wong, R. K. S., and Prince, D. A., 1977, Burst generation and cal-
       cium mediated spikes in hippocampal CA3 neurons, Neurosci.
       Abstr., 3, 148 (Abstr. 465).
Wong, R. K. S., and Prince, D. A., 1978, Participation of calcium
       spikes during intrinsic burst firing in hippocampal neur-
       ones, Brain Res., 159, No. 2, 385–390.
Wong, R. K. S., and Schwartzkroin, P. A., 1982, Pacemaker neurons in
       the mammalian brain: mechanisms and function, in: "Cellular
       Pacemakers, Vol. 1, Mechanisms of Pacemaker Generation," D.
       Carpenter, ed., John Wiley & Sons, New York, pp. 237–254.
Woodrow, K. M., Reifman, A., and Wyatt, R. J., 1978, Amphetamine
       psychosis – a model for paranoid schizophrenia?, in: "Neuro-
       pharmacology and Behavior," B. Haber and M. N. Aprison, eds.,
       Plenum Press, New York – London, pp. 1–21.
Woodruff, M. L., 1975, Midbrain and callosal influences on the
       spread of focal cortical epileptic activity, Brain Res., 85,
       No. 1, 53.

Wu, J.-Y., Matsuda, T., and Roberts, E., 1973, Purification and
    characterization of glutamate decarboxylase from mouse
    brain, J.Biol.Chem., 248, 3029-3034.
Wyler, A. R., 1974, Epileptic neurons during sleep and wakefulness,
    Exp.Neurol., 42, No. 4, 593-608.
Wyler, A. R., and Fetz, E. E., 1974, Behavioral control of firing
    patterns of normal and abnormal neurons in chronic epileptic
    cortex, Exp.Neurol., 42, No. 3, 448-464.
Wylerk, A. R., and Ward, A. A., 1980, Epileptic neurons, in: "Epil-
    epsy: A Window to Brain Mechanisms, J. S. Lockard and A. A.
    Ward, eds., Raven Press, New York, pp. 51-68.
Wyler, A. R., Fetz, E. E., and Ward, A. A., 1973, Spontaneous firing
    patterns of epileptic neurons in the monkey motor cortex,
    Exp.Neurol., 40, No. 6, 567-585.
Yamaguchi, Y., 1980, A possible rat model for chronic pain, J.Phys-
    iol.Soc.Japan, 42, 306,
Yarborough, G., 1975, Supersensitivity of caudate neurons after
    repeated administration of haloperidol, Eur.J.Pharmacol.,
    31, No. 4, 367-369.
Yerokhina, L. G., 1973, "Facial Nerves (Trigeminal Neuralgia and
    Other Forms of Prosopalgia)," (in Russian), Izd. Meditsina,
    Moscow.
Yerzina, G. A., 1961, Influence of the gamma-neuronal system on the
    electric activity of muscles in local tetanus in rats,
    Fiziol.Zh., 67, No. 8, 971-975.
Yessaian, N. H., Armenian, A. R., and Chiflikian, M. D., 1973,
    Effect of gamma-aminobutyric acid on release and metabolism
    of serotonine, Voprosy biokhimii mozga (Yerevan), 8,
    203-207.
Yokoi, K., Rose, S. E., and Yanagihara, T., 1981, Benzodiazepine
    receptor: heterogeneity in rabbit brain, Life Sci., 28, No.
    14, 1591-1595.
Yokota, T., and Nishikawa, N., 1977, Somatotopic organization of
    trigeminal neurons with caudal medulla oblongata, in: "Pain
    in the Trigeminal Region," D. J. Anderson and B. Matthew,
    eds., Elsevier, Amsterdam, pp. 243-257.
Yokota, T., Nishikawa, N., and Nishikawa, Y., 1979, Trigeminal
    nociceptive neurons in the trigeminal subnucleus caudalis and
    bulbar lateral reticular formation, in: "Advances in Pain
    Research and Therapy," J. J. Bonica et al., eds., vol. 3, Raven
    Press, New York.
Young, R. F., and King, R. B., 1972, Excitability changes in trige-
    minal primary afferent fibers in response to noxious and
    nonnoxious stimuli, J.Neurophysiol., 35, No. 1, 87-95.
Yasa, H., Iwata, K., Tasaki, f., Kajeyama, N., Miyake, A., and
    Watanabe, S., 1981, Threshold of penicillin-induced epilepsy
    in brain slices of guinea-pig, Electroenceph.Clin.Neuro-
    physiol., 52, No. 1, 98.
Zachar, J., and Zacharová D., 1963, Mechanisms Vzniku Siziacy sa
    Kortikalnej Depresie. Vyd.Slow.Acad.Ved., Bratislava.

Zakusov, V. V., Ostrovskaya, R. U., Markovitch, V. V., Moldavkin, G.
      M., and Bulayev, V. M., 1975, Electrophysiological evidence
      for an inhibitory action of diazepam upon cat brain cortex,
      Arch.Int.Pharmacodyn., 214, No. 2, 188-205.
Zakusov, V. V., Ostrovskaya, R. U., 1977, Kozhechkin, S. W.,
      Markovich, V. V., Molodavkin, G. M., and Voronina, T. A.,
      1977, Further evidence for GABA-ergic mechanisms in the
      action of benzodiazepines, Arch.Int.Pharmacodyn., 229, No.
      2, 313-326.
Zambrzhitsky, I. A., 1972, "Limbic Region of the Brain," (in
      Russian), Izd. Meditsina, Moscow.
Zancetti, A., 1967, Brain stem mechanisms of sleep, Anesthesiology,
      28, No. 1, 81-99.
Zimmermann, M., 1979, Peripheral and central nervous mechanisms of
      nociception, pain and pain therapy: facts and hypotheses,
      in: "Advances in Pain Research and Therapy," J. J. Bonica,
      J. C. Liebeskind, and D. G. Albe-Fessard, eds., vol. 3,
      Raven Press, New York, pp. 3-32.
Zisper, B., Crain, S. M., and Bornstein, M. B., 1973, Directly
      evoked 'paroxysmal' depolarizations of mouse hippocampal
      neurons in synaptically organized explants in long-term
      culture, Brain Res., 60, No. 3, 489-495.
Zorman, G., Hentall, I. D., Adams, J. E., and Fields, H. L., 1981,
      Naloxone-reversible analgesia produced by microstimulation
      in the rat medulla, Brain Res., 219, No. 1, 137-148.
Zorman, Y., Belcher, J. E. G., Adams, J. E., and Fields, H. L.,
      1982, Lumbar intrathecal naloxone blocks analgesia produced
      by microstimulation of the ventromedial medulla in the rat,
      Brain Res., 236, No. 1, 77-84.

Index

411